Bio-aggregate-based Building Materials

Bio-aggregate-based Building Materials

Applications to Hemp Concretes

Edited by
Sofiane Amziane
Laurent Arnaud

Series Editor
Noël Challamel

iSTE

WILEY

First published 2013 in Great Britain and the United States by ISTE Ltd and John Wiley & Sons, Inc.

ISTE Ltd
27-37 St George's Road
London SW19 4EU
UK

www.iste.co.uk

John Wiley & Sons, Inc.
111 River Street
Hoboken, NJ 07030
USA

www.wiley.com

Library of Congress Control Number: 2012954575

British Library Cataloguing-in-Publication Data
A CIP record for this book is available from the British Library
ISBN: 978-1-84821-404-0

MIX
Paper from
responsible sources
FSC® C013604
FSC
www.fsc.org

Table of Contents

Foreword xi

Chapter 1. Environmental, Economic and Social Context of Agro-Concretes 1
Vincent NOZAHIC and Sofiane AMZIANE

1.1. Sustainable development, construction and materials 1
1.1.1. Environmental impacts of the construction sector. 2
1.2. Standardization and regulation: toward a global approach 3
1.2.1. Standardization and regulation in force 3
1.2.2. Limitations of the normative and regulatory framework 5
1.3. The materials: an increasingly crucial element 7
1.3.1. Role of the materials in energy consumption. 7
1.3.2. What is a low-environmental-impact material? 7
1.3.3. Constantly-changing regulations 8
1.4. The specific case of concretes made from lignocellular particles. 9
1.4.1. Development of agro-concretes in the context of France 10
1.5. What does the term "Agro-concrete" mean? 13
1.5.1. General definition. 13
1.5.2. Lignocellular resources 13
1.5.3. General characteristics of lignocellular agro-resources. 15
1.6. Conclusions. 19
1.7. Bibliography 19

Chapter 2. Characterization of Plant-Based Aggregates 27
Vincent PICANDET

2.1. Microstructure of the shiv particles. 28
2.1.1. Structure of the stem of fibrous plants. 28
2.1.2. SEM observation of hemp shiv particles 30
2.1.3. Chemistry of the cell walls 31
2.1.4. Density and porosity, in the case of hemp shiv 35
2.2. Particle Size Distribution (PSD) 36
2.2.1. General characteristics of aggregates made from fibrous plants 36
2.2.2. Fiber content. 37
2.2.3. Methods for characterizing the PSD. 38
2.2.4. PSD analyses 48
2.2.5. Comparison of the results obtained by image analysis 52
2.2.6. Characterization of the geometry of the particles 57
2.2.7. Characterization of the PSD. 58
2.2.8. Conclusions 65
2.3. Compactness and compressibility. 66
2.4. Water absorption capacity 68
2.5. Bibliography 69

Chapter 3. Binders 75
Gilles ESCADEILLAS, Camille MAGNIONT, Sofiane AMZIANE and Vincent NOZAHIC

3.1. Portland cements. 75
3.1.1. General. 75
3.1.2. Production 76
3.1.3. Chemical and mineral composition 77
3.1.4. Properties 77
3.1.5. Environmental impacts 84
3.2. Lime 84
3.2.1. General. 84
3.2.2. Aerial lime 86
3.2.3. Natural hydraulic limes 89
3.3. Lime-pozzolan mixtures 92
3.3.1. Natural pozzolans. 93
3.3.2. Calcined natural pozzolans: metakaolin. 96
3.3.3. Fly ash 101
3.3.4. Blast furnace slag 103

3.4. Plaster 106
3.4.1. General 106
3.4.2. Production 106
3.4.3. Chemical and mineralogical composition. 108
3.4.4. Properties 108
3.4.5. Environmental impacts 110
3.5. Summary 110
3.6. Bibliography 111

Chapter 4. Formulation and Implementation 117
Christophe LANOS, Florence COLLET, Gérard LENAIN and Yves HUSTACHE

4.1. Objectives. 117
4.1.1. Preamble 117
4.1.2. Traditional applications 119
4.1.3. Constituents and mixture 120
4.1.4. Methods of implementation. 121
4.2. Rules of formulation 122
4.2.1. Basis of usual formulations 122
4.2.2. Influence of the proportion of paste in the mixture 124
4.2.3. Quality of the paste and water content 128
4.2.4. Homogeneity of the paste 135
4.2.5. The relationship between formulation and strength. 137
4.2.6. The relationship between formulation and thermo-hydric properties 141
4.3. Examples of formulations 141
4.3.1. Origin of the data 141
4.3.2. Walling application. 141
4.3.3. Flooring application 142
4.3.4. Roofing application. 142
4.3.5. Other applications 142
4.4. Installation techniques 143
4.4.1. Building a wall using formwork 143
4.4.2. Application by spraying 143
4.4.3. Laying of a floor 144
4.4.4. Creating a roof 144
4.4.5. Other uses 145
4.5. Professional rules for buildings using hempcrete and hemp mortars 145

4.5.1. History 145
4.5.2. Principles and content of the professional regulations 146
4.6. Bibliography 152

Chapter 5. Mechanical Behavior 153
Laurent ARNAUD, Sofiane AMZIANE, Vincent NOZAHIC and Etienne GOURLAY

5.1. Composite material 153
5.1.1. Making of the test tubes 154
5.1.2. Mechanical behavior 154
5.1.3. Effect of initial compression 157
5.1.4. Effect of the nature of the binder 159
5.1.5. Influence of the binder content 162
5.1.6. Influence of the particle size 164
5.1.7. Influence of the curing conditions 165
5.1.8. Evolution over time 166
5.1.9. Interaction between particles and binder 167
5.1.10. Anisotropic behavior 170
5.2. Modeling of the mechanical behavior 171
5.2.1. Empirical approach 171
5.2.2. Self-consistent homogenization approach 173
5.3. Toward the study of a stratified composite 174
5.4. Conclusion 175
5.5. Bibliography 176

Chapter 6. Hygrothermal Behavior of Hempcrete 179
Laurent ARNAUD, Driss SAMRI and Étienne GOURLAY

6.1. Introduction 179
6.2. Heat conductivity 180
6.2.1. Measurement of the conductivity 181
6.2.2. Modeling of the heat conductivity in dry and humid conditions 182
6.2.3. Heat transfers 185
6.3. Hygrothermal transfers 186
6.3.1. Experimental device 186
6.3.2. Stresses 189
6.3.3. Phase changes 191
6.3.4. Hygrothermal transfers 194
6.3.5. Role of coating products applied to hempcrete 196
6.3.6. Conclusions 200

6.4. Thermal characterization of various construction materials 201
6.4.1. Autoclaved aerated concrete 202
6.4.2. Vertically perforated brick 204
6.4.3. Hempcrete 205
6.4.4. Conclusions 210
6.5. Modeling of coupled heat- and mass transfers 211
6.5.1. Introduction 211
6.5.2. Transfer laws 212
6.5.3. Transfer model: the Künzel model. 216
6.5.4. Determination of the transfer coefficients 217
6.5.5. Numerical modeling 222
6.6. Conclusions. 235
6.7. Bibliography 238

Chapter 7. Acoustical Properties of Hemp Concretes. 243
Philippe GLÉ, Emmanuel GOURDON and Laurent ARNAUD

7.1. Introduction. 243
7.2. Acoustical properties of the material on the basis of the main mechanisms 244
7.2.1. Influence of the components 244
7.2.2. Influence of the casting method 249
7.3. Modeling the acoustical properties 252
7.3.1. Physical analysis of the acoustical properties being measured 253
7.3.2. The adapted double porosity model and its parameters. 255
7.3.3. Experimental validation of the model 257
7.4. Application of the model to the acoustical characterization of shiv 258
7.4.1. Porosity of shiv 258
7.4.2. Resistivity 262
7.5. Conclusion 264
7.6. Bibliography 264

Chapter 8. Plant-Based Concretes in Structures: Structural Aspect – Addition of a Wooden Support to Absorb the Strain 267
Philippe MUNOZ and Didier PIPET

8.1. Introduction. 267
8.2. Preliminary test 269
8.2.1. Description of the panel 269

8.2.2. Putting the panel in place on the bracing bank 270
8.2.3. Longitudinal loading and measurement of the movements 271
8.2.4. Behavior of the test bank 273
8.2.5. Behavior of the wooden panel 274
8.3. Test on a composite panel of a wooden skeleton and hempcrete 276
8.3.1. Description of the panel 276
8.3.2. Emplacement of the panel on the bracing bank 276
8.3.3. Vertical loading 279
8.3.4. Longitudinal loading and measurement of the movements 280
8.3.5. Running of the test 281
8.3.6. Feature of the ruin of the panel 283
8.4. Results and comparative analysis 285
8.5. Conclusions and reflections 287
8.6. Acknowledgements 288
8.7. Bibliography 288

Chapter 9. Examination of the Environmental Characteristics of a Banked Hempcrete Wall on a Wooden Skeleton, by Lifecycle Analysis: Feedback on the LCA Experiment from 2005 289
Marie-Pierre BOUTIN and Cyril FLAMIN

9.1. Introduction 289
9.2. Description of the products studied 291
9.3. Method for environmental evaluation of bio-sourced materials 292
9.4. Lifecycle Analysis on hempcrete – methodology, working hypotheses and results 294
9.4.1. Delimitation of the system under study 294
9.4.2. Inventory analysis 298
9.4.3. Impact evaluation 303
9.4.4. Results and interpretation of the lifecycle 305
9.5. Interpretations of the lifecycle, conclusions and reflections 306
9.6. Bibliography 310

List of Authors 313

Index 315

Foreword

I am writing this foreword soon after finishing reading *Sur la route du papier* (On the Paper Trail) by Erik Orsenna: a wondrous journey, a lesson in history but also, and above all, a revelation about the workings of globalization. Paper is a harmless fibrous pulp – originally created from old rags, and later on and to date, from wood – which, filtered in the form of a thin layer, has enabled the most abstract creations of the human mind to be promulgated and become immortal.[1] Though it has long been decried for the environmental consequences of its production, paper has now acquired a stamp of eco-friendliness thanks to the constant improvement of forestry and forest management, the manufacturing procedures and recycling. Is there any other "bio-sourced" material that has had a more profound impact on the development of civilization than paper? Assuredly not.

In another context, is there any other material that has made a greater contribution to the human race's economic development for over a century, and to the mass development of infrastructures that that development has required, than concrete? Along with petroleum (for mobility) and silicon (for information and communication technology (ICT)), this artificial rock is one of the material foundations of our so-called developed societies. Concrete has enabled us to harness the energy from rivers, to build ports and airports, levees, sewage systems, roads, bridges, tunnels and more buildings than any other material.

Yet this material, a mixture of cement, sand and gravel which we simply call "*concrete*", is in fact only one of the representatives of a broad category. After all, what is a concrete, if not a composite material made up of granular particles and a "glue" or binder holding everything together? According to that logic, bitumen concrete – that of the asphalt paving that now covers our roads in an almost

1 And, recently – but this is less noble – to package practically all the goods that we produce, in the form of cartons, boxes and bags.

monopolistic fashion – may be a serious challenger to cement concrete for the title of kingpin material in our infrastructures. Between them, these two materials alone constitute almost the totality of the "skeleton" of our lands. However, the intensive usage of these materials is not without consequences, either through greenhouse gas emissions or by the exhaustion of natural resources, be they fossil or mineral.

It might be tempting to leave the topic at that, unless for anecdotal purposes. However, there are at least two other concretes which merit our attention. The first is at least as widespread as cement and bitumen concretes on a worldwide scale; yet it is largely overlooked in our societies. Quite simply it is crude earth (not fired or baked like terracotta), which should, more correctly, be dubbed "clay concrete", because it is its fine-grained constituent – clay – which, upon interaction with water as in the case of cement concrete, ensures the cohesion of the larger grains. In various forms – compacted, molded, pasted or plastered – it provides shelter to over a quarter of the world's population. In France alone, the patrimony built of crude earth represents over a million houses. In the hands of master craftsmen and expert architects, the use of earth is fully capable of delivering on our desires for comfort and aesthetic beauty, whilst also satisfying our desire for eco-modernity.

This book is devoted to a fourth concrete, or rather a fourth family of concretes; original and promising from more than one point of view, which would seem to exhibit all the advantages of paper and earth, whilst still offering the convenience of use of our major industrial concretes. Contrary to popular opinion, sand and other granular (particulate) minerals are not an inexhaustible resource. Unless we wish to inflict irreparable damage on the environment, the time has come for recycling, or for using bio-sourced particulates, which is essentially the same thing. This is the path adopted by agro-concretes and, in particular, *hemp* concretes. France is the largest producer in Europe of *Cannabis Sativa*, whose fibers have been used to make rope for centuries. Yet this fast-growing plant, well adapted to temperate climates, harbors many other resources. Its stem, of a highly porous and therefore very lightweight wood, when ground up makes a surprising aggregate. Surprising, not on a mechanical level – the only level which truly counts for mineral aggregates, with cleanliness and shape in joint second place – but surprising, primarily, on a functional level: the level of hygrothermal equilibrium and acoustic properties.

Looking at the proliferation of synthetic materials available on the market, one might think that thermal, hydral and acoustic comfort is a domain in which the materials available – particularly when these materials are used in combination – have nearly reached the optimum desired. Polymer foams and organic and inorganic aerogels have extremely low thermal diffusivity and air permeability, which are difficult to better in the race toward very low values. Yet they lack inertia. When combined with other materials – or, even better, when a solid-to-liquid "phase-changing" material such as paraffin or a salt is added into the mixture – they (apparently) acquire the thermal inertia that they lack, by absorbing and reflecting the latent heat of fusion. In spite of their remarkable performances, these insulating materials still lack the "breathability" of certain natural materials, related to the capacity for absorption, transfer and phase-change of water in vapor and liquid form – all properties which depend on the characteristics of the porous space of the material and the thermal and hydric coupling which manifests itself in that space.

In the face of the complexity of combinations of synthetic materials employed to ensure an acceptable degree of comfort, agro-concretes and hemp concretes in particular offer a simple solution, which draws upon the exceptional porous texture – nearly always hierarchical – of certain plant structures. However, in order to take advantage of this property in terms of hygrothermal exchanges, the binder used must be able to work with the granular material rather than counteracting its properties.

The authors of this book present us with the elements, drawn directly from research, that help us comprehend the properties, function and formulation of agro-concretes. It is undoubtedly true that such concretes can never stand up to high- or ultra-high-performance mineral concretes, but it is not their intention to do so. First and foremost, they are intended to be insulating materials. Therefore, their primary intention is to sustainably ensure the comfort and durability of the dwelling, including the moderately dense dwellings towards which we are now tending.

This book has another virtue. It leads us to reflect on the physical bases for our criteria of "high environmental quality", which are still largely founded upon the segmentation and selection of a few physical properties. The very least that can be said is that this manner of proceeding is not hugely well-adapted to materials in which there is extensive coupling between properties. This is precisely the case where bio-sourced materials are concerned. Let us hope that this book is distributed as widely as possible, so that a global, "performance-oriented" approach can finally emerge.

Henri VAN DAMME
IFSTTAR
December 2012

Chapter 1

Environmental, Economic and Social Context of Agro-Concretes

1.1. Sustainable development, construction and materials

After decades of virtuous and limitless consumption, the evidence is incontrovertible: human activities are not without impact on the environment and on humans themselves. It was not until 1987, with the Brundtland Commission [UNI 87], that this observation gave rise to a new concept: sustainable development. The report published by this commission, *Our Common Future*, defines the term as follows:

> "Sustainable development is development that meets the needs of the present without compromising the ability of future generations to meet their own needs." [UNI 87]

Thereafter, this concept has pervaded modern societies, ultimately becoming a political and economic issue, and an issue of the very survival of the human race… All human activities – industry, construction, agriculture, energy, transport, etc. – now have to deal with so-called "sustainable development" issues. The report unveiled by the United Nations Environment Program (UNEP) [UNE 09] constitutes an overview of the evolution of our societies since the publication of the Brundtland Report. The following quote, taken from that text, highlights the enormity of the challenge:

> "There are no major issues raised in *Our Common Future* for which the foreseeable trends are favourable." [UNE 07]

Chapter written by Vincent NOZAHIC and Sofiane AMZIANE.

1.1.1. *Environmental impacts of the construction sector*

Above all, we must remember that the concept of sustainable development dealt with locally is often linked to problems on a worldwide scale, such as global warming or the gradual exhaustion of resources. These two criteria constitute the points of no return for our civilization.

As regards the climate, the scientific works of the IPCC[1] serve as a referential framework. The second assessment report (SAR) published by this organization in 1995 [IPC 95] concludes that the "the balance of evidence suggests a discernible human influence on global climate". A mere two years later, on the basis of this report and the UN Framework Convention on Climate Change [UNI 92], the international political debates culminated in the Kyoto Protocol [UNI 98]. This text commits the countries which have ratified it to reduce their GHG[2] emissions by 5.2% in comparison to their level in 1990 over the period 2008–2012. The protocol came into force in 2005 and therefore will conclude in 2012. Owing to its use of nuclear and hydroelectric energy, which do not produce much GHG, France is committed to maintaining these levels of emissions.

For its part, the construction sector (residential and tertiary), much like the agricultural or industrial sectors, finds itself facing significant challenges in terms of reducing GHG emissions and energy consumption. The figures speak for themselves, but they must be analyzed seriously. Indeed, it is not always entirely clear what data have been taken into account when producing the figures, particularly in terms of drawing the distinction between a building's function and its construction:

– **Total GHG emissions from both energetic and non-energetic sources (e.g. agriculture, forestry, etc.):** 7.9% on a global scale [IPC 07], 40% on the scale of the US [USD 11] and 18% on the scale of France in 2007 [CGD 10] for all residential/tertiary, institutional and commercial consumption (heating, specific electricity, hot water, cooking, etc.);

– **Final electrical energy consumption[3]:** 41% on the scale of the US in 2010 [USD 11] and 43.4% on the scale of France in 2008 [RIL 06; CGD 09] for all residential/tertiary, institutional and commercial use. These figures do not include the fossil energy required to produce the electricity.

---

1 IPCC: Intergovernmental Panel on Climate Change

2 GHG: Greenhouse Gas

3 Final energy is directly usable energy, bought and charged by each customer – e.g. liters of gasoline at the pumps.

However, while climate change represents an alarming phenomenon, it is not the only point that needs to be taken into account. The natural resources needed for the perpetuation of human activities and societies are, for the most part, finite and exhaustible. Similar to the threat of global warming, exhaustion of resources – be they minerals or arable land – is a major point of concern which will inevitably lead us to change our ways before the current century is out [OEC 08]. The activities relating to construction and to public projects, while they do not necessarily require materials to be used which come from exhaustible sources (with the exception of road infrastructures, which consume bitumen), constitute the single greatest cause of *consumption of natural resources* (31% in Europe [SER 09]). Furthermore, this consumption causes a large amount of *waste production*, even though 97% of the waste produced by construction and public works in France are inert [IFE 08] and are subject to a policy of value-creation. In France, the amount of waste generated by this domain equaled 343 million tons in 2004, 44% of the total mass [PEU 08].

In summary, the construction sector battles four main impacts on the environment:

– Its GHG emissions;

– Its energy consumption;

– Its consumption of natural resources;

– Its waste production.

1.2. Standardization and regulation: toward a global approach

1.2.1. *Standardization and regulation in force*

The legislation in charge of regulating these major impacts of the sector is a relatively recent phenomenon. The European framework was solidified in 2002 with the publication of the Energy Performance of Buildings Directive (EPBD). In the context of France, the successive "réglementations thermiques" (thermal regulations) RT 2000 and RT 2005 [FFB 09] follow this document. More recently, the *loi Grenelle 1* (First Conference Law) of 3 August 2009 defined the *Plan Bâtiment Grenelle*[4] (Conference Building Plan), launched in January 2009 (see Figure 1.1):

4 www.plan-batiment.legrenelle-environnement.fr (02/09/10)

> "The aim of the Plan Bâtiment Grenelle is to guide the implementation and deployment of the measures prescribed in the program to reduce buildings' energy consumption and greenhouse gas emissions" [GRE 09].

The approach deals with buildings' performances with a view to satisfying two main objectives:

– **38% goal**: to achieve a 38% reduction, by 2020, of the energy consumption of the existing residential/tertiary sector in comparison to its level in 2008;

– **Factor of 4 goal**: to achieve a fourfold reduction, by 2050, of GHG emissions by the existing residential/tertiary sector in comparison to its level in 1990. This commitment was made in 2003 as part of the National Climate Plan.

– The foundations for the *new thermal regulation RT 2012*, which is currently coming into force, were laid by Article 4 of the First Conference Law. This new regulation, which applies to new buildings in the residential and tertiary sectors, includes three main objectives:

– a numerical objective (C_{MAX} coefficient) regarding a reduction in energy consumption by 50 kWh/m^2/year, with a variety of criteria modulating that figure such as geographical location (see *Figure 1.1*). In terms of heating, this corresponds to an average of 15 kWh/m^2/year. In terms of controls, the emphasis is placed on the measure of the building's air-tightness, and on the monitoring of consumption;

– a significant technological and industrial change in the design and construction of buildings, for all areas of energy expenditure (heating regulators, lighting or water heating systems, etc.). However, air conditioning should be avoided, as the design of the building should take account of comfort in the summer months (T_{IC} coefficient – Interior Conventional Temperature);

– a balanced energy provision, which emits little in terms of GHGs and contributes to the country's energy independence (incentive to use renewable energies).

The primary goals of these new thermal regulations are to reduce GHG emissions by the construction sector, but also that sector's energy bill from a socio-economic standpoint and in terms of exhausting fossil resources. With the RT 2012, a *global design of the function of the building*, including consumption in terms of heating as well as air conditioning or lighting, is prescribed. This is the first step towards a "cradle to the grave"-type approach taken from the process put forward by the *Analyses de Cycle de Vie* (LifeCycle Analyses – LCA). The publication, between 2010 and 2012, of the European norms NF EN 15643 [AFN 10] also moves in this direction, proposing a system for evaluating buildings' contribution to sustainable development, based on an LCA approach. Design and diagnostics tools have been developed in the same vein by the CSTB (such as Team, Cocon or Elodie).

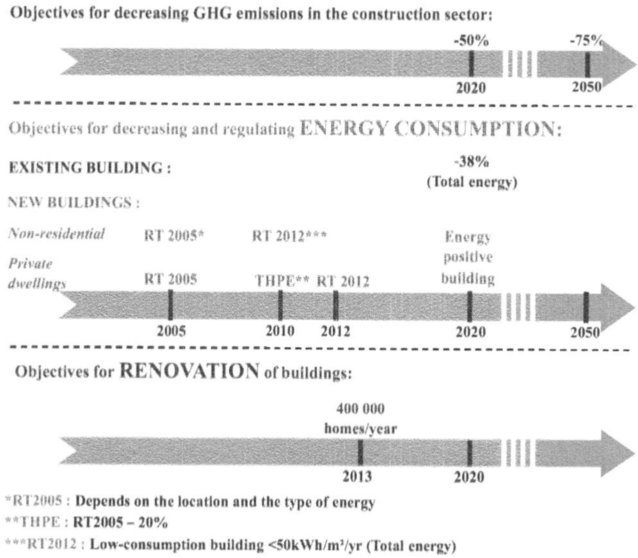

Figure 1.1. *Construction plan from the environmental conference, relating to energy consumption and reduction of greenhouse gas emissions by the construction sector*

Note also that in the context of the project's dwellings plan, the R&D program PREBAT[5] conducted a very exhaustive study which offers an overview of the international initiatives to reduce the consumption of construction [PRE 07].

1.2.2. *Limitations of the normative and regulatory framework*

As we saw above with regard to European and French legislation, the issue of a building's ecological impacts is, for now, centered on the period of time for which it functions. *The impacts of the material or the waste products (from the building site, or from demolition) do not enter into the debate.* Approaches do exist which encapsulate more criteria than this, but on a voluntary basis.

5 PREBAT: Programme de Recherche et d'Expérimentation sur l'énergie dans le BATiment (Program of Research and Experimentation on Energy in Buildings): www.Prebat.net (03/09/10)

Restricting the impacts on the external environment	Creating a satisfactory interior environment
ECO-CONSTRUCTION	COMFORT
1. Relation of the buildings with their immediate environment	8. Hygrothermic comfort
2. Integrated choice of construction procedures and products	9. Acoustic comfort
3. Low nuisance building sites	10. Visual comfort
	11. Olfactory comfort
ECO-MANAGEMENT	HEALTH
4. Energy management	12. Sanitary quality of the spaces
5. Water management	13. Sanitary quality of the air
6. Waste product management	14. Sanitary quality of the water
7. Maintenance management	

Choice of construction materials *crucially* important

Choice of construction materials important

Table 1.1. *The 14 targets of the HQE approach (High Environmental Quality) [HQE 01] and the importance of the construction materials in this approach*

In France, many associations bringing together professionals in construction, public institutions or regional bodies have given rise to labels certifying a building's environmental quality in accordance with a *global standard*. Such is the case, in particular, with the HQE approach[6] (see Table 1.1), which involves three applications: tertiary building, individual home or collective/group accommodation. Other certifications exist in France and offer similar approaches, such as *Habitat et Environnement* (Habitat and Environment)·[7] All are based on a top-down approach – i.e. working from the project's aims to the materials to be used [JUL 09]. Yet the HQE approach [HQE 01], while it is more complete than the regulatory framework detailed in the previous point, does not include a target as regards the choice of primary materials and their sustainability.

6 HQE: Haute Qualité Environnementale (High Environmental Quality), www.assohqe.org (02/09/10)

7 http://www.cerqual.fr/cerqual/habitat_environnement (02/09/10)

1.3. The materials: an increasingly crucial element

1.3.1. *Role of the materials in energy consumption*

The distribution of the energy consumption between the heating post and that devoted to the materials and construction of a conventional house (see Figure 1.2) reveals the greater relative weight that this second post as the building plan is gradually applied [PEU 08]. Indeed, we see a rise from 8% to 60% in the relative proportion of the materials and the construction in the energy consumption for a building with a lifespan of 100 years, while the consumption from heating plummets from 200 kWh/m²/yr to 15 kWh/m²/yr (RT 2012).

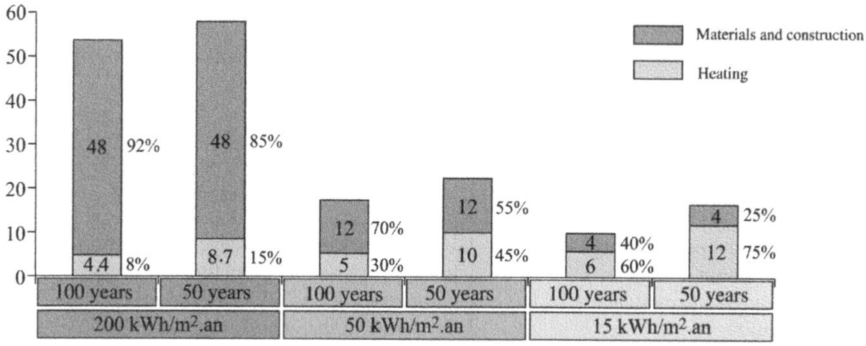

Figure 1.2. *Distribution of the energy consumption of buildings, due to heating and to materials and construction respectively, depending on their energy performances and their lifespan [MAG 10]*

In addition, the durability of the construction material used is also of crucial importance. If the lifespan drops from 100 to 50 years for a building that consumes 15 kWh/m²/yr, the relative proportion of the materials and construction in the building's energy consumption jumps from 60 to 75%. Thus, we can understand the important issues which will affect the construction materials market in a not-too-distant future, be it in terms of new construction or renovation of old buildings.

1.3.2. *What is a low-environmental-impact material?*

The points touched upon in the previous section outline a new set of specifications as regards the elaboration and choice of construction materials. Thus, we can define a new category: low-environmental-impact materials, or "eco-materials". At present, there are no clearly-defined criteria and even fewer norms that can be used to classify a material as an eco-material [PEU 08; ESC 06]. In fact, in addition to the technical characteristics typically required of a product for the home, we look for it to satisfy the specifications of sustainable construction in terms of respect for the environment, the comfort of the dwelling and the health of the users (see Table 1.2).

1.3.3. *Constantly-changing regulations*

The new criteria defined above are all points which need to be taken into account when creating new materials in the laboratory or research center, and when choosing a material for a particular construction project. Currently, lifecycle analysis (LCA) is the most widely-used method for approximating a product's environmental impact. This tool was standardized in the ISO 14040 series [AFN 06] on environmental management. There are other methods to define the impact of a manufactured product – particularly the carbon balance put in place by ADEME – but the application of these methods is limited.

Typical technical criteria	**Environmental quality**
Mechanical performances Architectural quality Fire- and heat-resistance Durability Easy maintenance Affordability	Use of renewable and recyclable primary materials (wood, plant matter, etc.); embodied energy; GHG emissions "from the cradle to the grave"; respect of the environment when exploiting the primary materials (water, landscape, biodiversity, etc.).
Sanitary quality	**Comfort of the dwelling and of implementation**
Limited VOC emissions in the dwelling and during construction. Limited emissions of volatile minerals (fibers, dust, etc.), particularly during construction.	Thermal qualities: thermal conductivity, inertia, thermal effusivity. Hygroscopic qualities: vapor permeability, diffusivity in gaseous and liquid phases, isotherms of absorption/desorption. Acoustic qualities: sound absorption coefficient, sound reduction index.

Table 1.2. *Qualities sought when creating an eco-material for construction. Overview of different definitions [AMI 09; MAG 10]*

In France, the implementation of FDESs[8] (type III environmental declaration) by the AIMCC[9] requires material manufacturers to carry out a LCA on their products, based on the aforementioned standard and to publish information about the hygiene, safety and health aspects of their products. They are the subject of the French norm NF P 01 010, and confer ISO 14025 certification [AFN 10]. Currently, FDES are not a legal obligation. For the time being, the introduction in 2008 of the European regulation REACH (Registration, Evaluation and Authorization of Chemicals – [REA 06]) is the main regulation which directly or indirectly affects construction materials. A more targeted legislative move was the French government's decree no 2011-321 of 23 March 2011, relating to the "labeling of products for construction or wall/floor coating and paints and varnishes as regards their emissions of volatile pollutants". This legislation requires manufacturers, importers, distributors of construction and decorating products, building contractors and buyers of such products to "indicate on a label, placed on the product or on its packaging, the characteristics of its volatile pollutant emissions once it is used".

A non-obligatory form of labeling (Type I environmental declaration) on construction materials is also available, but again its use is limited. The ACERMI[10] label [ACE 09] also relates to industrial insulating products delivered in rolls or in slabs. Products commercialized on the European market can also be granted the Natureplus[11] label, which is a reference point in terms of requirements because it relates only to products made up of at least 85% of renewable primary materials or materials of practically-inexhaustible mineral origin. To aid in the making of technical choices, the databases of software packages such as Cocon provide technical information above and beyond that provided on the FDESs for many construction materials (thermal conductivity, specific heat, resistance to water vapor diffusion, embodied energy and carbon impact, on the basis of a LCA).

1.4. The specific case of concretes made from lignocellular particles

The marriage of vegetable or animal materials and mineral binders is by no means a recent phenomenon. There are many vestiges in the past that bear witness to the durability of this type of mixture. The centuries, or even millennia,

---

8 FDES: *Fiches de Déclaration Environnementales et Sanitaires* (Environmental and Sanitary Declaration Certificates), fdes.fr (31/08/10)

9 AIMCC: *Association des Industries de Produits de construction* (Association of Construction Product Industries), www.aimcc.org (31/08/10)

10 ACERMI: *Association pour la CERtification des Matériaux Isolants* (Association for the Certification of Insulating Materials), http://acermi.cstb.fr (02/09/10)

11 NaturePlus: http://www.natureplus.org (02/09/10)

offer us an observation: it is possible to locally create construction materials that will stand the test of time.

Figure 1.3. *Ksar d'Aït Ben Haddou, Morocco, 13th Century. A constructive mix combining stones, earth and lignocellular vegetable matter*

Currently, nearly 60% of dwellings in the world are built of earth or a mixture and earth and plant matter [HEG 10]. Such is the case of the traditional Berber habitat, built of rammed earth, in southern Morocco (see Figure 1.3).

In this particular domain, countries' industrialization has caused locally-created materials to be gradually replaced with industrial materials. Yet, if we look at the goal set by the *plan bâtiment* of 400,000 renovations a year, two million tons of straw a year would serve this purpose, which is around 4% of France's annual production [AMI 09].

1.4.1. *Development of agro-concretes in the context of France*

It is certain that with the now-omnipresent "sustainable development", the use of so-called renewable materials (if they are managed correctly) and local materials presents a growing advantage in the world of construction materials, in France [ALC 07] and in the rest of the world [OEC 04]. Around the main markets generated by a cereal or petroleum culture, there are a great many secondary markets springing up, which will facilitate as complete a value creation as possible. Such is the case, in particular, for hemp [BOU 06], for which the areas of the market are as varied as the automobile industry, for the fibers, foodstuffs for the grain or indeed the wood of the stem (known as shiv) for construction. The quantities and the sources available are abundant [FRD 11]. *Hence, vegetable biomass has a bright future.*

1.4.1.1. *Environmental and socio-economic issues*

In the area of construction, France finds itself facing a fairly complex problem. For decades, in response to the poorly-constructed buildings of the post-war period [AMI 09], architects have congregated around a new design of a building: bioclimatic design. They were the first to reuse local materials to erect their buildings [PEU 08]. This new school of thought became progressively more popular in the circles of ecologists, and gave rise to numerous associations of self-constructors and a new generation of artisans [LAU 07]. For them, the use of eco-materials coupled with an eco-construction approach to building must, first and foremost, be a factor in local development and social links [AMI 10]. Thus, when these topics are taken up by industrial and scientific actors, they are understandably uncertain.

> "Eco-materials are usually produced from local resources, employing a local workforce, mobilizing local skills and *savoir-faire*, integrating themselves into the local art of construction and stimulating an economy that protects workers' social rights and redistributes the wealth that it creates. This approach flies in the face of industrial production of standardized materials for standard homes, which form a uniform landscape which does not adapt to regional architectural or climatic peculiarities" [AMI 09].

1.4.1.2. *The pitfall of novelty: technical opinions*

While this approach is possible in certain European countries – particularly the Scandinavian countries, where the certification system is more favorable [PRE 07] – it remains complicated in France. The conventional construction materials, such as concrete or brick, have clearly-defined techniques for their use which are set down in *Documents Techniques Unifiés* (DTUs – Unified Technical Documents), which often fit in well with a normative framework. Thus, it is easy to commercialize a product whose technical framework for use is the topic of one or more DTUs.

In order to have a hope of being used, innovative construction materials, such as lignocellular concretes, must be subjected to a more complex certification approach. This is the condition *sine qua non* for master craftsmen and artisans, who are subject to a review every ten years, to be able to safeguard their work. Therefore, from the very start, manufacturers and suppliers of innovative materials are constrained to provide guarantees in terms of performance and implementation. In the case of France, it is only after obtaining a *technical assessment* about a construction product, delivered by the CSTB and renewable every five years, that insurers will usually agree to a ten-year cover. It is clear that without this insurability, project managers cannot run the risk of using an innovative material. In the case of a product or system of construction whose goal is to satisfy a local demand, the financial burden of having a technical assessment carried out often proves too great to bear.

In this particular context, the role of associations such as *Construire en Chanvre* (Build with Hemp) is very advantageous, because they lend credibility to the methods by bringing together all the actors, from the grower to the researcher. As has been the case for hemp construction methods, the work of these groups may lead to the implementation of *professional rules of execution* which constitute an intermediary step before the DTU.

1.4.1.3. *Training professionals and offering incentives to the general public*

Beyond the issues of the ten-year guarantee and the securing of technical appraisals for innovative products, there is a lack of training for professional building contractors, who continue to employ conventional solutions [LAU 07]. One of the major problems for an artisan lies in the seasonal nature of plant-based concretes, which cannot be put in place in wintry conditions. Hence, year-round outdoor use requires industrial molding which enables the concretes to be dried beforehand. This type of molding could also be a significant limiting factor as regards the variability of the finished products generated by the plant-based primary material in combination with manual installation. Also, the use of plant matter from agriculture and whose yields vary from year to year is a further source of variability in terms of the users' supply. This is even more so in the absence of a specific process in charge of creating the distribution network.

Often, the development of such processes, at one time or another, requires the provision of help to private actors in order to get the market off the ground by financially guiding people's choices. Such assistance may be provided by the state itself or by local collectives at all levels (region, district, commune, etc.) [LAU 07]. The use of eco-materials may carry with it fiscal incentives such as tax credits, much like the purchase of any other insulating material [DGF 09]. Depending on the intended application, the material must satisfy the set specifications in terms of thermal conductive resistance (i.e. insulation).[12] We shall see later on (Chapter 2) that a lignocellular concrete, such as hemp concrete, possesses intrinsic properties which cannot be boiled down to its thermal performances alone. Thus, these dispositions do not favor the use of a material like hemp concrete.

It is also clear that the lack of accumulated scientific knowledge about this type of composite materials is a deterrent for decision-makers, project managers and master craftsmen. The most critical point relates to their durability and the

---

12 The thermal conductive resistance (R_{cd}) of a material denotes its capacity to resist the passage of a heat flux by conduction for a given thickness. It is expressed in $m^2.°C.W^{-1}$.

sustainability of the procedures for their implementation. Although archaeological observations stand in evidence of the durability of earth/organic mixtures, such as wattle and daub, (see section 1.2.1), they merely attest to their potential in terms of durability.

1.5. What does the term "Agro-concrete" mean?

1.5.1. *General definition*

A concrete in the conventional sense of the word consists of a heterogeneous mix between a mineral binder and aggregates (also mineral in origin) of graduated dimensions. Similarly, that which we define as *agro-concrete* will therefore consist of:

> "A mix between aggregates from lignocellular plant matter coming directly or indirectly from agriculture or forestry, which form the bulk of the volume, and a mineral binder".

This definition will not cover mixtures including:

– a low proportion of lignocellular aggregates;

– lignocellular plant fibers to reinforce conventional concrete.

Indeed, many projects aim to create construction materials using one or more forms of lignocellular matter as a *reinforcement to the structure* rather than as a lightweight aggregate with an insulating purpose. The materials used are generally fibers which serve to improve the traction resistance, ductility and post-fracture behavior of composite concretes made in this way. The scientific study of fiber-reinforced concrete (FRC) created from mineral or synthetic fibers began at the start of the 20th Century [BRA 08]. More recently, projects have been carried out to enhance the value of organic fibers to substitute industrial fibers. They are drawn from various sources, such as wood [COU 05; TON 10], coconut [GHA 99], sisal [LI 00; TOL 03], palm [KRI 05], bamboo [SUD 06], bagasse [AGG 95; BIL 08] or indeed diss [MER 07]. It is interesting to note that countries such as Brazil which have an exceptional range of flora have a wide range of fibers to experiment with, and research in this domain is very active [SAV 00; AGO 05].

1.5.2. *Lignocellular resources*

Many lignocellular substances have been the subject of research, with the aim of integrating them with mineral binders as a *lightweight aggregate.* Table 1.3 offers an overview, by country and by material, of the research carried out hitherto on what has been defined as agro-concrete. The table only includes studies performed on lightweight aggregates, which yield concretes with a dry density of less than 1000 kg/m^3. It is interesting to note that France, where agriculture and forestry are prevalent, finances research on numerous resources such as hemp, flax, wood, sunflower or beetroot. For concretes reinforced with natural fibers, the first scientific studies date from the turn of the century [ARN 00; BOU 98], although experiments had been carried out previously. The trajectory followed by the development of these materials is said to be *bottom-up*, i.e. from the building site to the laboratory.

Plant	Valuable material	Source(s)	Binder(s) used	Countries	References
Hemp	Hemp shiv	Agricultural co-product	Tradical® PF70, hydraulic lime, methacholine/ lime mix	France	[ARN 00; CER 05; CHA 08; ELF 08; COL 08; SAM 08; NGU 09; MAG 10; ARN 12]
				Elsewhere	[EVR 08; BUT 04; EIR 06; DEB 09; HIR 10]
Flax	Shiv, tow	Agricultural co-product	Portland cement, Cement + Sucrose	France	[KHA 08; AAM 08; ELH 10]
Wood (all types)	Sawdust, shavings	Sawmill waste	Portland cement, Cement/clay mix	France	[BOU 98; GOV 04; COA 06]
			Portland cement	Elsewhere	[SEM 02; TUR 07]
Sunflower	Stem	Agricultural by-product	Methacholine/ lime mix	France	[MAG 10; NOZ 12]
Beetroot	Dried beetroot pulp	Food industry waste	Portland cement	France	[MON 11]
Coconut	Shell	Food industry waste	Portland cement	Elsewhere	[KHE 01]
Durian	Shell	Food industry waste	Portland cement	Elsewhere	[KHE 01; CHA 05]
Cork	Wood	Industrial waste	Portland cement	Elsewhere	[KAR 06]
Miscanthus	Stem	Co-product (ethanol)	Portland cement	France	[LEN 11]

Table 1.3. *Overview of research into materials mixing mineral binders and lignocellular products for the making of lightweight concretes*

All these resources have a common point: they are either co-products or by-products, or industrial waste. This is not an exhaustive list, because industrial projects are conducted with other plant matter such as miscanthus or wheat straw [FRD 11]. These materials, readily and cheaply available, therefore logically hold a growing interest for many uses – particularly for creating agro-concretes. Their increasing value also facilitates a reduction of the environmental impacts as opposed to traditional building insulation systems. Indeed, these materials are renewable, biodegradable, neutral in terms of GHG emissions and require little energy to be transformed [BAL 05]. However, not all of them can be used, and it is necessary to define a set of specifications to guide their selection. A recent study carried out jointly between ADEME and FRD[13] makes the point about sources and worldwide availability of plant fibers [FRD 11].

1.5.3. *General characteristics of lignocellular agro-resources*

1.5.3.1. *Chemical composition*

The resources chosen as a matter of preference to create agro-concretes are said to be lignocellular. The etymology of the word reflects their composition, primarily, of *cellulose* and *lignin*, which are the two most common compounds in plant biomass (≈ 70%). Two other major molecular compounds are to be found in the stems of these plants: *hemicelluloses* and *pectins*. All these substances are made up of organic macromolecular chains which constitute polysaccharides. A few minor elements such as waxes and proteins are also present.

Cellulose is a polymer of glucose which is one of the main components of the plant cell wall. This biopolymer is responsible for most of the mechanical resistance in plants which have no secondary tissues. Its organization, which is mainly crystalline, renders cellulose *insoluble in most solvents, and particularly water*, although the compound is highly hydrophilic.

Lignins manifest themselves in the form of three-dimensional polymers. Their complex structure varies from one species to another, but so too do morphological elements (fibers, vessels, etc.). They lend *rigidity and impermeability* to plants containing them, as lignins are highly hydrophobic compounds. Finally, they are involved in the cohesion of the fibers in the lignocellular woody parts of the xylem and provide them with significant compression resistance.

---

13 FRD: Fibres Recherche et Développement (Fiber R&D).

Hemicelluloses are shorter-chain polysaccharides than cellulose, with an amorphous structure. They are hydrophilic, and notably they are *able to swell when they come into contact with water*. It is this swelling that means wood cannot be relied upon to retain the same size. In addition, hemicelluloses are water-soluble and can be removed from the wall, notably by soaking in an alkaline substance.

Pectins are acidic polysaccharides. They are present in large numbers in the middle lamella, where they hold adjoining cells together. Similarly to hemicelluloses, pectin is a water-soluble compound. It should also be noted that the composition and structure of the hemicelluloses and pectins varies from one plant species to another, and therefore is very complex to apprehend fully.

1.5.3.2. *Structural organization*

Lignocellular plants can be described at three levels: *macroscopic, microscopic and molecular*. In the case of stems or stalks, which represent the greatest potential for use in agro-concretes, the macroscopic structure can be generally characterized as shown in Figure 1.4. The stems of lignocellular plants may be of two types: monocotyledons or dicotyledons. In the case of perennial plants such as those used for straw, these two types of stems have the same composition but are organized differently. It is only at the start of the plant's second year that secondary tissues begin to be formed, constituting the age rings [WIE 05].

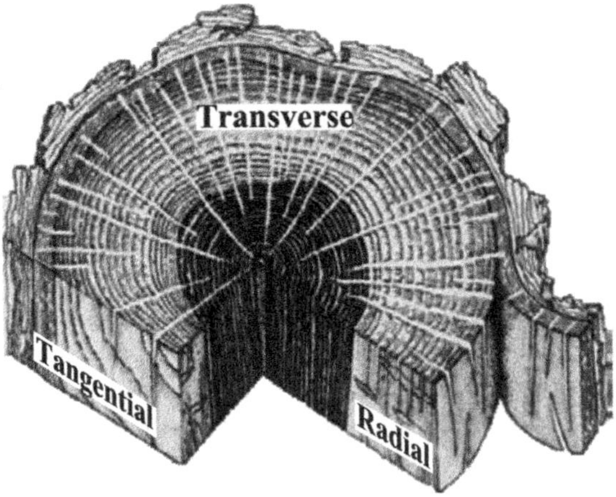

Figure 1.4. *Planes of a cross-section characteristic of lignocellular stems [RAV 70]*

Depending on the resource available and the co-valuations, the stem may be valued either partly or entirely for the making of agro-concretes. The use of the stem in its entirety, of course, increases the complexity the phenomena of interaction between the plant matter and any binder used. Indeed, the chemical composition of the different plant structures is subject to variation. In fact, the carminot-green de Mirande coloration on the cross-section shown in Figure 1.5 clearly distinguishes the zones with a high lignin content, with the other areas being richer in cellulose [GUY 10].

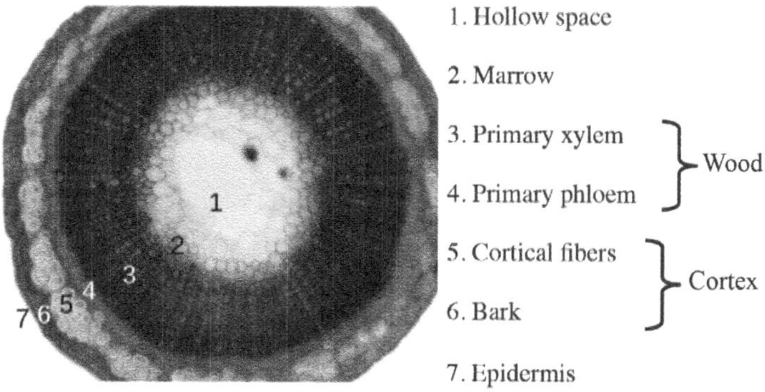

Figure 1.5. *Transversal cross-section of a young dicot stem (flax) colored with carminot-vert de Mirande and detail on the associated structures [GUY 10]*

On the microscopic scale, it is interesting to note the multiple porosity that characterizes these plants, the reason for which is to cater for the plant's physiological needs. For instance, the xylem and the phloem serve respectively for the circulation of raw sap and the elaborated sap [RAV 70].

A three-dimensional representation of the xylem of soft woods is given by River *et al.* [RIV 91] (see Figure 1.6). It stresses the organization into cells, mainly oriented in a longitudinal direction. Some cells do not have an end wall, and may constitute veritable tubes or tracheids that can transport sap (Figure 1.6). The lateral walls are dotted with punctuations: openings integrated into the walls, whose role is to transfer fluids between the different cells.

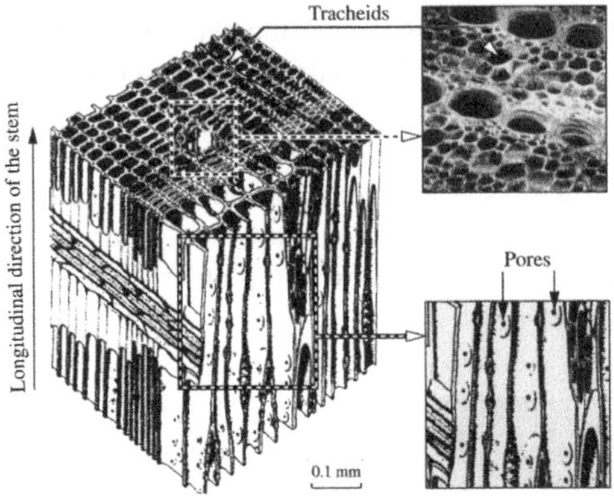

Figure 1.6. *Three-dimensional representation of the structure of a soft wood [RIV 91]*

1.5.3.3. *Cellular organization*

On the cellular scale, lignocellular plants are made up of long cells with complex organization, with the exception of cellular marrow with a beehive structure (see Figure 1.7). The middle lamella is made mainly of pectins which play the part of binders between the cells. The primary walls are made of cellulose, surrounded by microfibrils of hemicelluloses, and are linked by a web of pectins and proteins [SED 07]. The secondary walls have the same constituents, but also contain lignin, which intervenes and blocks cellular expansion. This process is carried out in the case of trees when a new layer of soft wood is laid down to replace the previous one in the spring. However, in annual lignocellular plants, the secondary wall does not have time to form. Hence, the role of the cell is to channel nourishing fluids (sap, water, etc.) through a conduit: the lumen.

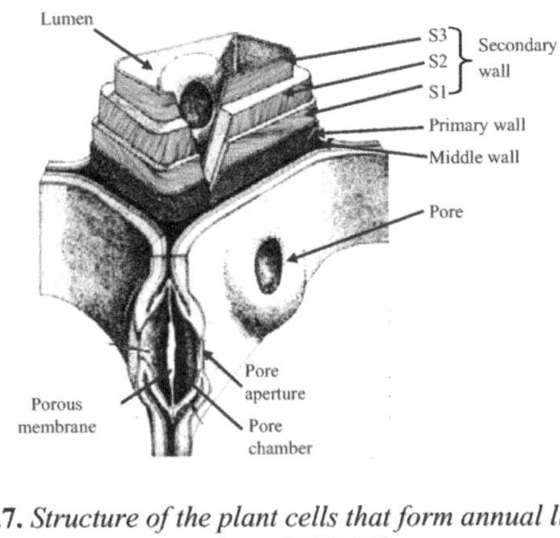

Figure 1.7. *Structure of the plant cells that form annual lignocellular plants [WIE 05]*

The different structure scales of plant matter described (see Table 1.4) are all parameters which need to be taken into account when making a technical choice [GAR 08]. The size of the structures, their organization and their molecular make-up will indeed exert an influence on the behavior of the plant in the alkaline milieu formed by the hydrated mineral binder. These parameters will be detailed more completely and comparatively between the plants used during the discussion later on in this book.

Type of structure	Characteristic dimension	Source
Diameter of pores in the primary wall	1.7–2 nm	[GAR 08]
Diameter of an open pore	200 nm	[GAR 08]
Diameter of the lumen of the tracheids	4–25 μm	[GAR 08]

Table 1.4. *Characteristic dimensions of the structures of stems of lignocellular plants*

1.6. Conclusions

The objectives are unambiguous, but there are genuine economic and scientific hurdles to overcome before we have at our disposal new products – often composites – that satisfy the same technical criteria but have enhanced qualities in terms of the environment, sanitation and comfort [ESC 06]. It is therefore necessary to create certifications and labeling to incite manufacturers to further develop their offering in terms of eco-materials, and thereby encourage construction professionals to use them. Undoubtedly one of the most promising avenues relates to lightweight concretes that combine mineral binders and plant particles.

1.7. Bibliography

[AAM 08] AAMR-DAYA E., LANGLET T., BENAZZOUK A., and QUENEUDEC M., "Feasibility study of lightweight cement composite containing flax by-product particles: Physico-mechanical properties", *Cement and Concrete Composites*, vol. 30, p. 957–963 Nov. 2008.

[ACE 09] ACERMI, *Règlement technique de la certification des matériaux et produits destinés à l'isolation thermique des bâtiments,* ACERMI, 2009.

[AGG 95] AGGARWAL L.K., "Bagasse-reinforced cement composites", *Cement and Concrete Composites*, vol. 17, p. 107–112, 1995.

[AGO 05] AGOPYAN V., SAVASTANO J., JOHN V. and CINCOTTO M., "Developments on vegetable fibre-cement based materials in Sao Paulo, Brazil: an overview", *Cement and Concrete Composites*, vol. 27, p. 527–536, May 2005.

[AFN 06] AFNOR, Norme NF EN ISO 14040:2006 - Management environnemental - Analyse du cycle de vie - Principes et cadre, AFNOR, 2006.

[AFN 10] AFNOR, Norme NF EN ISO 14025 : 2010 - Marquages et déclarations environnementaux - Déclarations environnementales de type III - Principes et modes opératoires, AFNOR, 2010.

[ALC 07] CABINET ALCIMED, Marché actuel des bioproduits industriels et des biocarburants & évolutions prévisibles à échéance 2015/2030, ADEME, 2007.

[AMI 09] CONTEVILLE L., DEN HARTIGH C., Les écomatériaux en France : État des lieux et enjeux dans la rénovation thermique des logements, Les Amis de la Terre, 2009.

[AMI 10] Amis de la Terre, Développer les filières courtes d'écomatériaux. Guide à destination des collectivités territoriales, Les Amis de la Terre, 2010.

[ARN 00] ARNAUD L., "Mechanical and thermal properties of hemp mortars and wools: Experimental and theoretical approaches", *Proceedings of 3rd International Symposium on Bioresource, Hemp and Other Fiber Crops*, Wolfsburg, Germany, 2000.

[ARN 12] ARNAUD L. and GOURLAY E., "Experimental study of parameters influencing mechanical properties of hemp concretes", *Construction Building and Materials,* vol. 28, p. 50–56, 2012.

[BAL 05] BALEY C., "Fibres naturelles de renfort pour matériaux composites", *Techniques de l'ingénieur*, April 2005.

[BIL 08] BILBA K. and ARSENE M., "Silane treatment of bagasse fiber for reinforcement of cementitious composites", *Composites Part A: Applied Science and Manufacturing*, vol. 39, p. 14881495, Sep. 2008.

[BOU 98] BOUGUERRA A., LEDHEM A., DE BARQUIN F., DHEILLY R.M. and QUÉNEUDEC M., "Effect of microstructure on the mechanical and thermal properties of lightweight concrete prepared from clay, cement, and wood aggregates", *Cement and Concrete Research*, vol. 28, p. 1179–1190, August 1998.

[BOU 02] BOUSTINGORRY P., Elaboration d'un matériau composite à matrice gypse et renfort bois fragmenté - Amélioration de la résistance au vissage de produits préfabriqués en gypse, Doctoral Thesis, ENSM St Etienne – INP Grenoble, 2002.

[BOU 06] BOULOC P., ALLEGRET S., ARNAUD L. and COLLECTIVE, *Le chanvre industriel: Production et utilisations*, Editions France Agricole, 2006.

[BUT 04] BUTSCHI P.Y., Utilisation du chanvre pour la préfabrication d'éléments de construction, Doctoral Thesis, University of Moncton, Canada, 2004.

[BRA 08] BRANDT A.M., "Fibre reinforced cement-based (FRC) composites after over 40 years of development in building and civil engineering," *Composite Structures*, vol. 86, p. 3–9, Nov. 2008.

[CER 05] CEREZO V., Propriétés mécaniques, thermiques et acoustiques d'un matériau à base de particules végétales, Doctoral Thesis, INSA de Lyon, 2005.

[CGD 09] COMMISSARIAT GÉNÉRAL AU DÉVELOPPEMENT DURABLE, Chiffres clés de l'énergie, Ministère de l'Ecologie, de l'Energie, du Développement durable et de la Mer, 2009.

[CGD 10] COMMISSARIAT GÉNÉRAL AU DÉVELOPPEMENT DURABLE, Chiffres clés du climat, Ministère de l'Ecologie, de l'Energie, du Développement durable et de la Mer, 2010.

[CHA 05] CHAROENVAI S., KHEDARI J., HIRUNLABH J., ASASUTJARIT C., ZEGHMATI B., QUENARD D. and PRATINTONG N., "Heat and moisture transport in durian fiber based lightweight construction materials", *Solar Energy*, vol. 78, p. 543–553, April 2005.

[CHA 08] CHAMOIN J., COLLET F., PRETOT S., "Optimisation de bétons de chanvre projetés et moulés – Caractérisation du matériau de référence", *Actes du 26ème congrès de l'AUGC*, Nancy, 4–6 June 2008.

[COA 06] COATANLEM P., JAUBERTHIE R. and RENDELL F., "Lightweight wood chipping concrete durability", *Construction and Building Materials*, vol. 20, p. 776–781, Nov. 2006.

[COL 08] COLLET F., BART M., SERRES L. and MIRIEL J., "Porous structure and water vapour sorption of hemp-based materials," *Construction and Building Materials*, vol. 22, p. 1271–1280, June 2008.

[COU 05] COUTTS R.S., "A review of Australian research into natural fibre cement composites", *Cement and Concrete Composites*, vol. 27, p. 518–526, May 2005.

[DEB 09] DE BRUIJN P.B., JEPPSSON K., SANDIN K. and NILSSON C., "Mechanical properties of lime-hemp concrete containing shives and fibres," *Biosystems Engineering*, vol. 103, p. 474–479, August 2009.

[DGF 09] DIRECTION GÉNÉRALE DES FINANCES PUBLIQUES, Crédit d'impôt pour dépenses d'équipement de l'habitation principale en faveur des économies d'énergie et du développement durable, Bulletin Officiel des Impôts, 2009.

[EIR 06] EIRES R., NUNES J.P., FANGUEIRO R., JALALI S. and CAMÕES A., "New eco-friendly hybrid composite materials for civil construction", *Proceedings of the 12th European Conference on Composite Materials*, Biarritz, 29 August-1 September 2006.

[ELF 08] ELFORDY S., LUCAS F., TANCRET F., SCUDELLER Y. and GOUDET L., "Mechanical and thermal properties of lime and hemp concrete ("hempcrete") manufactured by a projection process", *Construction and Building Materials*, vol. 22, p. 2116–2123, Oct. 2008.

[ELH 10] EL HAJJ N., DHEILLY R.M., ABOURA Z., BENZEGGAGH M.L. and QUENEUDEC M., "Procédé de fabrication des composites 100% végétaux : Effet de la granulométrie des étoupes de lin et de l'ajout des bios liants", *Actes des 16èmes JNC*, Toulouse, 10–12 June 2009.

[ESC 06] ESCADEILLAS G., "Les éco-matériaux dans la construction : enjeux et perspectives", *Actes des 7èmes (RF)²B*, Toulouse, 19–20 June 2006.

[EVR 08] EVRARD A., Transient hygrothermal behaviour of lime-hemp materials, Doctoral Thesis, UCL, Belgium, 2008.

[FFB 09] FFB, Construire en chanvre : Règles professionnelles d'exécution, Fédération Française du Bâtiment, Collection recherche développement métier, 2009.

[FRD 11] FRD, Evaluation de la disponibilité et de l'accessibilité de fibres végétales à usage matériau en France, ADEME, March 2011, 84p.

[GAR 08] GARDNER D.J., "Adhesion mechanisms of durable wood adhesive bonds", *Characterization of the Cellulosic Cell Wall*, Blackwell Publishing Professional, pp. 254–265, 2008.

[GHA 99] GHAVAMI K., TOLEDO FILHO R.D. and BARBOSA N.P., "Behaviour of composite soil reinforced with natural fibres", *Cement and Concrete Composites*, vol. 21, 1999, p. 39–48.

[GOV 04] GOVIN A., Aspects physico-chimiques de l'interaction bois-béton - Modification de l'hydratation du ciment par le bois, Doctoral thesis, University de Saint-Etienne, 2004.

[GRE 09] Ministère de l'Ecologie, de l'Energie, du Développement Durable et de la Mer, Dossier de Presse, Conférence sur les avancées du plan Bâtiment du grenelle de l'environnement, 28 October 2009.

[GUY 10] GUYON E., PEDREGOSA A. and SALVIAT B., *Matière et matériaux – De quoi est fait le monde ?*, Editions Belin, 2010, 336p.

[HEG 10] HEGGER M., AUCH-SCHWELK V., FUCHS M., ROSENKRANZ T., *Construire – Atlas des matériaux*, Presses Polytechniques et Universitaires Romandes, 2010.

[HIR 10] HIRST E., WALKER P., PAINE K. and YATES T., "Characterisation of low density hemp-lime composite building materials under compression loading", *2nd Conference on Sustainable Construction Materials and Technologies (SCMT)*, Ancona, Italy, 28–30 June 2010.

[HQE 01] VALICOURT D., HETZEL J., NAGY L., NIBEL S., OLIVE G., TROADEC P., *Référentiel HQE*, Association HQE, 2001.

[IFE 08] INSTITUT FRANÇAIS DE L'ENVIRONNEMENT, Les quantités de déchets produits et éliminés en France en 2004, Dossiers de l'IFEN, 2008.

[IPC 95] INTERGOVERNMENTAL PANEL ON CLIMATE CHANGE, Second Assessment Report (SAR) – Climate Change 1995, United Nations Environmental Program, 1995.

[IPC 07] INTERGOVERNMENTAL PANEL ON CLIMATE CHANGE, Second Assessment Report (SAR) – Climate Change 2007, United Nations Environmental Program, 2007.

[JUL 09] JULLIEN A., "A propos de développement durable en génie civil – Méthodes et applications du laboratoire à l'ouvrage", *Actes du 27ème congrès de l'AUGC*, Saint-Malo, 3–5 June 2009.

[KAR 06] KARADE S.R., IRLE M. and MAHER K., "Influence of granule properties and concentration on cork-cement compatibility", *European Journal of Wood and Wood Products*, vol. 64, 2006, p. 281–286.

[KHA 08] KHAZMA M., EL HAJJ N., GOULLIEUX A., DHEILLY R.M. and QUENEUDEC M., "Influence of sucrose addition on the performance of a lignocellulosic composite with a cementious matrix", *Composites Part A: Applied Science and Manufacturing*, vol. 39, December 2008, p. 1901–1908.

[KHE 01] KHEDARI J., SUTTISONK B., PRATINTHONG N. and HIRUNLABH J., "New lightweight composite construction materials with low thermal conductivity," *Cement and Concrete Composites*, vol. 23, February 2001, p. 65–70.

[KRI 05] KRIKER A., DEBICKI G., BALI A., KHENFER M. and CHABANNET M., "Mechanical properties of date palm fibres and concrete reinforced with date palm fibres in hot-dry climate," *Cement and Concrete Composites*, vol. 27, May 2005, p. 554–564.

[LAU 07] LAUMONIER C., ROUDIL N. and COLOMBARD-PROUT M., *Le rôle des artisans dans la diffusion des meilleures techniques énergétiques possibles*, CSTB, 2007.

[LEN 11] LE NGOC HUYEN T., QUENEUDEC T'KINT M., REMOND C., CHABBERT B. and DHEILLY R.-M., "Saccharification of Miscanthus x giganteus, incorporation of lignocellulosic by-product in cementitious matrix", *Comptes Rendus Biologies,* vol. 334, 2011, p. 837.e1–837.e11.

[LI 00] LI Y., MAI Y. and YE L., "Sisal fibre and its composites: a review of recent developments", *Composites Science and Technology*, vol. 60, August 2000, p. 2037–2055.

[MAG 10] MAGNIONT C., Contribution à la formulation et à la caractérisation d'un écomatériau de construction à base d'agroressources, Doctoral Thesis, University Toulouse III, 2010.

[MER 07] MERZOUD M., Elaboration et caractérisation d'un matériau composite à base de fibres de diss dans la fabrication de la maçonnerie, Doctoral Thesis, University Badgi Mohktar Annaba, Algeria, 2007.

[MON 11] MONREAL P., MBOUMBA-MAMBOUNDOU L.B., DHEILLY R.M., and QUÉNEUDEC M., "Effects of aggregate coating on the hygral properties of lignocellulosic composites", *Cement and Concrete Composites*, vol. 33, 2011, p. 301–308.

[NOZ 12] NOZAHIC V., AMZIANE S., TORRENT G., SAÏDI K. and DE BAYNAST H., "Design of green concrete made of plant-derived aggregates and a pumice-lime binder", *Cement and Concrete Composites*, vol. 34, 2012, p. 231–241.

[OEC 04] OECD/IEA, *Joint Workshop on Sustainable Buildings: Towards Sustainable Use of Building Stock*, OECD publications, 2004.

[OEC 08] OECD, *Household Behaviour and the Environment – Reviewing the Evidence*, OECD publications, 2008.

[PEU 08] PEUPORTIER B., *Eco-conception des bâtiments et des quartiers*, Presses de l'école des Mines, Collection Sciences de la Terre, 2008.

[PRE 07] PREBAT, *Comparaison internationale bâtiment et énergie*, ADEME-PUCA-CSTB, 2007.

[RAV 70] RAVEN P.H. and CURTIS H., *Biology of Plants*, Worth Publishers, 1970, 706p.

[REA 06] REACH, Regulation (EC) No 1907/2006 of the European Parliament, Official Journal of the European Union, 2006.

[RIL 06] RILLING J., *La construction intelligente en énergie ? Sciences, Technologies et Société*, Sciences et devenir de l'homme – Les Cahiers du M.U.R.S., 2006.

[RIV 91] RIVER B., VICK C. and GILLESPIE R., "Wood as an adherend," in *Treatise on adhesion and adhesives*. vol. 7, J. D. Minford, Ed., New York, 1991.

[SAM 08] SAMRI D., Analyse physique et caractérisation hygrothermique des matériaux de construction; approche expérimentale et modélisation numérique, Doctoral Thesis, INSA, Lyon, 2008.

[SEM 02] SEMPLE K.E., CUNNINGHAM R.B. and EVANS P.D., "The suitability of five Western Australian mallee eucalypt species for wood-cement composites", *Industrial Crops and Products*, vol. 16, September 2002, p. 89–100.

[SAV 00] SAVASTANO H., WARDEN P.G. and COUTTS R.S.P., "Brazilian waste fibres as reinforcement for cement-based composites," *Cement and Concrete Composites*, vol. 22, October 2000, p. 379–384.

[SED 07] SEDAN D., PAGNOUX C., CHOTARD T., SMITH A., LEJOLLY D., GLOAGUEN V. and KRAUSZ P., "Effect of calcium rich and alkaline solutions on the chemical behaviour of hemp fibres", *Journal of Materials Science*, vol. 42, November 2007, p. 9336–9342.

[SER 09] SUSTAINABLE EUROPE RESEARCH INSTITUTE (SERI), *Overconsumption? Our Use of the World's National Resources*, SERI, GLOBAL 2000, 2009.

[SUD 06] SUDIN R. and SWAMY N., "Bamboo and wood fibre cement composites for sustainable infrastructure regeneration", *Journal of Materials Science*, vol. 41, November 2006, p. 6917–6924.

[TOL 03] TOLEDO FILHO R.D., GHAVAMI K., ENGLAND G.L. and SCRIVENER K., "Development of vegetable fibre-mortar composites of improved durability", *Cement and Concrete Composites*, vol. 25, February 2003, p. 185–196.

[TON 10] TONOLI G., SAVASTANO JR. H., FUENTE E., NEGRO C., BLANCO A. and ROCCO LAHR F., "Eucalyptus pulp fibres as alternative reinforcement to engineered cement-based composites", *Industrial Crops and Products*, vol. 31, March 2010, p. 225–232.

[TUR 07] TURGUT P., "Cement composites with limestone dust and different grades of wood sawdust," *Building and Environment*, vol. 42, November 2007, p. 3801–3807.

[UNE 07] UNITED NATIONS ENVIRONMENTAL PROGRAM (UNEP), New science and developments in our changing environment, 2009.

[UNI 87] UNITED NATIONS, Bruntland Report – Our common future, World Commission on Environment and Development, 1987.

[UNI 92] UNITED NATIONS, Framework Convention on Climate Change, 1992.

[UNI 98] UNITED NATIONS, Kyoto Protocol, 1998.

[USD 11] US DEPARTMENT OF ENERGY, Buildings Energy Data Book 2010, D&R International, 2011.

[WIE 05] WIEDENHOEFT A. and MILLER R., "Structure and Function of Wood", in *Handbook of Wood Chemistry and Wood Composites*, CRC Press, 2005.

Chapter 2

Characterization of Plant-Based Aggregates

Bio-based aggregates present characteristics which are very different from the mineral aggregates typically used in concretes, for which there are standardized tools and techniques for characterization.

In this chapter, the aggregates examined come from the stem of plants cultivated either for their fibers (hemp, flax, etc.) or for their seeds oleaginous flax, sunflower, etc. In all cases, our aim is to enhance the value, as aggregates for construction materials, of co-products from the stem which have, hitherto, hardly been used (if at all). Cultivation of these plants solely for the purposes of production of aggregates would not be advisable either from an economic or an environmental point of view: the cost of such materials would prove significant in relation to, for example, mineral aggregates – the price of which is steadily increasing as resources become less readily available – and their production would mobilize agricultural land for non-food purposes. In the case of fibrous plants, it should be noted that this latter point is compensated by the fact that cultivation of such plants contributes to balanced land management and constitutes a beneficial component in cereal crop rotation.

Owing to the structure of the stem of the plant they are made from, such aggregates are generally malleable, elongated and highly porous with an apparent low density. Within the finished product, they do not play the part of a rigid skeleton, as do mineral aggregates in hydraulic concretes, but are instead very flexible, and for large quantities, their compactness in the material depends on the

---

Chapter written by Vincent PICANDET.

compacting performed during the casting stage. They can also absorb large amounts of water, and by water competition, prevent potential hydraulic reactions of the binder.

Hence, the characterization of these aggregates, which is crucial to a proper awareness of the quality of the materials in which they are incorporated, requires adaptations to be made to the techniques usually employed for mineral aggregates, or the devising of new characterization procedures.

Following a brief description of the microstructure of these particles, based on current studies aimed at enhancing the value of hemp shiv, flax shiv and sunflower stems, a study relating to the characterization of particle size distribution is detailed in the case of hemp shiv. Measurements of the bulk density and compressibility of the hemp shiv are also presented. Finally, the water-absorption capacity of these aggregates is illustrated by a few very simple tests.

2.1. Microstructure of the shiv particles

2.1.1. *Structure of the stem of fibrous plants*

In a transversal cross-section, going from the outside toward the center of the stem, which often forms a hollow cylinder (see Figures 2.1 and 2.2), the different cellular tissues making up the plant are composed as follows [CRO 05; BOU 06]:

– *Epidermis*: this constitutes the stem's protective layer, and is the area in which exchanges with the surrounding environment take place.

– *Primary fibers*: these are associated with the primary phloems, running from the primary meristem. The primary fibers are distinguished from the wood fibers by their length, a very slender cell wall and a particular chemical composition.

– *Phloem*: this is the tissue that channels elaborated sap, which is a solution rich in carbohydrates such as saccharose, sorbital and mannitol.

– *Secondary fibers*: these are generated from the cambium.

– *Cambium*: this enables the stem to grow thicker, which is referred to as secondary growth. The cambium originates from the procambium – a tissue derived from the primary meristems (creating the primary vascular bundle). Cambial activity, therefore, is responsible for the production of the secondary xylem which is the main element in the central part of the stem.

– *Meristem*: this is a biological tissue comprising non-differentiated (or slightly differentiated) cells forming an area of growth where cell division takes place. We usually distinguish the primary meristems, which ensure the growth of the stem in terms of length, from the leaves or roots and the secondary meristems, which are responsible for the transversal growth of the organs of certain plants.

– *Xylem*: this ensures the plant's uprightness, and the transport of minerals – functions performed respectively by the fibers and vessels.

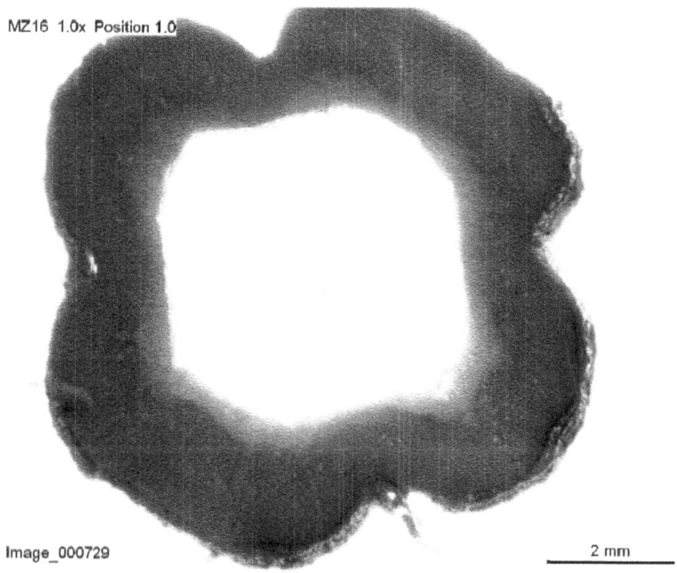

Figure 2.1. *Transversal cross-section of a hemp stem halfway up*

In the case of hemp, the part assimilated to the cortex containing the epidermis and the fibers represents barely 10% of the cross-section of the stem. Conversely, the secondary xylem, assimilated to the woody part, represents over 85% of the cross-section of the stem [FID 08]. Hemp shiv comes from this part of the stem.

The inner layer of the hollow stem is made of a material that is more translucent than the woody part. These cells located at the heart of the stem constitute the primary xylem – the site of cell multiplication in the process of the plant's growth.

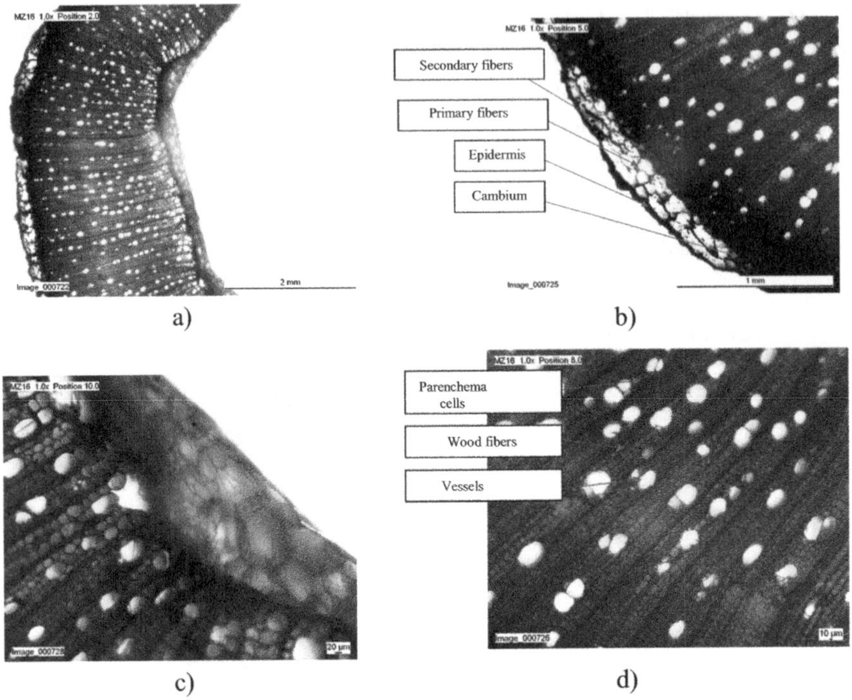

Figure 2.2. *View, under a trinocular microscope, of the transversal cross-section of thin slivers of hemp stem*

2.1.2. *SEM observation of hemp shiv particles*

Cross-sections were taken of the woody part of the stem of a hemp plant and observed under a SEM (Scanning Electron Microscope) at various levels of magnification, both along and across the stem (see Figure 2.3). These observations confirm the highly porous nature of this material, which accounts for its excellent capacity to absorb and retain water. It is made up of capillaries, formed by the cell walls and oriented longitudinally (i.e. in the direction of the stem). The width of these capillaries is variable, but on the transversal views they appear to be connected, end-to-end, to one another.

The process of defibration produces hemp shiv particles from the woody part of the stem, named as shiv, which are elongated along its axis. The porosity of these particles is also mainly oriented along the same axis. These pores have a diameter that essentially varies between 10 and 50 µm and a length of around 80 µm (see Figure 2.3d).

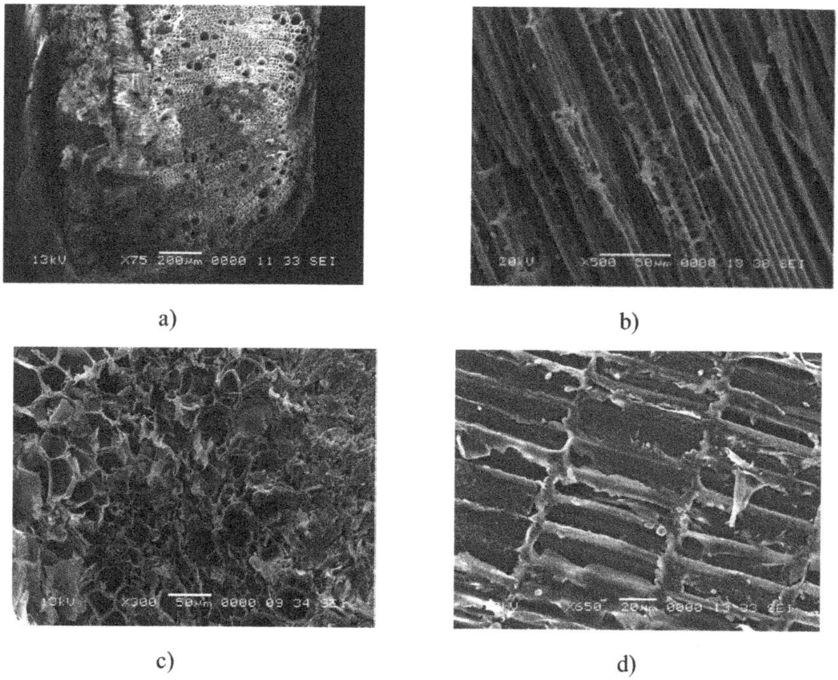

Figure 2.3. *SEM observations of hemp shiv: transversal (a and c) and longitudinal (b and d) views*

2.1.3. *Chemistry of the cell walls*

Plant aggregates are made up of cell walls – remnants of the cells that make up the plant from which they are taken. These lignocellular walls are composed primarily of cellulose, hemi-celluloses and lignin in varying proportions depending on the species of plant and the position or function of the cell within the plant's structure. For instance, a fiber is created by a cell which is elongated and whose metabolism has focused entirely on the creation of the cell wall.

Cell walls also contain various aromatic compounds associated with lignins, pectins, waxes, fats or lipids and ash or mineral compounds which can be extracted by calcination [AKI 10]. Cellulose, hemi-celluloses and lignin, in that order, are the three most abundant types of natural polymers [BUR 08].

Table 2.1 gives a non-exhaustive overview of some values found in the existing body of literature for the distributions of the primary constituents of the cell walls in plants likely to produce aggregates for use in construction materials, which have now been studied to a relatively advanced level. This category includes two fibrous plants – hemp and flax – and an oily plant – sunflower – for which the data were obtained by Vincent Nozahic [NOZ 12].

2.1.3.1. *Cellulose*

Cellulose is primarily a linear polymer of glucose [NEL 00]. The way in which glucose is linked or arranged to form this linear polymer determines the properties of the particular cellulose. Generally, the glucose can be arranged in a crystalline manner, giving rise to a stable, hydrophobic polymer with excellent mechanical stress resistance. Cellulose is present in the cell walls in the form of microfibrils (between 2 and 20 nm in diameter and 100–400 nm in length), constituting a mechanically-resistant linear structure [AKI 10]. A number of different models of the arrangement of these microfibrils in the cellulosic fiber may be envisaged, while in other parts of the cell wall, cellulose may also be present in a less ordered form than the crystalline state mentioned above, with significant differences in terms of physical properties and functions.

Mass %	Cellulose	Hemi-celluloses	Lignins	Pectins	Ash	Wax
Hemp shiv [GAR 98]	48	12	28	6	2	4
Hemp shiv[VIG 96]	44	18	28	4	2	1
Hemp fibers [GAR 98]	55	16	4	18	4	3
Hemp fibers [VIG 96]	55	16	4	14	4	1
Hemp fibers [SED 08]	56.1	10.9	6	20.1	-	7.9
Hemp fibers [TRO 08]	58.7	14.2	6	16.8		4.3
Flax shiv [FEN 89]	53	13	24	-	>2	-
Flax shiv [BUR 07]	34.2	21.3	30.2	-	1.2	-
Flax shiv [COX 99]	46	26.2	23.1	-	3.1	
Flax fibers [SAI 02]	78	6	5	-	2	-
Sunflower stalk [JIM 93]	42.1	29.7	13.4	5.9	7.9	1
Marrow-less sunflower [JIM 93]	38.6	22.8	16.2	-	12.2	-
Marrow-less sunflower [KHR 96]	41.4	30	18.3	-	8.9	-
Sunflower marrow [YIN 07]	47.4	9.4	3.5	6	20.4	-

Table 2.1. *Mass fractions of the main categories of constituents of plant cell walls*

In the case of fibrous plants, cellulose is essentially present in the cortical fibers, the role of which is to improve the rigidity of the stem. These long fibers constitute the main added value of the plant.

2.1.3.2. *Hemicellulose*

After cellulose, hemicelluloses are the second most omnipresent carbohydrate in plant cell walls. The term "hemicelluloses" covers many different polysaccharides which are rather heterogeneous because of their origin and therefore their composition and their structure or arrangement. Hemicelluloses are not linear polymers, and in cell walls they are generally linked with pectins, aromatic compounds or cellulose [AKI 10]. They are often compared to a matrix component, which can be present in the lamellae that hold the cell walls in fibrous tissues together, and in the primary and secondary cell wall, which is finer and rich in cellulose, where they serve as the link between cellulose and lignin [FOC 92].

It should be noted that hemicelluloses are relatively hydrophilic, and contain carbohydrates which are potentially water-soluble. If included in an aqueous solution, the quantities and solubility of the polysaccharides likely to interact with, e.g. a paste of mineral binder, altering the kinetics of its binding and the binding itself, are highly variable. However, care should be taken when envisaging this option.

2.1.3.3. *Lignin*

The aromatic ring is the basic chemical component in lignin and other aromatics. These compounds are extremely diverse and are present in various forms within the plant and the cell walls. Three main groups of lignins can be distinguished [BUR 08]: softwood lignin (gymnosperms), e.g. conifers; hardwood lignin (angiosperms), and lignin from herbaceous plants. This latter group is beginning to attract an increasing amount of interest in R&D, because it is a renewable material produced in an annual cycle, and can be provided, primarily by the annual production of biomass. The lignin largely conditions the properties and treatment of bio-sourced materials. Hence, lignin is the compound which, indirectly, plays a capital role in the development of the market for these materials [BUR 08]. The estimated quantity of lignin is often relative to the measuring method used, and there may be an appreciable difference in the amounts reported in the literature, even for the same material [AKI 10].

Lignin also provides protection against the development of bacteria and pathogenic microbes that are harmful to the cell walls [AKI 08]. The nature and amount of the lignin present in the different parts of the plant have a significant impact on the effectiveness of retting procedures applied to fibrous plants, and also, more generally, on the durability or biodegradability of materials made from the plant. In partnership with cellulose, lignin gives the plant a rigid structure, enabling it to stand upright. In the cell walls, lignin is closely associated with hemicelluloses and cellulose. Covalent bonds are formed between lignin and hemicelluloses, and it is associated with the cellulose by the intermediary of the hemicelluloses [AKI 10].

In the case of fibrous plants – mainly flax or hemp – most of the lignin is to be found in the tissues in the center of the stem, containing the xylem and the other cell walls serving to channel water and sap [AKI 96]. Flax or hemp shiv obtained by defibration comes from this "lignified" part of the plant (see Table 2.1).

2.1.3.4. *Pectin*

Pectins, much like hemicelluloses, are water-soluble and include many different components which are present in cell walls. Of these, galactose and rhamnose are the most representative of pectins. In fibrous plants they are often present in small quantities, but they are located strategically in the plant's tissues. Pectins, in juxtaposition to hemicelluloses, constitute the polysaccharide matrix in the different tissues of the plant and in the fibers [AKI 10]. It should be noted that retting of fibrous plants causes degradation of the pectins by the action of bacteria and mold, thereby enabling the fibers to be separated from the non-fibrous part of the plant.

2.1.3.5. *Waxes, fats and lipids*

These hydrocarbons are of different sorts, but share the peculiarity of being insoluble in water [NEL 00]. The biological functions they perform are also diverse. Fats and oils are the main forms of energy storage for many living organisms. Phospholipids and sterols are structural components in membranes. Other lipids play various roles, including enzymatic functions, pigmentation, etc. Biological waxes are esters, comprising long chains of alcohols. The proportion of these constituents is relatively low in fibrous plants, but may be higher in grasses (bagasse and cereals) [AKI 10].

Lipids are particularly important on the external wall of plants, and around the fibers. In addition, the accumulation of wax on the cuticle forms a protective barrier against dehydration and the entry of infection agents into the plant. During the retting of fibrous plants, the epidermis of the stem separates from the fibers and the lignified central part. Flax shiv, like hemp shiv, contains only very small amounts of wax.

2.1.3.6. *Ash*

The quantity of insoluble mineral matter can be determined by a number of methods, the results of which are fairly congruous, overall. The quantity of ash in fibrous plants is generally low, whereas in grasses it may be significantly higher – particularly in rice or wheat straw, whose silica (SiO_2) content is higher. Yet fibrous plants – particularly flax and hemp – contain greater quantities of heavy metals such as lead (Pb), copper (Cu), zinc (Zn) and cadmium (Cd) [AKI 10]. The capacity of these plants to accumulate these heavy metals can, furthermore, be exploited to de-pollute some soils [LIN 02].

2.1.4. *Density and porosity, in the case of hemp shiv*

The densities and porosities measured on two types of hemp shiv [NGU 10a], "HS" (Hemp Shiv) and "FHS" (Fibrous Hemp Shiv), presented in Figure 2.4 (see section 2.2), are summarized in Table 2.2.

The apparent density when loose and dry is measured on the basis of a cylindrical volume 160 mm in diameter and 320 mm in height, in which the loose dry hemp shiv is poured.

The apparent density of the particles was measured on the basis of a straight section of stem, the area of which was determined by image analysis and the measured height. This figure, given as an indicative value, is underestimated in that the hemp shiv particles probably have a greater density [CEY 08; CER 05], owing to the stresses undergone during the defibration process and the stress of confinement when they were conditioned and kept in a 20 kg sack. A value of 300 kg.m^{-3} would, at first glance, seem to constitute a more meaningful density of the particles used. It would represent a decrease of the intra-granular porosity but a slight increase in the inter-granular porosity as reported in Table 2.2.

The apparent density of the solid phase is determined by a pycnometer, using toluene as filling fluid.

From these measurements, we deduce the following porosity values:

– total porosity: $\phi_{total} = 1 - \rho_V/\rho_S$

– intra-granular porosity: $\phi_{intra} = 1 - \rho_P/\rho_S$

– inter-granular porosity: $\phi_{inter} = 1 - \rho_P/\rho_V$

It should be noted that when long fibers are present in hemp shiv, as is the case of fibrous shiv, these fibers make up a significant proportion of the loose volume, and largely contribute to the increased inter-granular porosity.

	HS	*FHS*
ρ_L apparent density loose and dry [kg.m^{-3}]	112	71
ρ_P apparent density of the dry particles [kg.m^{-3}]	256	256
ρ_S apparent density of the solid phase [kg.m^{-3}]	1460	1440
ϕ_{total}, total porosity	92%	95%
ϕ_{intra}, intra-granular porosity	82%	82%
ϕ_{inter}, inter-granular porosity	56%	72%

Table 2.2. *Densities and porosity of the hemp shiv under examination*

Although the apparent density of the dry particles is probably underestimated, the intra-granular porosity proves to be very high, which accounts for the excellent absorbent quality of this material.

2.2. Particle Size Distribution (PSD)

At present, no norm exists to cover the PSD of bio-sourced aggregates. They are different in many respects from the mineral aggregates traditionally employed in hydraulic concretes – which rounder, very unyielding with low porosity and considerably denser – for which methods of characterization, mainly by sieving, have been defined and are employed in the published standards. Yet industrial application, either on-site or in a precast factory, calls for a more in-depth examination of the properties of these aggregates, in order to stay abreast of the quality of the finished materials.

2.2.1. *General characteristics of aggregates made from fibrous plants*

Hemp straw and flax straw is composed of very long and not heavily lignified cortical fibers surrounding a woody part (very heavily lignified short fibers) at the center of the stem [CRO 05], corresponding to the part which, while the plant was growing, carried the sap. The cortical fiber, rich in cellulose, represents the main value of this agricultural product [BOU 06]. During the process of defibration, the straw is ground, usually using hammer mills. The woody part is detached from the fibers, and shredded into small pieces to form hemp or flax shiv. In the case of hemp, 100 kg of straw, when ground, yield around 30 kg of fiber, 60 kg of hemp shiv and 10 kg of dust [BOU 06; BEV 09; BRU 09]. Although it is the main constituent, hemp shiv is merely a co-product of the exploitation of hemp. Its main use up until now has been in animal litter or horticultural straw.

The term "hemp shiv" is currently used to denote aggregates from the stem of the hemp plant which may be very varied, as they come from agricultural products that are subject to weather hazards and obtained using various post-harvest processes. Flax shiv, for its part, present still more disparate characteristics, particularly due to the larger variety of species that can be grown. In addition, the implementation processes employed generally tend to give preference to these particles, and therefore – depending on the overall shape of the particles – give rise to a tangible anisotropy of the material, particularly in terms of its thermal characteristics [ELF 08; NGU 10b].

2.2.2. *Fiber content*

When the plant is felled, it may be left on the ground for a variable length of time so that retting will facilitate the process of defibration. The advance of this process, the dampness of the straw and the regulations of the grinders used affect the size of the particles obtained when the straw is ground [MAN 04; MIA 11]. Also, the defibration performed may be more or less vigorous, and the fiber content in the hemp shiv or the flax shiv may be variable.

a) HS Hemp shiv b) FHS Fibrous Hemp Shiv

Figure 2.4. *Hemp shiv under investigation, laid out in a spread 7 cm in diameter*

The study presented below is based on the example of two types of hemp shiv: a hemp shiv gained from an advanced process of defibration with very few residual fibers, denoted HS, and another gained from a partial process of defibration, containing a significant amount of short cortical fibers (shorter than 4 cm), denoted FHS. It is very easy to distinguish these two hemp shiv with the naked eye, as shown in Figure 2.4.

2.2.3. *Methods for characterizing the PSD*

Two methods can be easily employed to study the PSD of the hemp shiv, each with advantages and drawbacks:

The conventional method of sieving with dry particles enables us to take measurements directly on a sample of a few hundred grams. The limitations of this method are due to the elongated shape of hemp shiv particles, and their low density, which render sieving less appropriate and unreliable [IGA 09].

2D image analysis of particles spread out over a flat surface gives us access to more information. Thereby, the width and length of each particle detected can be measured. However, this method is more complex and requires samples no larger than a few grams. The precision of the results produced is therefore limited by the representativeness of the sample and by the only dimensions obtained for the aggregates by projection onto a plane.

2.2.3.1. *Sieving method*

Sieving was performed on dry material, so that the finest particles could separate from the others. Sieves with standardized square mesh were used, as was a mechanical siever for the study of soils and mineral aggregates (NF ISO 3310.1 – ASTM E-11-95). In order to obtain repetitive results, the vibration time was extended to an hour for a 200-gram sample and for five consecutive sieves. The apertures of the sieves used ranged from 10 to 0.315 mm as follows: 10; 8; 6.3; 5; 4; 3.15; 2.5; 2; 1.25; 1; 0.63 and 0.315 mm.

PSD analysis by sieving assumes that all the particles are practically spherical in shape, and pass through a square aperture when their diameter is less than the side of the square. For flat or elongated particles, such as hemp shiv particles, this point is developed in further detail in section 2.2.5.3. The particles may either pass through the sieve in the direction of their length (see Figure 2.20) or be retained if they are positioned across the aperture. In the latter case, these particles may also block the passage of particles located above them.

Generally speaking, increasing the time taken over sieving helps to reduce the relative differences of the refused particles obtained for each sieve. On the basis of several tests, the precision of the results, i.e. the evaluation of the particles retained in each sieve, was evaluated at ±15% with the series of sieves used. These uncertainties are illustrated in Figure 2.6. Globally we observe that the PSD of the two hemp shiv under study (apart from fibers), are relatively close.

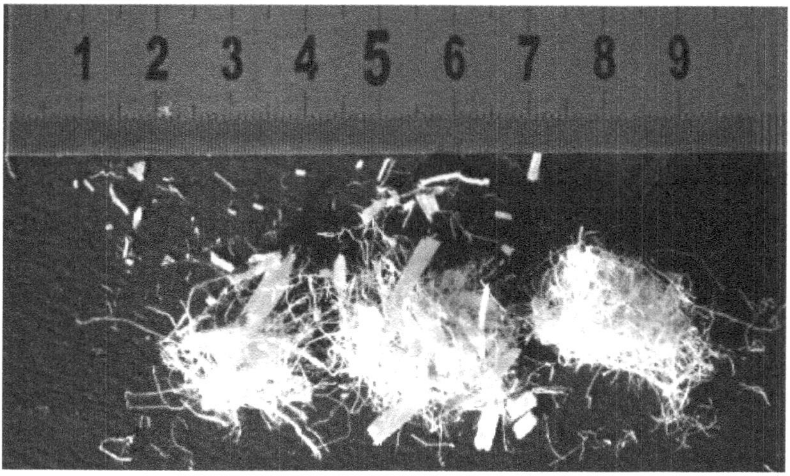

Figure 2.5. *Pellets of fibers formed with the first sieves (4 and 5 mm)*

When fibers are present, sieving can be used to complete the separation of the fibers from the hemp shiv. Indeed, the fibers tend to form pellets in the first sieves – see Figure 2.5. Also, when the fiber and the hemp shiv are still linked, the vibrations of the sieve are enough to detach them. Generally, the amount of fibers in the hemp shiv tested varies between 1 and 15% of the mass. For instance, this fiber content was evaluated as being between 12% and 15% in the case of the fibrous hemp shiv (FHS) as shown in Figure 2.4. The PSDs presented below do not take account of these free fibers. Yet even in the case of hemp shiv obtained by an advanced process of defibration, a detailed examination of the aggregates shows that a small amount of short fibers may still be found attached to a few hemp shiv particles, even after the stages of sieving performed, for at least 30 minutes.

2.2.3.2. *Image-analysis method*

This method requires good-quality sampling, in that only a finite amount of material can reasonably be analyzed. A classic method of quartering can be applied to a 20 kg sack of hemp shiv and repeated as many times as need be. The representativeness of the sample selected is the key element guaranteeing the relevance of the results produced.

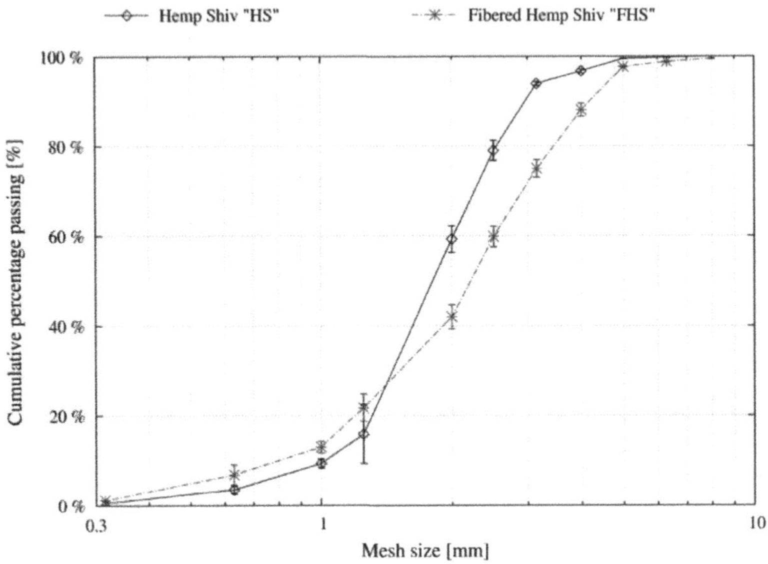

Figure 2.6. *Cumulative size distribution obtained by sieving*

2.2.3.2.1. Acquisition of a digital image

Image analysis is based on a two-dimensional observation of particles spread out over a flat surface. In this study, the images were obtained with a conventional scanner generally used to digitize documents. This technique offers the advantage of avoiding any distortion of the image which might occur if a camera were used. The scanner can acquire a color image or an image converted into 8-bit grayscale that can be processing by an image-analysis program such as "Image Tool" or "ImageJ", which are freely accessible online.

The particles are spread out so that they do not overlap or touch. This requires very detailed attention, because the particles are fine and in these conditions it is difficult to avoid some overlap. Certain algorithms can be used to deal with this problem at the image-processing level [SHA 06], but with the aim of simplifying the procedure of analysis and increasing the precision of our measurements [IGA 09a] [IGA 09b], we limited ourselves to a procedure of rather dispersed distribution, requiring a greater number of images to be analyzed and a case-by-case verification of potential overlaps of particles.

As the hemp shiv under investigation is light in color, a dark background was used in order to obtain a maximum degree of contrast. The grayscale image was processed at a resolution of 600 DPI (dots per inch) on both the vertical and horizontal axes. This corresponds to a constant scale factor of 0.04233 mm per pixel. This scale factor can be verified by calibration. On an A4 surface, i.e. 210 × 297 mm^2, the image produced in an uncompressed format occupies around 35MB of memory. The precision of the measurements can further be improved by increasing the resolution, but this is limiting because of the time needed for the scanner to transfer the data and for the storage and processing of the multiple images.

Image analysis requires a binarized image, which necessitates prior thresholding of the grayscale image. This is the trickiest step in this method, particularly if a large number of fine particles (of less than 0.5 mm width) are present. Automatic thresholding procedures may be available, but there will be a halo effect, to a greater or lesser degree of severity, around the particles identified. The halo effect tends to decrease the level of gray of the pixels on the outer boundary of the lightest objects. Incorrectly adapted thresholding may therefore contribute to an artificial increase of the size of these objects and consequently, noticeably over- or underestimate (depending on whether the background is dark- or light-colored) the relative size of the smallest objects detected [NGU 10a; IGA 09b]. Yet this problem can be greatly assuaged by manual thresholding to process the image. The lower bound of the threshold can be adjusted so that this halo effect is contained in a band approximately one pixel in width around each object. Appropriate thresholding should cover the surface of the objects needing to be detected as precisely as possible. The quality of this thresholding can be verified using small "standard" objects the same color as the particles we wish to detect.

Generally, in the case of light-colored particles on a dark background, the upper threshold is fixed at the maximum, i.e. 255, and the lower threshold is between 60 and 90. In our case, in order to create the most apt binarized image, the value of this lower threshold was set at 80 for all the images processed.

Hemp shiv also contains dust of organic or mineral origin, usually less than 2% of the weight of the particles passing through the 0.315 mm sieve. The tail of distribution toward the finest particles is difficult to quantify using image analysis. Such a task would require additional observation, on a microscopic scale, of finest sieved sample.

Next, therefore, it is useful to define a detection threshold, i.e. a minimum projected area of the objects to be processed, so as to take account only of the hemp shiv particles and avoid dust particles represented by merely a few pixels. In the case presented here, only the particles whose area is greater than 0.08 mm^2 (i.e. objects represented by at least 45 pixels) and width greater than 0.1 mm (i.e. over two pixels across) were taken into account for the PSD analysis.

Other processing operations can also be carried out with a view to preparing the images before analysis. One such operation, which involves producing an erosion of a given number of pixels followed by an operation of expansion of the object by adding the same number of pixels onto its boundary, can help eliminate dust particles and fibers which are not representative of the hemp shiv particles needing to be identified. This operation is referred to as an opening operation, because it may lead to the de-compartmentalization of cavities, separated by thin boundaries, which may be contained in the objects. By way of example, its effect is illustrated in Figure 2.10 for a thickness of two pixels.

2.2.3.2.2. Measurements of length and parameters of shape of the objects

Image analysis gives us access to far more information than does sieving. For each particle detected, its projected area and the perimeter of that projected area are directly measured and recorded.

Other more elaborate parameters, such as the minimum convex area surrounding the object (see Figure 2.7), can be used to define different formal parameters to help better characterize the particles. Of these, the convexity ratio, χ (also called "solidity") [MOR 00], defined as the object's projected area over the convex area surrounding that object, reveals the form of the particles. Perfectly convex particles have a convexity ratio of 1. In this study, we labeled particles with a convexity ratio of less than 2/3 as non-convex (see section 2.5.1).

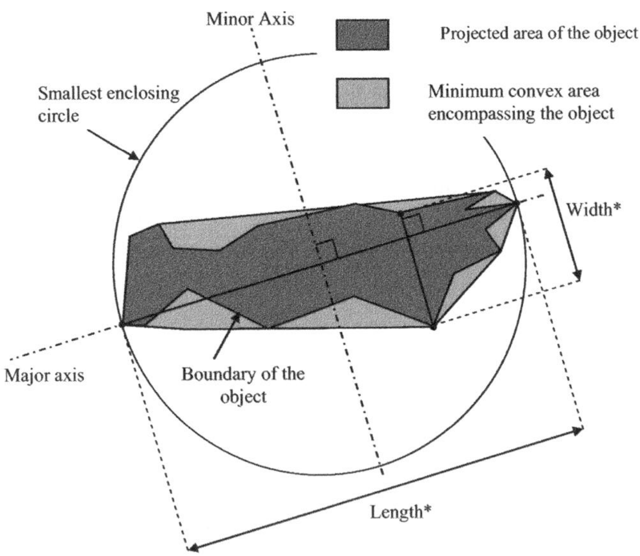

Figure 2.7. *Evaluation of the convex area and the maximum Féret diameter*

Similarly, the convexity of the particles can also be estimated by looking at the ratio of the projected area to the product of the length by the width of the identified object (Area/(Length × width)) when this measurement is not embedded in the software tool being used.

Observation of different types of ground-up straw shows that the resulting particles have irregular and angular forms due to the microstructure of the plant, oriented along the axis of the stem and to the shredding action that can sometimes be caused by hammer mills, for instance. In this scenario, the shapes of the finest particles therefore tend to be polygonal and convex, whereas the shapes of the coarsest particles tend to diversify to include non-convex particles [BIT 09a].

Overall, in the case of convex and non-convex particles gleaned from ground straw, the method for determining the length based on the diameter of the smallest enclosing circle or maximal caliper (which some writers also refer to as the Féret diameter) is fairly representative of the length of the object [IGA 09a]. Hence, the length can be directly quantified using this maximum diameter, defining the major axis of the projected area (see Figure 2.7).

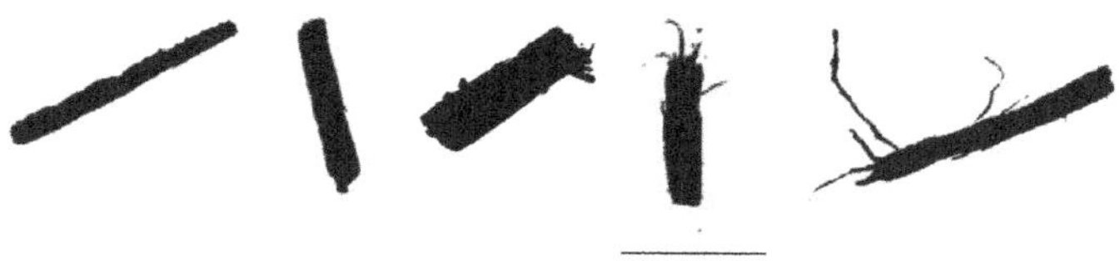

Figure 2.8. *Hemp shiv particles after binarization of the image, classified on the basis of biases in the analysis which will cause protuberances due to shredding and to remaining connected fibers*

2.2.3.2.3. Determination of the particle size

The measurement of the length and width of these particles may be subject to different definitions, depending on the representativeness of these dimensions in the case of the type of object needing to be analyzed.

The width can be defined and measured using an Image Tool, as the maximum length along the minor axis, perpendicularly to the major axis. This is denoted as "width*" in Figure 2.7 and Figure 2.9. For around 2600 analyzed particles contained in a 4 g sample of HS hemp shiv, Figure 2.11 gives a view of the logarithmic scale of the widths in comparison to the lengths analyzed. This method leads to a slight overestimation with rectangular shapes, because it is the diagonals that are identified as the lengths. However, the overestimation of the lengths therefore also applies, to the same degree, to the widths, so that the elongation of the particles, ε (the ratio of length to width of the particles) – a parameter which we shall study later on – remains the same. Overestimation is also important in the case of non-convex particles, because the measured width starts from the major axis even if this axis lies outside the object.

The width can also be defined, and measured using ImageJ, as the minimal Féret diameter or minimal caliper, i.e. the minimum distance between two parallel straight lines (or planes) encompassing the object, or indeed as the width of the narrowest rectangle (or parallelepiped) containing the object – see Figure 2.9. This method could lead us to suppose that the estimation of the width is correct in the case of rectangular particles. However, in the case of hemp wood, the short fibers still connected and the particles shredded by milling give rise to outcrops from the projected areas and ultimately cause an overestimation of the widths obtained by way of this method.

In order to iron out some of the outgrowths of the objects studied (see Figure 2.8) and analyze them using geometric forms deemed to be representative, other methods exist. Such methods consist of adjusting the basic geometric shapes (rectangles, ellipses, triangles, polygons, etc.) to the objects detected [IGA 08; BIT 09a] so as to determine their length, and above all their representative width. Of these, in the case of hemp shiv, an ellipse can be adjusted so that its center of gravity corresponds to that of the object and its projected area is identical to that of the object. The lengths and widths of the object are therefore defined respectively in accordance with the large and small radii of the adjusted ellipses (see Figure 2.9).

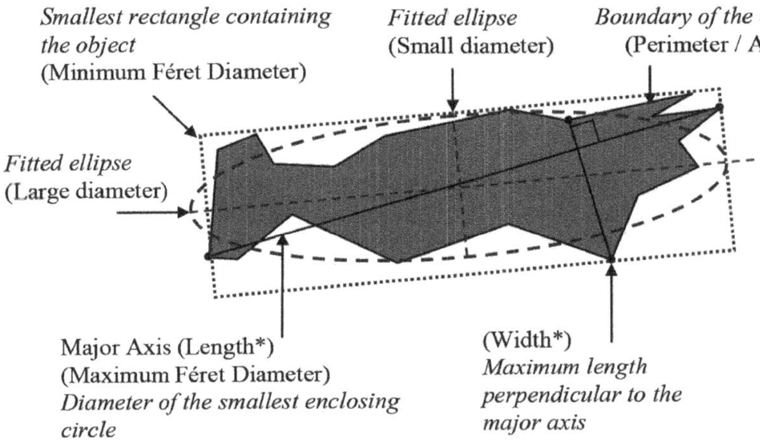

Figure 2.9. *Lengths and widths analyzed*

a) b) c)

Figure 2.10. *Analysis of a hemp shiv particle approximately 20 mm in length: a) grayscale scanned image; b) image after thresholding; c) image after operation of opening of one pixel followed by adjustment of an ellipse, facilitating the evaluation of its length and width*

It should be noted that with rectangular shapes, the adjustment of an ellipse also leads to an overestimation of the lengths and widths in identical proportions, so that the elongation of the particles, ε, again remains unchanged.

For the same images obtained with a 4 g sample of CP hemp shiv, the widths on the basis of the lengths analyzed by adjustment of an ellipse on each object detected (around 2600 – see section 2.5.1) are represented in Figure 2.12. Of these point clouds, two general categories of particles may appear: the particles representative of hemp shiv, for which the corresponding point cloud is centered around a width of around 2 mm; and dust or micro-fibers, for which the point cloud seems to be truncated by the threshold selected.

Of the various characteristics that are measured, the elongation seems critically important, because it will condition the orientation of the granular arrangement and the anisotropy of the finished materials. For the same sample, it is represented as a function of the projected area of each particle when the lengths are measured from the major and minor axes in Figure 2.14 and when they are evaluated on the basis of the diameters of the adjusted ellipses in Figure 2.13. The point cloud is primarily clustered around a straight line denoting a constant length-to-width ratio. For an elongation on a logarithmic scale, the points appear to be distributed in accordance with a normal law. In these figures, therefore, we have considered the geometric mean of the elongation of each particle weighted by its area, notated as $\overline{\varepsilon}_{gm}$, and the corresponding standard deviation, notated as σ_{gm}, in order to calculate the confidence intervals for the different segments of the projected area in question.

In Figure 2.14, it appears that the smallest particles, with a projected area of less than 0.4 mm^2 may contain short fibers (shorter than 5 mm) which are not straight.

These particles noticeably increase the average elongation of the particles contained in the smallest intervals of projected area when the lengths and widths are measured in accordance with the major and minor axes.

Overall, for these two methods of analysis, there does not appear to be any significant change in elongation with the projected area of the particles. Over all the particles identified, the average value of this elongation is near to four, with a geometric standard deviation of around 1.5. The impact of the dimensional approach to particle analysis on the results is presented in greater detail in section 2.6.2.

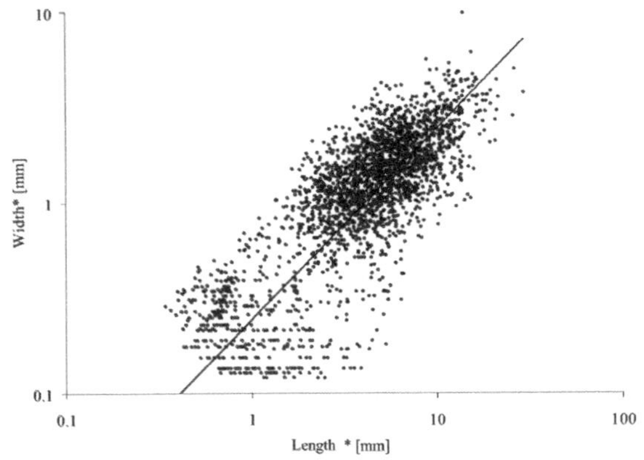

Figure 2.11. *Length and width of hemp shiv particles evaluated by the major and minor axes of the projected area*

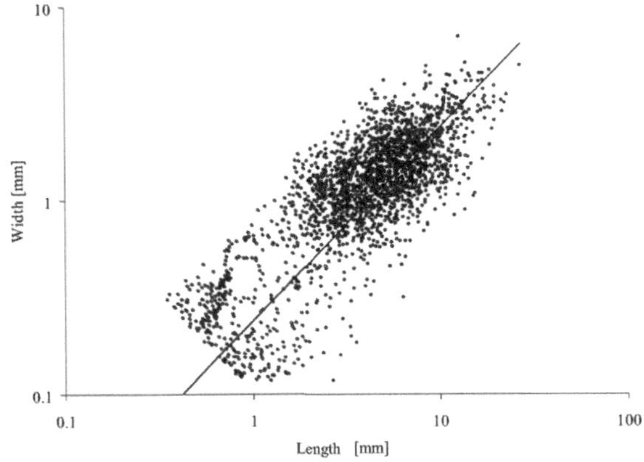

Figure 2.12. *Length and width of hemp shiv particles evaluated by adjustment of an ellipse over the projected area*

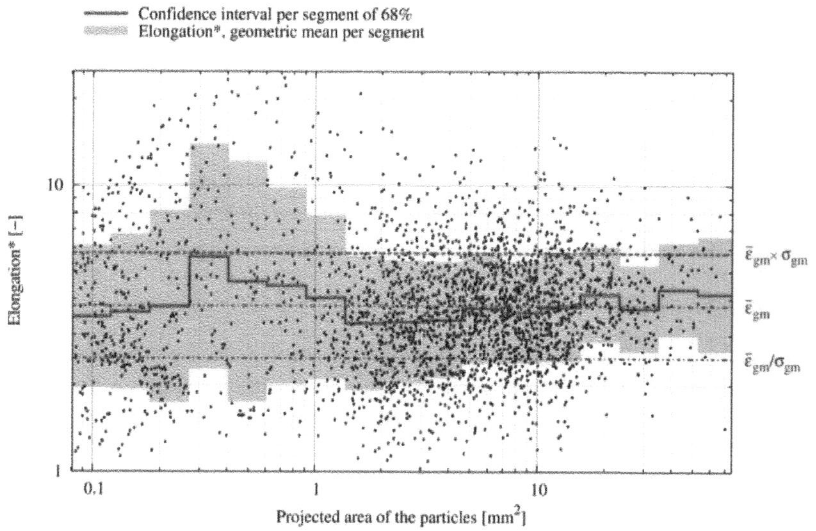

Figure 2.13. *Elongations of hemp shiv particles evaluated along the major and minor axes of the projected area*

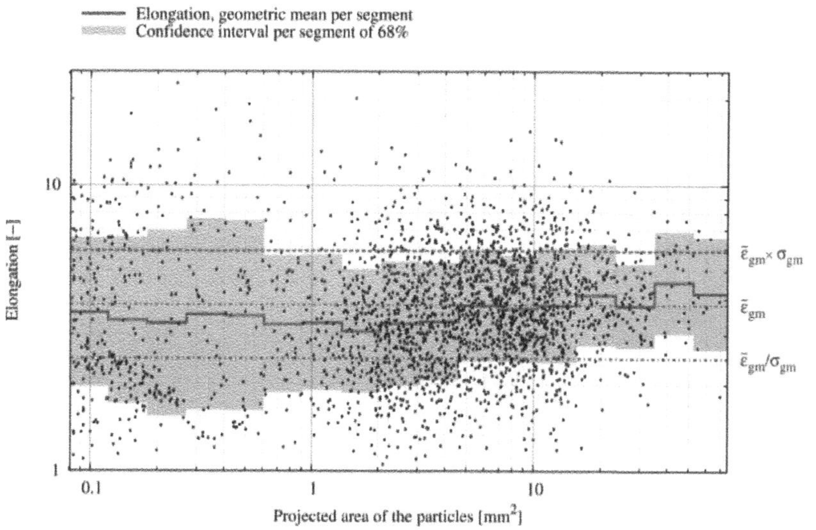

Figure 2.14. *Elongations of the hemp shiv particles evaluated by adjustment of an ellipse over the projected area*

2.2.4. *PSD analyses*

Various types of distribution, or probability density, of size of particles may be deduced from these tests, depending on whether they are distributed in accordance with their number or with their projected area. However, so as to be meaningfully representative of potential granular packing and to be comparable with the analyses obtained by sieving, this analysis should be performed on the basis of a distribution of the volume of the particles.

2.2.4.1. *Frequency distribution*

Distributions on the basis of the number of particles can be performed directly from the raw data from the image analysis step. However, this type of distribution is very sensitive to the number of the smallest particles, and particularly to their detection threshold. In the study presented, we set a detection threshold based on the projected area of the particles equal to 0.08 mm^2. Yet clearly, the material contains even smaller particles which will not be taken into account when counting the particles. Such a distribution based on the relative cumulative number of particles passing through a sieve of a given size, denoted as N% in Figure 2.15, therefore depends heavily on this detection threshold.

2.2.4.2. *Area fraction distribution (projected area)*

Image analysis also gives us access to the projected area of each particle detected: A_i. The influence of the finest particles on the PSD can therefore be weighted by this criterion, so as to consider a distribution by the cumulative projected area of the particles whose considered size is less than a given value. The distribution of cumulative passing can be directly calculated on the basis of the sum of the projected areas of *n* particles, arranged in order of increasing size, for a total number of *N* particles detected, whose total area is A_T. The cumulative distribution by increasing size, $P_A(X \le x_n)$, similar to the "cumulative passing" obtained by sieving (see Figure 2.6), can then be written as:

$$P_A(X \le x_n) = \frac{\sum_{i=1}^{n} A_i}{\sum_{i=1}^{N} A_i} = \frac{1}{A_T} \sum_{i=1}^{n} A_i \qquad [2.1]$$

where X is the considered size of the particles and $P_A(X \le x_n)$ the proportion of the projected area accounted for by particles smaller than the nth particle of size x_n. The cumulative distributions $P_A(X \le x_n)$ and the distributions of size based on the area of the particles are therefore annotated ($A\%$) in the figures.

2.2.4.3. *Mass fraction distribution*

The PSD curves are usually plotted on the basis of the results obtained by sieving, i.e. on the basis of a mass distribution. If M_i is the mass of the particle i, the distribution $P_M(X \leq x_n)$ is directly calculated from the cumulative mass of the n smallest particles passing through a sieve of given size out of a total of N particles of mass M_T:

$$P_M(X \leq x_n) = \frac{1}{M_T} \sum_{i=1}^{n} M_i \qquad [2.2]$$

where X is the considered size of the particles and $P_M(X \leq x_n)$ the proportion of the mass of particles smaller than the nth particle of size x_n.

2.2.4.4. *Relation between area fraction- and mass distributions*

In order to be able to compare the results obtained by sieving and by image analysis, a distribution of the size of the particles in relation to their mass must be considered – that is, we must consider a distribution of type $P_M(X \leq x_n)$ for both the width and length.

If the apparent density of the particles is independent of their dimensions, $P_M(X \leq x_n)$ can also be written in accordance with the volume V_i of each particle and their total volume V_T. Supposing that the particles are similar in shape, e_i denotes the average thickness over the whole of the projected area of each particle, i.e. their third dimension (inaccessible by 2D image analysis), and the volume of each particle can be considered to be the product of its projected area, A_i, by its average thickness, e_i: $V_i = e_i A_i$.

$$P_M(X \leq x_n) = \frac{1}{V_T} \sum_{i=1}^{n} V_i \cong \frac{\sum_{i=1}^{n} e_i.A_i}{\sum_{i=1}^{N} e_i.A_i} \qquad [2.3]$$

From hereon in, many complementary tests can be performed; yet it is not easy to approximate the thickness e_i of each particle. It can only be supposed that since the particles are simply and freely spread out over a plane, this average thickness is less than the width of each particle. The thickness of the woody part in the stem varies noticeably, depending on the climatic conditions, the date of harvesting and the density of the plantation [SCH 06] on the one hand, and then depending on the height of the particular section within the stem [RAH 10], with the thickness decreasing toward the apex. However, the process of defibration is applied to all the cropped straw, giving rise to multi-directional grinding. As observed for different types of ground straw, the general shape of the particles does not seem to be affected by the diameter of the stems being ground [NGU 10a; IGA 09a] whereas the size of the particles produced depends essentially on the process of grinding itself and on the settings used [BIT 09a]. Figure 2.11 shows that in this case, the average elongation ratio of the particles is, overall, independent of the particles' projected area. This observation can be supposed to extend into the third dimension: the average ratio of width to average thickness, $\bar{\Phi} = e_i/l_i$, may also be reasonably assumed to be constant.

If the density of the particles is identical, with a similar shape irrespective of their size – i.e. if they are generally homothetic and their volume V_i can be approximated by $\bar{\Phi} A_i l_i$, the mass distribution of the particles can be deduced from the projected area A_i and the width l_i of the particles by the following relation:

$$P_M(X \leq x_n) \cong \frac{\sum_{i=1}^{n} l_i.A_i}{\sum_{i=1}^{N} l_i.A_i} \left(\text{if } \frac{e_i}{l_i} \cong \bar{\Phi} \text{ (const.)} \right) \qquad [2.4]$$

It is in view of this hypothesis that the cumulative distributions $P_M(X \leq x_n)$ and distributions of size based on the mass of the particles are annotated as (M%) in the figures hereafter.

It should be noted that the cumulative distribution $P_M(X \leq x_n)$ depends neither on the value of the apparent density of the particles, ρ_a, nor on $\bar{\Phi}$.

Because $P_M(X \leq x_n)$ is sensitive to the largest particles, this distribution may be considerably different from $P_A(X \leq x_n)$ in the case of a spread distribution. It is interesting to note that if all the particles had the same thickness, the approximation of the cumulative distribution by mass, $P_M(X \leq x_n)$, defined in equation [2.3], would be equivalent to $P_A(X \leq x_n)$ defined in equation [2.2].

Figures 2.15 and 2.16 illustrate the difference observed between the cumulative distributions based on the number of particles, the projected area and the supposed mass of the particles, for both the width and length of the particles.

In addition, the standard distributions of particle size represented in Figure 2.17 confirm the uni-modal nature of the area fraction (A%) and mass fraction (M%) distributions for the width and length of the particles.

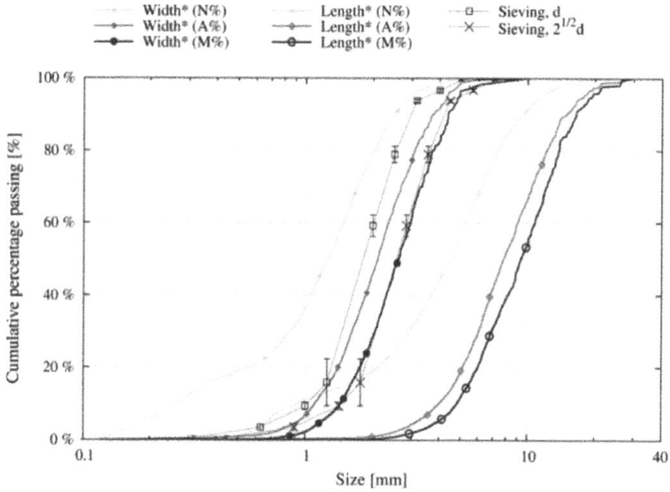

Figure 2.15. *Cumulative size distribution of HS hemp shiv obtained by sieving, considering a mesh-size measuring d and $2^{1/2}$d in the sieves, and frequency distributions (N%), in area fraction (A%) and mass fraction (M%) obtained by image analysis from the lengths and widths of the hemp shiv particles evaluated along the major and minor axes of the projected area of the particles*

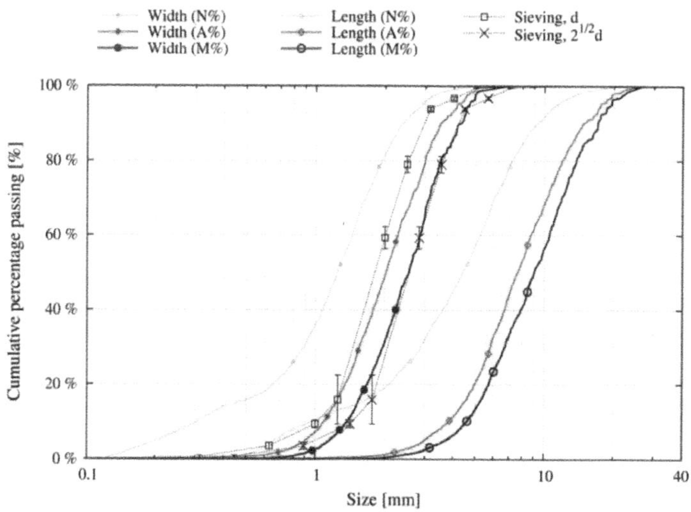

Figure 2.16. *Cumulative size distribution of "CP" hemp shiv obtained by sieving, considering a mesh-size measuring d and $2^{1/2}$d in the sieves, and frequency distributions (N%), in area fraction (A%) and mass fraction (M%) obtained by image analysis from the lengths and widths of the hemp shiv particles evaluated by adjustment of ellipses on the projected areas of the particles*

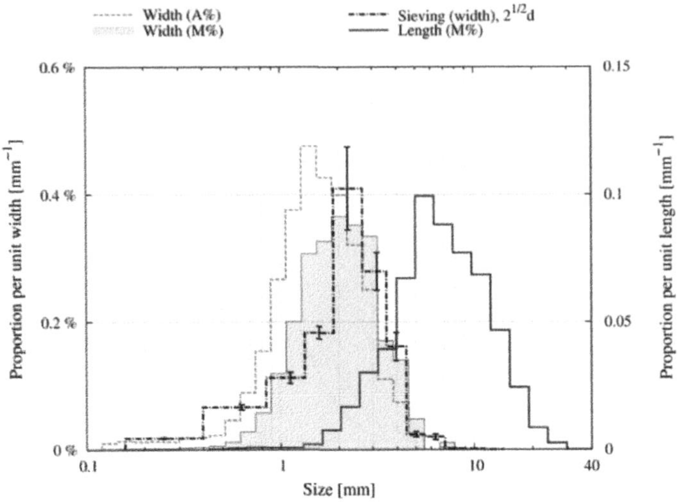

Figure 2.17. *PSD in mass fraction of "CP" hemp shiv obtained by sieving, considering a mesh-size measuring d and $2^{1/2}$d in the sieves, and density of distributions in area fraction (A%) of the widths, and mass fraction (M%) of the widths and lengths obtained by image analysis from the dimensions evaluated by adjustment of ellipses on the projected areas of the particles*

2.2.5. *Comparison of the results obtained by image analysis*

2.2.5.1. *Impact of the selection of the particles to be analyzed*

The detected area and minimal width of the particles needing to be analyzed constitute the first parameter when selecting which particles to analyze. In the wake of the defibration process, the hemp shiv particles may have varied shapes, which deviate from the ellipsoidal or parallelepipedic shapes to which it is possible to compare them. Short fibers may still be attached, and the shredding action to which the stem is subjected leads to very varied shapes that are sometimes difficult to qualify in terms of representative length and width. Thus a minimum convexity ratio can be set in order to discount particles whose shape differs too greatly from the conventional shapes which can be correctly analyzed. Figure 2.18 therefore presents the effect of this selection on the distribution of the widths of the same sample containing: 2600 particles whose area is greater than 0.08 mm^2 with a convexity coefficient χ, greater than 0; 2460 particles with $\chi > 1/2$; 2300 particles with $\chi > 2/3$; and 1980 particles with $\chi > 3/4$. There is a noticeable effect on frequency distributions, for the finest particles, while the differences become very small for area fraction distributions, and insignificant for mass fraction distributions. It should be noted that the same observation can be made as regards the distributions of the lengths, with differences reduced still further.

The results presented in this study relate to particles whose area is greater than 0.08 mm^2, width greater than 0.1 mm and convexity ratio greater than 2/3. It should be noted that this final condition, here applied operationally in order to shed light on the results presented, disqualifies less than 12% of particles detected (300 in this case) out of the sample tested, and therefore has little significant influence on the progression of the analyses.

The influence of the threshold concerning the minimum area of the particles taken into consideration in the analyses has also been studied. At the resolution selected – 23.6 pixels/mm (600 DPI) – it does not seem representative to consider objects whose surface area is less than or equal to 0.08 mm^2, i.e. we look at particles represented by at least 45 pixels – notably to study the overall shape of the particles, for instance. When this threshold is increased in iterative steps, up to ten times this initial value (i.e. up to 0.8 mm^2), the number of particles taken into account decreases significantly, but we do not see a significant difference in the distributions of observed sizes in area fraction and mass fraction. Increase it still further, and slight differences appear, primarily on the distributions of width of the particles. In reality, for the hemp shiv being studied, it still seems possible to consider only those particles whose area is greater than or equal to 1 mm^2 to perform a simplified but representative PSD study. However, a lower threshold is preferable, wherever possible.

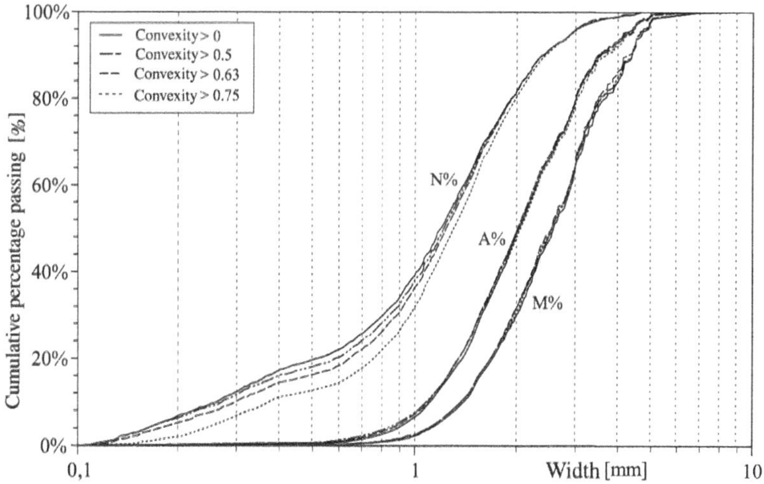

Figure 2.18. *Threshold effect of the convexity-to-width ratio of the ellipses adjusted to the projected areas of the particles being analyzed*

2.2.5.2. *Impact of the method of analysis*

The measurement of the length and width of these particles may be subject to different definitions depending on the representativeness of these dimensions in the case of the type of object needing to be analyzed – see Figure 2.9. Three different analytical methods have been applied to the same images, and the distributions of lengths and widths are presented and compared in Figure 2.19.

1) Maximum length along the major axis (or maximum Féret diameter) and maximum length along the minor axis, perpendicular to the major axis, denoted as "Major/Minor axes".

2) Maximum and minimum Féret diameter, i.e. minimum distance between two parallel lines (or planes) encapsulating the object, denoted as "Féret".

3) Large and small diameter of an ellipse adjusted so that its center of gravity corresponds to that of the object and its projected area is identical to that of the object, denoted "Ellipse".

It is clear that the distribution of the length is hardly sensitive at all to the analytical method employed. It should be noted, however, that the first two methods lead to theoretically-identical results for the length, even if two different programs are used to measure it. Notably, there is no significant difference between the distributions of lengths analyzed by methods 2 and 3 using the same program.

The widths, for their part, are more sensitive to the analytical method used. This point develops differently depending on the nature of the particles studied, particularly in the case of different food grains [IGA 09c]. In the case of hemp shiv, however, the discernible difference is relatively small, and this difference is comparable for both types of distribution shown in Figure 2.19, in terms of the area fraction (A%) and mass fraction (M%).

As mentioned in section 2.2.3.2.c, adjustment of ellipses to the projected areas of the hemp shiv particles enables us to iron out some of the remaining short fibers, supple and slight in terms of volume in comparison to the woody particles to which they are still attached, which are more rigid and voluminous, that we are seeking to identify. This method, leading to the smallest estimated widths, was therefore selected for the continuation of the study presented here.

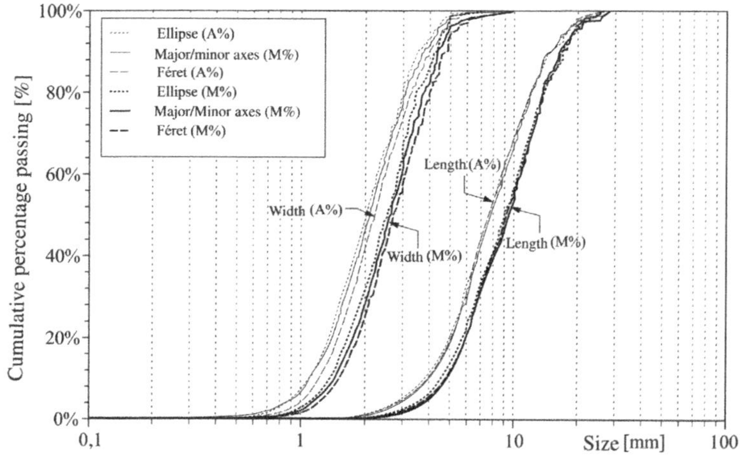

Figure 2.19. *Impact of the analytical method on the distribution of widths and lengths of the objects (projected areas of the particles) analyzed*

2.2.5.3. *Comparison with the results obtained by sieving*

As presented in Figures 2.15 and 2.16, the cumulative distributions of the particles are, overall, closer to the curve of cumulative percentage passing obtained by sieving, considering the width, rather than the length. This implies that the particles are capable of passing through the sieves in a direction perpendicular to the meshes, and that the sorting of the hemp shiv when the particles are sieved is done primarily on the basis of the width rather than the length of the particles.

If we assume that the particles are ellipsoidal, elongated and flat in shape, when these particles can pass lengthways through the sieves, their width may be oriented along the diagonals of the square openings (see Figure 2.20). In this case, only those particles whose width is greater than $2^{1/2}d$ are caught by a sieve whose mesh size measures d.

Figure 2.21 shows that the distribution of the cumulated mass widths (M%) is very similar to the distribution obtained by considering the diagonal of the square opening in the sieves, $2^{1/2}d$ mm instead of the mesh size i.e. the side of the square opening, measuring d mm. This indicates that the majority of the particles are oriented along the diagonal of the opening when they pass through the sieves.

As Figure 2.22 confirms, the size-per-mass distribution of the particles is also that which corresponds most closely to the results obtained by sieving. Yet this is slightly more skewed toward fine particles. This may be due to some of the finer particles being retained in the sieve, particularly in the last and finest sieves, whose meshes are finer than 1.25 mm.

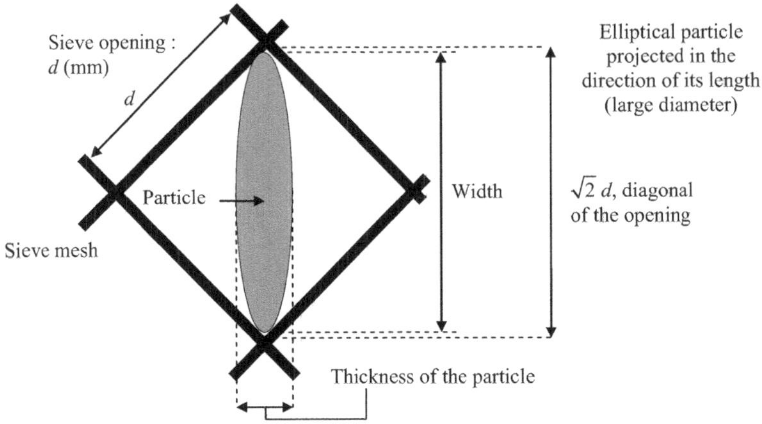

Figure 2.20. *Cross-section of the largest elliptical particle able to pass through the sieve, in the direction of its length and along the diagonal of the square opening*

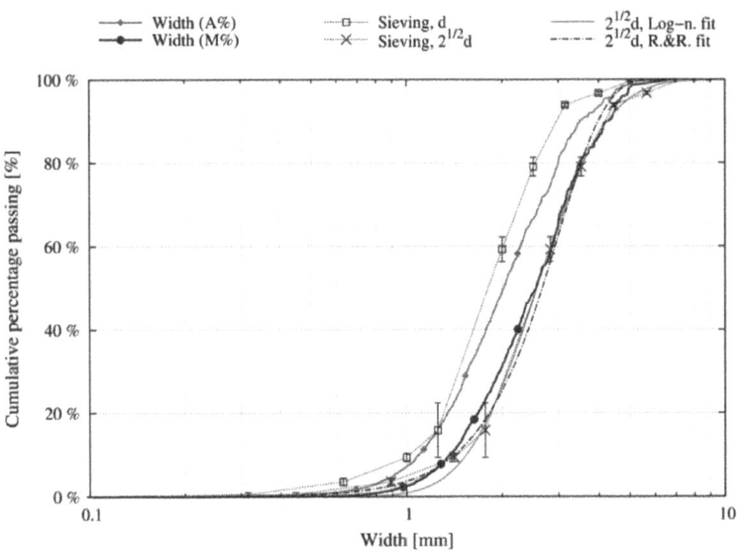

Figure 2.21. *Comparison of the width cumulative distributions in terms of area fraction and mass fraction obtained by image analysis and by sieving, considering the side of the square opening d and its diagonal $2^{1/2}d$*

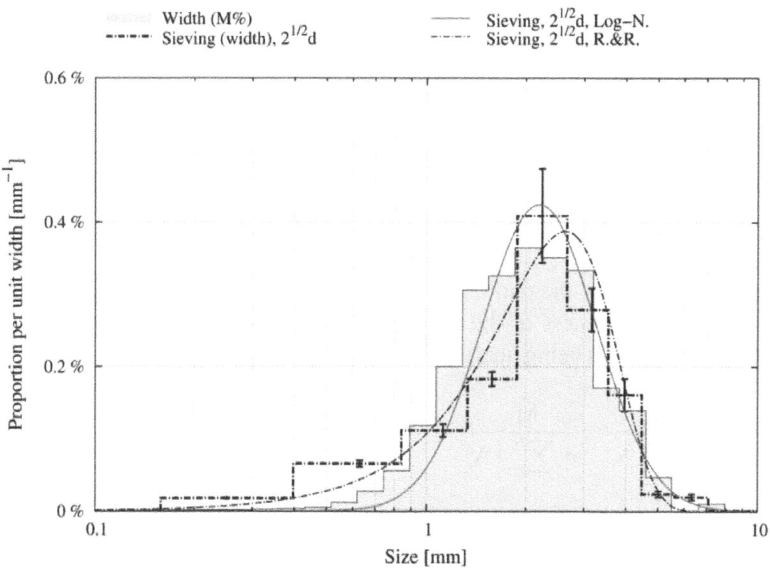

Figure 2.22. *Comparison of the width distributions in mass fractions by image analysis and by sieving, considering the diagonal of the square opening $2^{1/2}d$*

2.2.6. *Characterization of the geometry of the particles*

2.2.6.1. *Average elongation*

The elongation of the particles can lead to an anisotropy in the material, owing to the preferred orientation that their implementation usually causes, and to the orientation of the microstructure and porosity within these particles. In view of their respective volume, the elongation of the largest particles will have a greater impact on the general properties of the material than will the smaller particles. The average elongation of the particles, $\bar{\varepsilon}$, therefore needs to be weighted by the volume of each particle and can be written as follows:

$$\bar{\varepsilon} = \frac{\sum_{i=1}^{N} V_i \frac{L_i}{l_i}}{\sum_{i=1}^{N} V_i} \cong \frac{\sum_{i=1}^{N} L_i A_i}{\sum_{i=1}^{N} l_i A_i} \quad \left(\text{si } V_i \cong \bar{\Phi}\, l_i A_i\right) \qquad [2.5]$$

The estimations of lengths are essentially identical no matter what the analytical method, whereas for the width the estimations exhibit a slight difference (see Figure 2.19).

Consequently, for the sample of hemp shiv under consideration, the average elongation varies between 3.93 (adjustment of the ellipse) and 3.65 (Féret diameters), with 3.85 (major/minor axes).

The geometric means of the elongation are very similar to these values for all cases (see section 2.2.3.2.c), with an associated standard deviation of less than 1.54.

2.2.6.2. *Average flatness*

The average ratio of the particles' thickness to their width, $\bar{\Phi}$, can be considered an indicator of the flatness of the particles. It can be evaluated if the mass or total volume of the particles being analyzed is known [KWA 99].

$$\bar{\Phi} = \frac{V_T}{\sum_{i=1}^{N} l_i A_i} = \frac{M_T}{\rho_a \sum_{i=1}^{N} l_i A_i} \qquad [2.6]$$

In our case, the mass of the sample is known, and the apparent density of the particles ρ_a can be estimated [NGU 10a; CEY 08]. We shall consider the apparent density ρ_a of the particles to be approximately 300 kg/m^3, in the case at hand; $\bar{\Phi} \cong$ 1/3 if the dimensions are estimated on the basis of adjusted ellipses. In that the widths estimated differ slightly depending on the analytical method in question, the value of $\bar{\Phi}$ will be slightly smaller when the widths are estimated as being greater. For instance, $\bar{\Phi} \cong$ ¼ when length and width are measured in accordance with the major and minor axes.

2.2.7. *Characterization of the PSD*

2.2.7.1. *Means and standard deviations*

The arithmetic mean of the size (width or length) of the particles, weighted by their area or mass, and the associated standard deviation, can be used to gain an overall characterization of the PSD. The standardized degrees of skewness and kurtosis relative to the third- and fourth-order moments of distribution can also be calculated.

However, in that the distributions can be approximated by a normal law according to a scale of logarithmic size, the weighted geometric mean X_{gm} and its associated standard deviation σ_{gm}, defined in the two equations below, appear to be more relevant when seeking to characterize the distribution, as in the case of numerous distributions of particle sizes obtained from the comminution of the stems of different herbaceous plants [ASA 06; BIT 09a; BIT 09b; MIA 11] or organic dusts [IGA 09b].

$$X_{gm} = \exp\left(\frac{\sum M_i \ln(x_i)}{\sum M_i}\right) \quad [2.7]$$

$$\sigma_{gm} = \exp\left(\sqrt{\frac{\sum M_i \left(\ln(x_i) - \ln(X_{gm})\right)^2}{\sum M_i}}\right) \quad [2.8]$$

2.2.7.2. *Graphic methods*

Different sizes corresponding to representative fractions of cumulative percentage passing such as D_{95}, D_{90}, D_{84}, D_{75}, D_{60}, D_{50}, D_{30}, D_{25}, D_{16}, D_{10}, and D_5 are classically used to determine different parameters which are supposed to characterize the PSD. In particular, the mean, the standard deviation, the skewness and the kurtosis, on the basis of a normal law, can be approximated from these different values [FOL 74].

Other parameters can also be used, with the aim of evaluating the uniformity and the extent of the PSD at hand. Among the commonly-used dimensionless coefficients, the Hazen coefficient or coefficient of uniformity, $C_u = D_{60}/D_{10}$ and the coefficient of curvature $C_c = D_{30}^2/D_{10}D_{60}$ can be used, although the nomenclature employed to qualify mineral aggregates is not directly applicable to the current case.

2.2.7.3. *Distribution models*

The maximum amount of information can be obtained if a distribution model is adjusted to the PSD under investigation [DJA 97]. Different laws exist that are based on semi-infinite variables. Usually, basic models with two parameters – one relating to the average size and the other to the extent of the distribution – are used.

2.2.7.3.1. Log-normal distribution

Given the uni-modal nature of the distributions presented and apparently symmetrical according to a logarithmic scale of size, the Log-normal law seems obvious as a first approach [LIM 01]. Its distribution function of the lengths X is written as follows:

$$P_{Log.N}(X \leq x) = \frac{1}{2}\left[1 + \text{erf}\left(\frac{\ln(x) - \mu}{\sigma\sqrt{2}}\right)\right] \qquad [2.9]$$

where μ and σ are the parameters needing to be identified, and erf(x) denotes the Gauss error function. It should be noted that e^{μ} and e^{σ} reciprocally represent the weighted geometric mean and the associated standard deviation, which can both also be calculated on the basis of the whole dataset including each particle identified, and denoted as X_{gm} and σ_{gm} (see Table 2.3).

The probability density of this distribution function is written thus:

$$P_{Log.N}(x) = \frac{1}{x\sigma\sqrt{2\pi}} \exp\left[-\left(\frac{\ln(x) - \mu}{\sigma\sqrt{2}}\right)^2\right] \qquad [2.10]$$

The mode, $Mode_{Log.N}$, or dominant value (i.e. the value which is most represented in the size considered in the population of objects under investigation) is obtained on the basis of the maximum of this latter function: $Mode_{Log.N} = e^{(\mu - \sigma^2)}$mm.

The arithmetic mean of the size, or its expected value, $E_{Log.N}$, is written as:

$$E_{Log.N} = \exp\left(\mu + \frac{\sigma^2}{2}\right) \qquad [2.11]$$

and its associated standard deviation is given by:

$$\sigma_{Log.N} = E_{Log.N}\sqrt{\exp(\sigma^2) - 1} \qquad [2.12]$$

2.2.7.3.2. Rosin-Rammler distribution

When the size distribution relates to particles or fragments obtained by grinding, one of the other models used most frequently in the existing body of literature is the Rosin-Rammler model [DJA 97], particularly in the case of biomass, [ALL 03; ALL 04; BIT 09b; BIT 11]. The distribution function of this model, identical to the Weibull distribution, can better represent skewed distributions [ROS 33]:

$$P_{RR}(X \leq x) = 1 - \exp\left[-\left(\frac{x}{\lambda}\right)^k\right] \qquad [2.13]$$

where λ and k are constants relating respectively to the dimension of the 63.2th percentile of the distribution function and to the tightening of the distribution. This distribution function has the advantage of exhibiting a reciprocal function which can be used to directly calculate the dimensions corresponding to a given cumulative fraction.

$$x = \lambda\left[-\ln\left(1 - P_{RR}(X \leq x)\right)^{1/k}\right] \qquad [2.14]$$

The median size of the particles $D_{RR}(50)$ corresponds to $P(X \leq x) = 50\%$ and is equal to $\lambda.ln(2)^{1/k}$mm.

The probability density derived from the Rosin-Rammler distribution function can be written as follows:

$$P_{RR}(x) = \frac{k}{\lambda}\left(\frac{x}{\lambda}\right)^{k-1} \exp\left[-\left(\frac{x}{\lambda}\right)^{k}\right] \qquad [2.15]$$

The mode of this distribution, $Mode_{RR}$, is written:

$$Mode_{RR} = \lambda\left(\frac{k-1}{k}\right)^{1/k} \qquad [2.16]$$

The arithmetic mean of the size, or its expected value E_{RR}, is:

$$E_{RR} = \lambda\,\Gamma\left(1 + \frac{1}{k}\right) \qquad [2.17]$$

where $\Gamma(x)$ denotes the gamma function.

The associated arithmetic standard deviation is given by:

$$\sigma_{RR} = \sqrt{\lambda^2\,\Gamma\left(1 + \frac{2}{k}\right) - E_{RR}{}^2} \qquad [2.18]$$

2.2.7.3.3. Adjustment of the distribution laws

The aforementioned two distribution models can easily be adjusted to the area fraction and volume fraction distributions – see Figure 2.23. They encompass each of the distributions, in width or length. Overall, the distributions are better represented by a log-normal law – particularly for the smallest values. These two models can also be adjusted for data obtained by sieving, but with far fewer points (see Figure 2.21). Using a "least-squares" method, the correlation coefficient R^2 is mentioned in each of the cases studied in Table 2.3. The values deduced from the parameters of these models can thus be used to give a fairly accurate overall characterization of the PSD under study.

The standardized distributions and the superposition of probability density functions of these models are shown in Figure 2.22 in the case of sieving, considering the diagonal of the square opening, i.e. $2^{1/2}d$, and in Figures 2.24 and 2.25 for the distributions by area fraction and mass fraction respectively. Overall, these different figures show that these two models describe the distributions observed fairly well. The Rosin-Rammler model is better at describing the PSD obtained by sieving of the finer particles, because it takes account of greater skewness toward fine particles. The log-normal law, for its part, is better at describing the distributions obtained by image analysis and corresponds more closely to the modes observed.

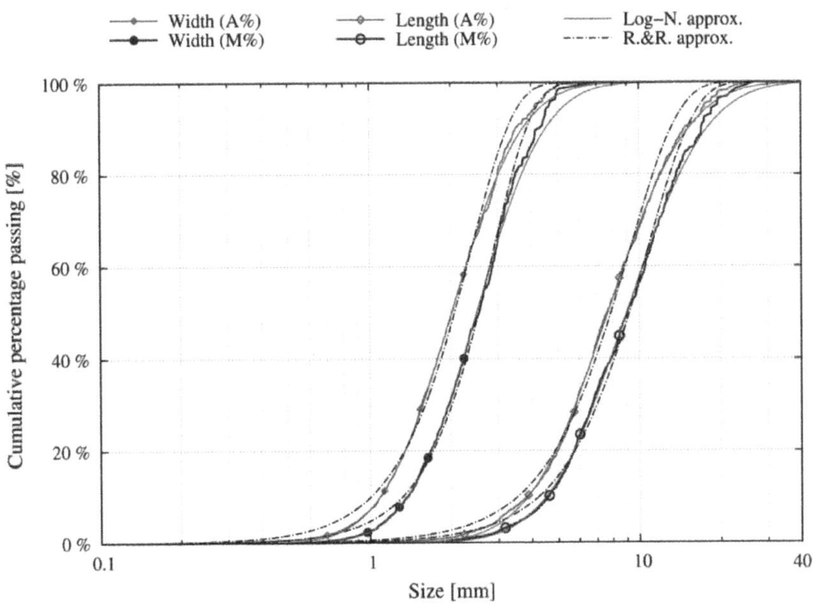

Figure 2.23. *Comparison of the models adjusted to the cumulative distributions of width and length in an area fraction and mass fraction*

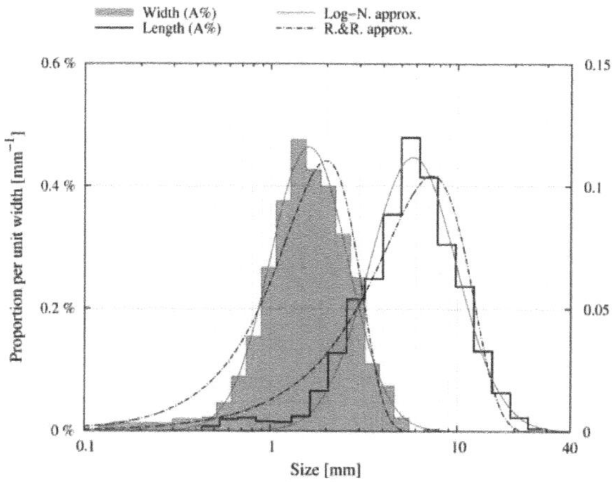

Figure 2.24. *Comparison of the models adjusted to the distributions of width and length in an area fraction*

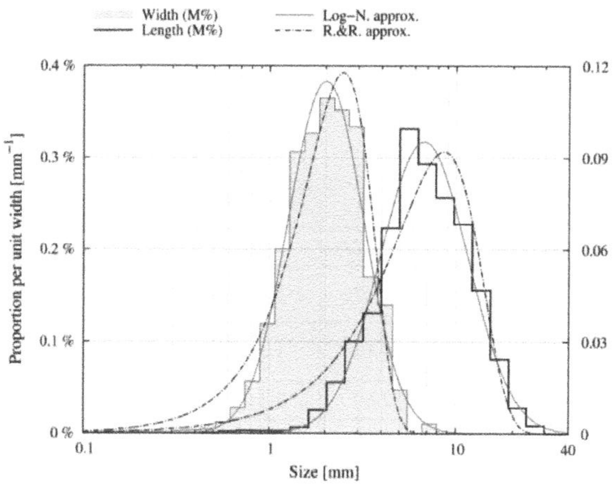

Figure 2.25. *Comparison of the models adjusted to the distributions of width and length in mass fraction*

2.2.7.3.4. Values characteristic of the present case

The values of the different parameters characteristic of the distributions are recapped in Table 2.3. Of these parameters selected and to qualify the distributions, in that these distributions are fairly well represented by a log-normal model, the geometric mean X_{gm} and its associated standard deviation σ_{gm}, which correspond respectively to the values of e^{μ} and e^{σ} for a long-normal distribution described in equation [2.9], seem fairly representative of the PSDs presented – particularly as regards the description of the granular packing that they are likely to cause. The comparison of these values, juxtaposed in Table 2.3, enables us to fully appreciate the representativeness of the adjustments made.

	Sieving		Width		Length	
	d	$2^{1/2}$d	(A%)	(M%)	(A%)	(M%)
Log-normal, see equation [2.9]						
R^2	0.99998	0.99998	0.99963	0.99902	0.99931	0.99916
μ	0.595	0.942	0.701	0.914	2.039	2.197
σ	0.396	0.396	0.480	0.465	0.537	0.539
$Mode_{Log.N} = e^{(\mu-\sigma^2)}$ [mm]	1.55	2.19	1.60	2.01	5.76	6.74
Rosin-Rammler, see equation [2.13]						
R^2	0.99997	0.99998	0.99342	0.99393	0.99385	0.99241
λ [mm]	2.13	3.01	2.38	2.88	9.22	10.71
k	2.98	2.98	2.62	2.87	2.35	2.42
D_{50} see equation [2.14] [mm]	1.86	2.63	1.98	2.48	7.28	8.60
ModeRR see equation [2.16] [mm]	1.86	2.63	2.07	2.54	7.89	9.21
Arithmetic mean and standard deviation [mm]						
E_{am}	1.65	2.33	2.21	2.68	8.69	10.04
σ_{am}	0.65	1.77	1.02	1.15	4.58	5.00
$E_{Log.N}$ see equation [2.11]	1.96	2.77	2.26	2.78	8.87	10.41
$\sigma_{Log.N}$ see equation [2.12]	0.81	1.14	1.15	1.37	5.13	6.04
E_{RR} see equation [2.17]	1.88	2.69	2.11	2.57	8.17	9.50
σ_{RR} see equation [2.18]	0.69	0.98	0.87	0.97	3.70	4.18
Geometric mean and standard deviation [mm]						
X_{gm}	1.45	2.06	1.99	2.45	7.58	8.85
$e^{\mu} = D_{50\ Log.N}$	1.81	2.56	2.02	2.50	7.68	9.00
σ_{gm}	1.71	1.71	1.61	1.55	1.72	1.67
e^{σ}	1.49	1.49	1.62	1.59	1.71	1.71
Dimensionless coefficients						
Cu	1.97	1.97	2.09	2.09	2.29	2.29
Cg	1.03	1.03	0.97	0.98	1.02	0.93

Table 2.3. *Calculated and adjusted parameters of the PSDs exhibited in the case of the hemp shiv under investigation*

The correspondence of these values and standard deviations obtained is excellent, while the geometric means obtained are in accordance with the distributions obtained by image analysis, and differ slightly for the distributions obtained by sieving. This difference may be due to some of the finer particles being retained in the sieve, particularly in the last and finest sieves, whose meshes are finer than 1.25 mm. It is also interesting to note that the geometric standard deviations, for the area fraction or mass fraction weighted variables, are very similar for both the widths and lengths. This confirms the generally homothetic nature of the geometry of the particles.

Finally, as regards geotechnical parameters using the dimensionless coefficients Cu and Cg, these PSDs (in mass fraction) would be qualified as uniform, and therefore incorrectly graded or too well sorted to be used in a concrete mix.

2.2.8. *Conclusions*

The image analysis method, based solely on 2D observations, can be used to gain a precise measurement of the length of the hemp shiv particles and, to a lesser extent, when these particles retain connected fibers, of their width. Various image analysis algorithms can be used to determine their width, such as the minimum Féret diameter or the measurement of the small diameter of an adjusted ellipse. Many different shape parameters are available, and a large number of particles can be identified to deliver statistically robust results.

The near-constancy of the average elongation of the particles for the different surface intervals considered, obtained on the basis of the 2D analysis, suggests that the particles generally have a homothetic shape in 3D, i.e. a quasi-constant average flatness. In addition, according to this basic hypothesis, image analysis allows us access to the average elongation $\overline{\varepsilon}$ and the average flatness $\overline{\Phi}$ of the particles which are likely to cause significant anisotropy in the materials, in that the modes of implementation or casting of the materials including hemp shiv tend to orient these particles. If we also consider this hypothesis, the comparison between sieving and image analysis shows that the square-mesh sieves conventionally used for mineral aggregates separate the hemp shiv particles, elongated and flat, essentially on the basis of their width. Furthermore, given how thin they are in comparison to their width, the hemp-shiv particles pass through the sieves if that width is less than the diagonal of the square openings.

Hence, conventional sieving offers an initial approach to the distribution of the width of the particles. Notably, it enables us, if need be, to separate the cortical fibers remaining after the defibration operations, and also to evaluate the amount of finer particles, or dust depending on the definition adopted, contained in the hemp shiv. Sieving can also be used to supplement the image analysis techniques presented herein.

From a practical point of view, therefore, image analysis can be performed very easily with basic office materials, a flatbed scanner (or digital camera) and a modern computer. The software packages, which are freely available online, can be used in their basic, default configurations. Experience shows us, in the case of the hemp shiv under investigation, that a detection threshold for the particles set between 0.08 and 0.8 mm^2, with no special discrimination between the particles in view of their convexity, yields similar results: with no significant differences in terms of the distributions either of area fraction or of mass fraction. In addition, it is sufficient to sample 4 g of material obtained by successive stages of quartering to provide reliable and reproducible results.

The distributions obtained by sieving and image analysis, for their width or length, cumulative in terms of area or mass, are uni-modal and can be accurately approximated using a classic log-normal law. The weighted geometric mean and the associated standard deviation therefore constitute the main representative parameters with regard to the granular packing and to the overall properties induced in the materials, to characterize this type of PSD.

2.3. Compactness and compressibility

Unlike mineral aggregates, usually employed in hydraulic concretes, aggregates made from plant particles and highly deformable and cannot provide the material with a rigid skeleton.

The apparent density ρ_V of loose dry hemp shiv (see Table 2.2 and section 2.1.4), can be significantly increased by the application of confining stress. The application of such stress is sometimes necessary for storage and/or transport. Applied upon implementation or casting, it can help greatly reduce the porosity in the elaborated material and increase its mechanical resistance [NGU 09].

A compacting test is performed in a steel cylinder with an internal diameter of 160 mm and a height of 320 mm. The stress is applied axially by a piston sliding into the cylinder. Compacting is performed by a hydraulic press with a capacity of 500 kN. The movement of the piston and the force applied are recorded. The force-displacement curve for both types of aggregates is recorded and expressed as Figure 2.26, plotting axial applied stress against equivalent bulk density of dry hemp shiv.

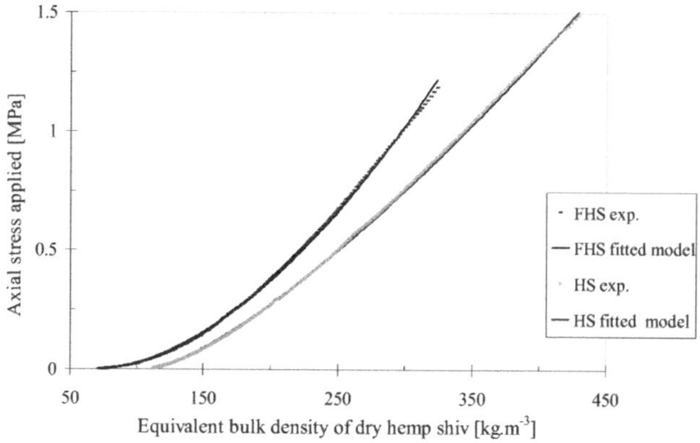

Figure 2.26. *Change in the apparent density of loose dry hemp shiv as axial stress is applied in the cylinder*

For this test, the hemp shiv was humidified so that its water content was 100%, so that the test would be representative of the mechanical behavior of the hemp shiv at the time when it is cast. This in fact corresponds to its water content at the moment of mixing during the manufacture of blocks destined for prefabrication by controlled compacting in the fresh state of hemp concrete [NGU 09; NGU 10a].

Figure 2.26 demonstrates that the apparent density of loose hemp shiv becomes greater than that of the uncompressed shiv once stress of 0.7 MPa is applied. The inter-granular porosity is then probably reduced, but, in view of the deformability of the shiv particles, no information can be gleaned from these tests about the inter-granular porosity in the compressed bulk.

In order to characterize the compressibility of hemp shiv in bulk, a basic model with two parameters, σ_o and k, described by equation [2.19], inspired by models frequently used in the study of compacting of different biomasses [EMA 07], was used. It directly links the stress applied to the hemp shiv, σ, to the equivalent bulk density of dry hemp shiv in its initial state, ρ_{Vo}, and to that in the stressed state ρ_V.

$$\sigma = \sigma_o \left(\frac{\rho_V - \rho_{Vo}}{\rho_{Vo}} \right)^k \qquad [2.19]$$

This model fairly accurately describes the experimental behavior observed within the range of stresses applied. The values of the parameters σ_o and k are recapped in Table 2.4. The parameter k, which relates to compressibility, indeed appears greater in the case of fibrous hemp shiv, FHS, because it exhibits greater relative deformation under equal stress.

	ρ_{V_0} [kg.m^{-3}]	σ_0 [MPa]	k (compressibility) [–]
HS	112	0.38	1.3
FHS	71	0.13	1.8

Table 2.4. *Fitted parameters of the compressibility model to the cases under investigation here*

2.4. Water absorption capacity

In order to determine the water absorption capacity of the hemp shiv and its absorption kinetics, measurements were taken from 100 g samples of completely dry shiv. They were dried beforehand at 60 °C until a uniform weight was reached, before being immersed in water for different periods of time: one, two, five, 10, 30 and 60 minutes, and then 24 hours and 48 hours. The mass of the sample at a given time enables us to determine the degree of water absorption, which is expressed as the ratio between the sample's weight gain at that particular moment and its initial dry weight.

The absorption kinetics, measured by sequential weighing after immersion in water, is similar for both aggregates, HS and FHS: very fast. The loose dry aggregates absorb 50% of their maximum water capacity in less than a minute [NGU 10a], which suggests significant competitiveness in terms of water mobilization between the loose aggregates and the lime during casting and setting. As an indication, we note that after 48 hours of immersion, the absorption of the HS aggregates reaches 400% and that of the FHS aggregates 350%.

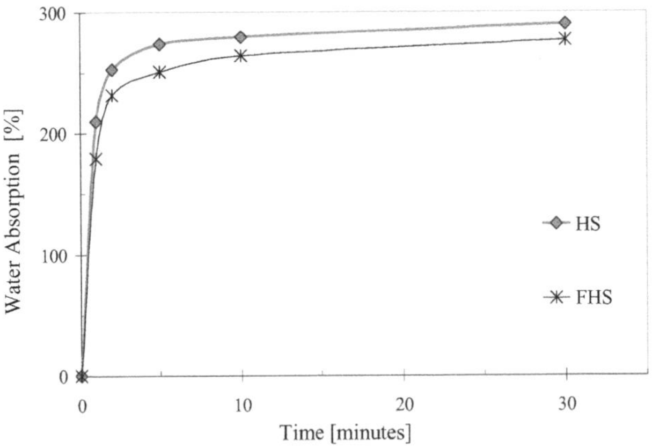

Figure 2.27. *Change in water content of hemp shiv over increasing time of immersion*

The fibers contained in the Fibrous Hemp Shiv are probably less hydrophilic because of their less porous structure. Hence, although the inter-granular porosity is greater in the case of FHS, the kinetics and absorption capacity are less than in the case of HS, which contains very few fibers.

Water absorption capacity plays a very important role for the formulation and casting of hemp concrete. This hydrophilic property of the shiv conditions the use of hydraulic binders. Indeed, it means that hemp shiv will be in competition with the binder to mobilize water, thereby altering the hydration of the binder which may need to react with the water. The function of the binder may therefore be profoundly altered, leading to serious problems such as incomplete setting, "crumbling" and loss of resistance and cohesion.

2.5. Bibliography

[AKI 10] AKIN D.E., "Chemistry of plant fibres" in J. MÜSSIG, *Industrial Applications of Natural Fibres, Structure, Properties and Technical Application*, Wiley, p. 13–22, 2010.

[AKI 96] AKIN D.E., GAMBLE G.R., MORRISON III W.H., RIGSBY L.L., DODD R.B., "Chemical and structural analysis of fibre and core tissues from flax", *J. Sci. Food Agric.*, n° 72, p. 155–165, 1996.

[AKI 08] AKIN D.E., "Plant cell wall aromatics: influence on degradation of biomass" *Biofuels, Bioproducts, and Bioprocess,* n° 2, p. 288–303, 2008.

[ALL 03] ALLAIRE S.E., PARENT L.E., "Size Guide Number and Rosin–Rammler approaches to describe particle size distribution of granular organic-based fertilisers" *Biosystems Engineering,* n° 86, p. 503–509, 2003.

[ALL 04] ALLAIRE S.E., PARENT L.E., "Physical Properties of Granular Organic-based Fertilisers, Part 1: Static Properties" *Biosystems Engineering,* n° 87, p. 79–87, 2004.

[ASA 06] ASABE, Standards. Method of determining and expressing particle size of chopped forage materials by screening, ANSI/ASAE S424.1., St. Joseph, MI, p. 619–662, 2006.

[BEV 08] BEVAN R., WOOLLEY T., "Hemp Lime Construction, A guide to building with hemp lime composites" *HIS Bre Press,* 2008.

[BIT 09a] BITRA V.S.P., WOMAC A.R, CHEVANAN N., MIU P.I., IGATHINATHANE C., SOKHANSANJ S., SMITH D.R., "Direct mechanical energy measures of hammer mill comminution of switchgrass, wheat straw, and corn stover and analysis of their particle size distributions", *Powder Technology*, n° 193, p. 32–45, 2009.

[BIT 09b] BITRA V.S.P., WOMAC A.R., CHEVANAN N., YANG Y.T., MIU P.I., IGATHINATHANE C., SOKHANSANJ S., "Mathematical model parameters for describing the particle size spectra of knife-milled corn stover", *Biosystems Engineering,* n° 104, p. 369–383, 2009.

[BIT 11] BITRA V.S.P., WOMAC A.R., YANG Y.T., MIU P.I., IGATHINATHANE C., CHEVANAN N., SOKHANSANJ S., "Characterization of wheat straw particle size distributions as affected by knife mill operating factors", *Biomass and Bioenergy*, vol. 35, n° 8, p. 3674–3686, 2011.

[BOU 06] BOULOC P., ALLEGRET S., ARNAUD L., "Le chanvre industriel : production et utilisations", *France agricole*, Paris, 2006.

[BRU 09] De BRUIJN P.B., JEPPSSON K.-H., SANDIN K., NILSSON C., "Mechanical properties of lime-hemp concrete containing shives and fibres", *Biosystems Engineering,* n° 103, p. 474–479, 2009.

[BUR 08] BURANOV A.U., MAZZA G., "Lignin in straw of herbaceous crops", *Industrial Crops and Products*, n°28, p. 237–259, 2008.

[BUR 07] BURANOV A.U., MAZZA G., "Fractionation of flax shives by water and aqueous ammonia treatment in a pressurized low polarity water (PLPW) extractor", *J. Agric. Food Chem,* n° 55, p. 8548–8555, 2007.

[CER 05] CEREZO V., Propriétés mécaniques, thermiques et acoustiques d'un matériau à base de particules végétales : approche expérimentale et modélisation théorique, Doctoral Thesis, Ecole Doctorale MEGA, Lyon, 2005.

[CEY 08] CEYTE I., Béton de chanvre, définition des caractéristiques mécaniques de la chènevotte, Travail de Fin d'Etudes, ENTPE, 2008.

[COX 99] COX M., EL-SHAFEY E., PICHUGIN A.A., APPLETON Q., "Preparation and characterization of a carbon adsorbent from flax shive by dehydration with sulfuric acid", *J. Chem. Technol. Biotechnol,* n° 74,p. 1019–1029, 1999.

[CRÔ 05] CRÔNIER D., MONTIES B., CHABBERT B., "Structure and chemical composition of bast fibers isolated from developing hemp stem", *Journal of agricultural and food chemistry*, n° 53, p. 8279–8289, 2005.

[DJA 97] DJAMARANI K.M., CLARK I.M., "Characterization of particle size based on fine and coarse fractions", *Powder Technology*, n° 93, p. 101–108, 1997.

[ELF 08] ELFORDY S., LUCAS F., TANCRET F., SCUDELLER Y., GOUDET L., "Mechanical and thermal properties of lime and hemp concrete ("hempcrete") manufactured by a projection process", *Construction and Building Materials*, n° 22, p. 2116–2123, 2008.

[EMA 07] EMAMI S., TABIL L.G., "Friction and compression characteristics of chickpea flour and components", *Powder Technology*, n°182, p. 119–126, 2008.

[FEN 89] FENGEL D., WEGENER G., *Wood Chemistry, Ultrastructure, Reactions*, Walter de Gruyter Press, New York, p. 56–58, 1989.

[FID 09] FIDELIUS S., Etude de la mise en œuvre du béton de chanvre pour la construction, Research Masters Dissertation, Lorient, 2008.

[FOC 92] FOCHER B., "Physical characteristics of flax fibre", SHARMA, H.S.S. and VAN SUMERE C.F.M., *The Biology and Processing of Flax*, Belfast, p. 11–32, 1992.

[FOL 74] FOLK R.L., *Petrology of Sedimentary Rocks*, Hemphill Publishing Co., Austin, Texas, 1974.

[GAR 98] GARCIA-JALDON C., DUPEYRE D., VIGNON M.R., "Fibres from semi-retted hemp bundles by steam explosion treatment", *Biomass and Bioenergy*, n° 14, p. 251–260, 1998.

[IGA 08] IGATHINATHANE C., PORDESIMO L.O., COLUMBUS E.P., BATCHELOR W.D., METHUKU S.R., "Shape identification and particles size distribution from basic shape parameters using ImageJ", *Computers and Electronics in Agriculture*, n 63, p. 168–182, 2008.

[IGA 09a] IGATHINATHANE C., PORDESIMO L.O., COLUMBUS E.P., BATCHELOR W.D., SOKHANSANJ S., "Sieveless particle size distribution analysis of particulate materials through computer vision" *Computers and Electronics in Agriculture*, n° 66, p. 147–158, 2009.

[IGA 09b] IGATHINATHANE C., MELIN S., SOKHANSANJ S., BI X., LIM C.J., PORDESIMO L.O., "Columbus E.P., Machine vision based particle size and size distribution determination of airborne dust particles of wood and bark pellets", *Powder Technology*, n° 196, p. 202–212, 2009.

[IGA 09c] IGATHINATHANE C., PORDESIMO L.O., BATCHELOR W.D., "Major orthogonal dimensions measurement of food grains by machine vision using ImageJ", *Food Research International*, n° 42, p. 76–84, 2009.

[JIM 93] JIMENEZ L., LOPEZ F., "Characterization of paper sheets from agricultural residues" *Wood and Science Technology*, n°27, p. 468–474, 1993.

[KAR 04] KARUS M., VOGT D., "European hemp industry: Cultivation, processing and product lines", *Euphytican*, n° 140, p. 7–12, 2004.

[KHR 96] KHRISTOVA P., YOSSIFOV N., GABIR S., "Particle board from sunflower stalks: Preliminary trials", *Bioresource Technology*, n° 58, p. 319–321, 1996.

[KWA 99] KWAN A.K.H., MORA C.F., CHAN H.C., "Particle shape analysis of coarse aggregate using digital image processing", *Cement and Concrete Research*, n° 29, p. 1403–1410, 1999.

[LIM 01] LIMPERT E, STAHEL W.A., ABBT M., "Log-normal Distributions across the sciences: keys and clues", *BioScience*, vol. 51, n° 5, p. 341–352, 2001.

[LIN 02] LINGER P, MÜSSIG J, FISCHER H, KOBERT J., "Industrial hemp (Cannabis sativa L.) growing on heavy metal contaminated soil: fibre quality and phytoremediation potential", *Industrial Crops and Products*, n°16, p. 33–42, 2002.

[MAN 04] MANI S., TABIL L.G., Sokhansanj S., "Grinding performance and physical properties of wheat and barley straws, corn stover and switchgrass", *Biomass and Bioenergy*, n° 27, p. 339–352, 2004.

[MIA 11] MIAO Z., GRIFT T.E., HANSEN A.C., TING K.C., "Energy requirement for comminution of biomass in relation to particle physical properties", *Industrial Crops and Products*, n° 33, p. 504–513, 2011.

[MOR 00] MORA C.F., KWAN A.K.H., "Sphericity, shape factor, and convexity measurement of coarse aggregate for concrete using digital image processing", *Cement and Concrete Research*, n° 30, p. 351–358, 2000.

[NEL 00] NELSON, D.L., COX M.M., *Lehninger Principles of Biochemistry*, 3rd edition, Worth Publishers, New York, NY, 2000.

[NGU 09] NGUYEN T.T., PICANDET V., AMZIANE S., BALEY C., "Influence of compactness and hemp hurd characteristics on the mechanical properties of lime and hemp concrete" *European J. of Env. and Civil Eng.* n° 13, p. 1039–1050, 2009.

[NGU 10a] NGUYEN T.T., Contribution à l'étude de la formulation et du procédé de fabrication d'éléments de construction en béton de chanvre. Doctoral Thesis, University Bretagne Sud, Ecole Doctorale SICMA, Lorient, 2010 (Available online: http://web.univ-ubs.fr/limatb/lab/).

[NGU 10b] NGUYEN T.T., PICANDET V., CARRÉ P., LECOMPTE T., AMZIANE S., BALEY C., "Effect of compaction on mechanical and thermal properties of hemp concrete", *European J. of Env. and Civil Eng.* n° 14, p. 545–560, 2010.

[NOZ 12] NOZAHIC V., Vers une nouvelle démarche de conception des bétons de végétaux lignocellulosiques basée sur la compréhension et l'amélioration de l'interface liant/végétal. Doctoral Thesis, University Blaise Pascal, Clermont-Ferrand, 2012.

[RAH 10] RAHMAN KHAN MD. M., CHEN Y., LAGUË C., LANDRY H., PENG Q., ZHONG W., "Compressive properties of hemp (Canabis Sativa L.) stalks", *Biosystems Engineering*, n° 106, p. 315–323, 2010.

[ROS 33] ROSIN P., RAMMLER E., "The laws governing the fineness of powdered coal", *Journal of Instrument Fuel*, n° 7, p. 29–36, 1933.

[SAI 02] SAIN, M., FORTIER D., LAMPRON E., "Chemi-refiner mechanical pulping of flax shives: refining energy and fiber properties", *Biores. Technol.* n° 81, p. 193–200, 2002.

[SED 08] SEDAN D., PAGNOUX C., SMITH A., CHOTARD T., "Mechanical properties of hemp fibre reinforced cement: Influence of the fibre/matrix interaction", *Journal of the European Ceramic Society*, n° 28, p. 183–192, 2008.

[SCH 06] SCHÄFER T., HONERMEIER B., "Effect of sowing date and plant density on the cell morphology of hemp (Cannabis sativa L.)", *Industrial Crops and Products*, n°23, p. 88–98, 2006.

[SHA 06] SHAHIN M.A., SYMONS S.J., POYSA V.W., "Determining soya bean seed size uniformity with image analysis", *Biosystems Engineering*, vol. 94, n° 2, p. 191–198, 2006.

[TRO 08] LE TROEDEC M., SEDAN D., PEYRATOUT C., BONNET J.P., SMITH A., GUINEBRETIERE R., GLOAGUEN V., KRAUSZ P., "Influence of various chemical treatments on the composition and structure of hemp fibres", *Composites: Part A*, n° 39, p.514–522, 2008.

[VIG 96] VIGNON M.R., DUPEYRE D., GARCIA-JALDON C., "Morphological characterization of steam exploded hemp fibers and their utilization in polypropylene-based composites", *Bioresource Technology*, n° 58, p. 203–215, 1996.

[YIN 07] YIN Z., PAN Z., WANG C., DONG Y. and OU Y., "Composition, structure and mechanical properties of several natural cellular materials" *Chinese Science Bulletin*, n° 52, p. 2903–2908, 2007.

Chapter 3

Binders

Various binders may be used to make building materials based on aggregates and fibers of plant origin. The choice is essentially made on the basis of the main properties sought (mechanical properties, thermal properties, etc.), which will depend on the usage to be made of the material (support, insulation, etc.), on its manufacture (prefabricated in a factory, poured on site, projected ("shotcrete"), etc.) and on its location regarding the construction (indoors, outdoors, sheltered or otherwise). The choice may also be guided by financial considerations (e.g. local prices of materials) and environmental factors (CO2 balance, for instance).

Among the most commonly used materials, we find Portland cements and hydraulic or aerial lime, most often associated with pozzolanic admixtures such as fly ash, blast furnace slag and metakaolin. We also find plaster for applications where it is sheltered from humidity. Certain commercial binders developed specifically for hemp concrete (hempcrete) are also obtained by mixing these different compounds: for instance, 70% slaked lime, 15% hydraulic lime and 15% pozzolana.

3.1. Portland cements

3.1.1. *General*

By definition, cement is a hydraulic binder, i.e. a finely-ground mineral material which, when mixed with water, forms a paste which sets and hardens by a series of

---

Chapter written by Gilles ESCADEILLAS, Camille MAGNIONT, Sofiane AMZIANE and Vincent NOZAHIC.

reactions and hydration processes and which, once hardened, retains its strength and stability even when exposed to water.

From a mineralogical point of view, cement can be defined as a mixture of base oxide CaO and acid oxides such as SiO_2, Al_2O_3 or Fe_2O_3.[1]

3.1.2. *Production*

The main constituent in cements is clinker, which is obtained by sintering an appropriate mix of limestone (80%), clay (20%) and sometimes shale:

– the limestone provides calcium oxide;

– the clays provide silica, aluminum, iron oxides and alkalines;

– the shales provide all the oxides.

The main stages of manufacture are as follows (dry or semi-dry manufacture):

– preparation of the raw materials (conditions the hydraulicity and regularity of the clinker):

- extraction and crushing of the primary materials with pre-homogenization;

- grinding after possible addition of admixtures (bauxite, silica sand, fly ash) followed by homogenization.

– kiln-firing of the raw materials (conditions the clinker's reactivity):

- decarbonation of the limestone in a pre-heating cycle (between 700 and 950°C);

- calcination in a horizontal rotary kiln a few meters in diameter and tens of meters in length (continuous production of between 500 and over 1000 tons per day) and formation of new compounds in solid and liquid phase (between 950 and 1450°C);

- clinkerization (at 1450°C) by immersing the liquid solution in air.

– milling of the clinker (conditions the cement's reactivity) and elaboration of the cements:

---

1 It is common, in cement chemistry, to use the initial letter of the oxides instead of the conventional chemical notations: C for CaO, S for SiO_2, A for Al_2O_3, F for Fe_2O_3, $\underline{K}$ for K_2O, N for Na_2O and H for H_2O. For sulfate and carbonate the initial letter is overlined: $\overline{S}$ for SO_3 and $\overline{C}$ for CO_2.

- milling along with any other mineral compounds (cements with admixtures) and gypsum or anhydrite (set regulators) in ball grinders or rolling presses equipped with closed-circuit separators (only the finest grains come out).

It is important to note that every type of clinker will have different properties depending on the nature of the raw material (which depends on the sources of the primary materials) and the clinkerization (which depends on the procedure at the cement plant). Similarly, using the same clinker, the cement plant may produce numerous different types of Portland cement, playing on the fineness and the type and the content of other constituents (limestone admixtures, fly ash, blast furnace slag, etc.).

3.1.3. *Chemical and mineral composition*

From a chemical point of view, the clinker, which is the main component in Portland cement, is made primarily of calcium oxides (60–69%), silicon (18–24%), aluminum (4–8%) and iron (1–8%) and, in a minor degree, magnesium oxides (less than 5%), sodium and potassium (less than 2%). In cement, we also find sulfates (less than 3%) which are introduced in the form of gypsum or anhydrite after clinkerization so as to regulate its setting.

From a mineralogical point of view, the compounds in the clinker are grouped together in the form of calcium silicates ($3CaO.SiO_2$ or C_3S or alite, and $2CaO.SiO_2$ or C_2S or belite), which make up the main bulk of the reactive phases (60–65% and 10–20% respectively), and calcium aluminates ($3CaO.Al_2O_3$ or C_3A, and $4CaO.Al_2O_3.Fe_2O_3$ or C_4AF) which represent the rest (8–12% for each of these compounds).

3.1.4. *Properties*

3.1.4.1. *Applicable standard*

Portland cements have to conform to the standard NF EN 197-1 [NFE 01], which applies to all 27 common forms of cement, and prescribes:

– the properties of the components of these cements and the proportions in which they must be combined;

– the mechanical, physical and chemical specifications;

– the rules for evaluating conformity to the specifications.

Thus, the usual cements are subdivided into five main types:

– CEM I: Portland cement with at least 95% clinker;

– CEM II: Compound Portland cement which may contain between six and 35% of components other than clinker (blast furnace slag, silica fume, fly ash, limestone, etc.);

– CEM III: blast furnace cement with 36–95% blast furnace slag;

– CEM IV: pozzolanic cement which may contain between 11 and 55% of pozzolanic components;

– CEM V: cement composed of blast furnace slag (18–50%) and pozzolanic components (18–50%, mainly accounted for by fly ash).

For all cements, the specifications relate to the mechanical compressive strength (determined after 28 days in accordance with NF EN 196-1 [NFE 06a]) and to the set starting time and the stability:

– the typical classes of strength are 32.5, 42.5 and 52.5 (MPa) and correspond to the minimum strengths measured at 28 days on standardized mortar specimens;

– the set starting time must be greater than 45 minutes for 52.5 cements and 75 minutes for 32.5 cements;

– the stability (expansion) must be below 10 mm.

Chemical specifications are also defined, depending on the cements, and relate to loss on ignition, insoluble residue, sulfate and chloride content and pozzolanicity.

3.1.4.2. *Setting and hardening mechanisms*

The setting of cements was described in 1887 by Le Chatelier and obeys thermodynamic and kinetic laws. It takes place in two stages: a stage of dissolution followed by a stage of crystallization. Crystallization results from the difference in solubility between the anhydrous elements and the hydrous elements. In this case, the dormant period (where there is no rise in temperature) corresponds to the dissolution of the anhydrous constituents and the saturation of the solution. The setting (with temperature elevation) corresponds to the desupersaturation of the solution by crystallization of stable hydrates. This period of setting is followed by a period of hardening which may last several months (the rate of hydration is limited by the diffusion of water and ions through the first layer of hydrates).

3.1.4.2.1. Hydration of Portland cement alone

The hydration of Portland cement is usually broken down into the hydration of its different components, even if there is significant interaction between them.

For calcium silicates (C_3S and C_2S), the hydration reactions can be written diagrammatically (and in cement-science notation) as follows:

$$C_2S + H \longrightarrow C\text{-}S\text{-}H + 0.3CH$$

$$C_3S + H \longrightarrow C\text{-}S\text{-}H + 1.3CH$$

Although these reactions are chemically similar, they differ greatly by their kinetics of reaction (visible hydration front of around 8 µm at 28 days for C_3S and only 0.6 µm for C_2S) and their exotherm (respectively 502 J/g and 259 J/g at 180 days for C_3S and C_2S).

The characteristics of the products formed are as follows:

– calcium silicate hydrates, or C-S-H[2], represent 60–70% of the volume of a hydrated paste. It is an incompletely-crystallized, nanocrystalline hydrate, with a porous structure with small pores, which offers a large specific surface or active area (around 250 m^2/g). Its composition is variable (the C:S ratio varies around an average of 1.7) depending on its age, the hydration temperature, the E:C ratio, etc.

– Portlandite, $Ca(OH)_2$ or CH, represents 20–25% of the hydrated paste. It crystallizes in the form of hexagonal platelets, which in a saturated solution helps maintain a very high (alkaline) pH (12.45) which protects the reinforcing rods from corrosion. It is highly soluble in water (1.2 g/l at 20°C).

For calcium aluminates (C_3A and C_4AF), and in the presence of a sufficient quantity of gypsum, the hydration reactions can be written as follows:

$$C_3A + 3C\bar{S}H_2 + 26H \longrightarrow C_6A\bar{S}_3H_{32}$$

$$3C_4AF + 12C\bar{S}H_2 + 110H \longrightarrow 4C_6(A,F)\bar{S}_3H_{32} + 2(A,F)H_3$$

These two reactions lead to the formation of a tri-sulfoaluminate of calcium hydrate, generally called AFt (alumina, ferric oxide, tri-sulfate), with very different hydration kinetics and heats (respectively 857 and 209 J/g at 28 days for C_3A and C_4AF).

---

2 In cement notation, it is usual to write calcium silicate hydrates in the form C-S-H, with dashes between uppercase letters, denoting the fact that the molar ratio between the different chemical components is not 1:1:1. The same is true for certain calcium aluminate hydrates (C-A-H) and calcium alumino-silicate hydrates (C-A-S-H).

The hydration product of C_3A ($C_6A\overline{S}_3H_{32}$) is called ettringite. It crystallizes in the hexagonal system in the form of needles, is stable at ambient temperature but disappears above 60°C. It must not be confused with secondary or differed ettringite, which are potentially harmful for concrete durability.

When the concentration of sulfates in solution decreases, ettringite may change into calcium monosulfoaluminate, also called AFm, by the reaction:

$$2C_3A + C_6A\overline{S}_3H_{32} + 4H \longrightarrow 3C_4A\overline{S}H_{12}$$

When all the sulfates have been exhausted, calcium hydroaluminates such as C_4AH_{13} or C_3AH_6 begin to form.

These reactions may also occur with C_4AF and also yield AFm and calcium hydroaluminate phases in which some of the aluminum is substituted by iron.

3.1.4.2.2. Hydration of Portland cement in the presence of pozzolanic admixtures

Pozzolanic admixtures which are rich in silica (silica fume) and sometimes in aluminum (fly ash, natural or calcined pozzolans, etc.) and low in calcium, do not cause any chemical reaction on contact with water alone. However, they react in an alkaline environment with the hydrates already formed by the cement and thereby participate in the formation of new hydrate compounds. We speak of a pozzolanic reaction (in reference to Roman cement), which takes place in several stages.

– Step 1 – Rapid hydration reactions:

$$\text{clinker} + nH \longrightarrow C\text{-}S\text{-}H_I + C\text{-}A\text{-}H_I + CH$$

– Step 2 – Slow dissolution of the pozzolanic admixtures:

$$\text{additive} + H \longrightarrow S\text{-}H \ (+A\text{-}H)$$

– Step 3 – Combinations of the pozzolanic admixtures with the hydrates from the binder:

$$S\text{-}H + CH \longrightarrow C\text{-}S\text{-}H_{II}$$

$$A\text{-}H + CH \longrightarrow C\text{-}A\text{-}H_{II}$$

$$S\text{-}H + A\text{-}H + CH \longrightarrow C\text{-}A\text{-}S\text{-}H$$

In brief, the pozzolanic reaction progressively consumes portlandite, which improves the chemical resistance of the concretes in a hostile environment, and forms new hydrates such as C-S-H (with C:S ratios between 1.1 and 1.3), C-A-H or C-A-S-H, which play a part in mechanical strength. However, we must take care to ensure that a certain amount of portlandite is preserved, and therefore a sufficiently alkaline pH of the interstitial solution to protect the reinforcement rods against corrosion. In fact, pozzolanic admixtures are present in limited amounts in cements.

3.1.4.2.3. Hydration of Portland cement in the presence of potentially hydraulic admixtures

Granulated blast furnace slag, made up of calcium oxides (38–45%), silicon (32–37%), aluminum (10–16%), magnesium (3–9%) and iron (0–10%) present in a vitreous form, are potentially hydraulic but with too low water solubility to be used on their own. The sufficiently high pH of the cement paste is able to activate these ground slag particles to form new hydrates by the following overall reaction:

$$\text{clinker} + \text{slag} + n\text{H} \longrightarrow \text{C-S-H} + \text{C-A-H} + \text{CH}$$

Therefore, they can be substituted for clinker in large proportions (up to 95% in CEM III/C cements). Their properties will depend on the proportions of slag used (with more slag, the acid resistance increases, but the mechanical strength falls, particularly at an early age).

3.1.4.3. *Mechanical properties*

Portland cements, used as binders in mortars and concretes, offer great mechanical strength – mainly compressive strength (compressive strength 10 times greater than flexural strength). These strengths change rapidly over time, as 40% of the final strength is obtained after two days and 70% after seven days. Therefore, the formwork can be removed from these concretes at an early stage (generally less than one day) and they can begin to bear a load early (generally between three and seven days).

The essential parameters regarding the strength of a mortar or concrete are the effective water[3]-to-cement ratio (or W_{eff}/C), the type of cement and the dosage. These parameters are to be found in the methods of formulation with, among others, the Bolomey principle which expresses the strength of a concrete, *Rcn*, as a function of the strength of the cement, *Rc*, on the same day, and the dosage of the cement, *C*, of the quality of the aggregates, *G* (a value between 0.40 and 0.60 for mineral

3 Effective water = Total water – Water absorbed by the aggregates.

aggregates) and the content in terms of effective water, W_{eff}, and air, V (C, W_{eff} and V are expressed in kg/m^3 and Rcn and Rc in MPa):

$$R_{cn} = R_c \times G \times \left(\frac{C}{W_{eff} + V} - 0.5 \right)$$

Thus we can obtain materials that exhibit compressive strength, at 28 days, anywhere from 25 MPa (modern concretes with a W_{eff}/C ratio of 0.60) to over 100 MPa (very high-performance concretes with a W_{eff}/C ratio of 0.30–0.40) or even over 150 MPa (ultra-high-performance concretes with a W_{eff}/C ratio of 0.15–0.20). These very high-performance concretes are obtained by also adding ultra-fine particles into the concrete, to increase the compactness of the granular skeleton, and superplasticizers which disperse the ultra-fine particles, reduce the dosage of water and facilitate the use of the concrete.

The temperature also influences the acquisition of strength in concretes by activating chemical reactions, as described by the Arrhenius Equation.

Finally, a parameter which must not be overlooked is the curing of the concrete, i.e. its protection (mainly hydric protection) against the external conditions during the first few days after it is poured. Indeed, significant drying of the concrete at an early stage leads to a loss of performance or even cracking.

3.1.4.4. *Physical and chemical properties*

Hydrated cement, and consequently mortars and concretes, are porous materials. This porosity is the crucial factor in the durability of these materials. It depends essentially on the dosage of cement and the water content of the concrete, but also on the curing of the material. Therefore, it constitutes one of the main physical properties of the material. For instance, the porosity of an ordinary concrete is approximately 15–18% at 28 days, that of a high-performance concrete is 10–12% and that of a very-high-performance concrete is 7–9%.

The porosity of a cement paste takes account of two kinds of pores:

– capillary pores, which are the largest (on average 55 nm) and are the remnants of the inter-granular spaces in the fresh paste. They decrease with age but increase with the W/C ratio;

– micropores (5 nm on average), which constitute the internal porosity of the hydrates.

It is the capillary pores which have an impact on the durability, influencing the permeability and diffusivity of the material.

From a chemical standpoint, hydrated Portland cement is a material qualified as "durable" in a typical environment (building, engineering structures, etc.). Conversely, it cannot withstand harsh chemical ambiences such as a highly acidic environment (it degrades totally below a pH of four) or highly alkaline ones. Similarly, in the presence of renewed water, it will leach out (dissolution of the portlandite and then decalcification of the C-S-H), which will lead to a gradual increase in its porosity and a decrease in its mechanical performances.

The inclusion of pozzolanic materials (blast furnace slag, fly ash, etc.) improves the durability of cements in these chemically hostile environments.

3.1.4.5. *Thermal and hydric properties*

From the point of view of its thermal properties, traditional vibrated concrete based on Portland cement has a very poor insulating capacity, because its thermal conductivity λ[4] is 2 W/(m.K). Conversely, it exhibits excellent heat capacity of 2500 KJ/(m^3.K), which lends it good thermal inertia [THU 05].

As regards its hydric properties, the capillary porosity is crucially the important factor, because the permeability and diffusion depend on the interconnectedness of the capillary pores (this capillary porosity depends essentially on the W_{eff}/C ratio and the degree of hydration of the cement).

To begin with, concrete can be considered as an impermeable material, because its permeability to water is approximately 10^{-20} m^2, while its permeability to gas varies from 10^{-15} to 10^{-17} m^2. Cement paste's low permeability to water vapor is a disadvantage in the case of its association with plant particles to create a sweating composite material, participating in the regulation of the ambient relative humidity and in which hydric transfers play a part in the energy efficiency of the wall [SAM 08].

If we consider concrete in an ambient environment (e.g. a building), the free water present in the capillary porosity will find a state of equilibrium with the humidity of the ambient air, but with very slow kinetics. This can be expressed by tracing the isotherms of sorption (which takes several months!) and we often observe a phenomenon of hysteresis between the adsorption (with increasing hygrometry) and desorption (with decreasing hygrometry) – see Figure 3.1).

---

4 The insulating or conductive capacity of construction materials is generally characterized by their thermal conductivity, λ, expressed in W/(m.K). Indeed, in most cases, we consider that heat transfers within the material by convection and radiation are negligible in comparison to the transfer by conduction.

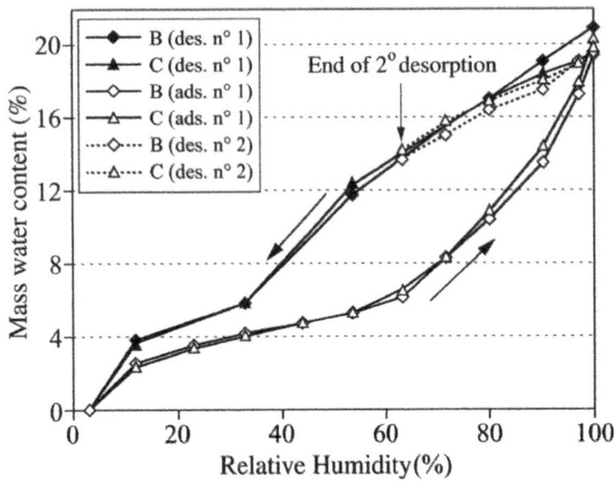

Figure 3.1. *Isotherms of adsorption/desorption of water vapor, determined at 23°C for concrete (Cn) and hardened cement paste (C) (W/C = 0.45) [BAR 07]*

3.1.5. *Environmental impacts*

For a CEM I-type cement (without admixtures), the consumption of energy resources is 5950 MJ/T and the potential for global warming (greenhouse gases), expressed in equivalent kilos of CO_2 per ton of product, is 866 [ATI 11].

These values may be significantly decreased by the mass introduction of constituents other than clinker (CEM II to CEM V cements) which only require one round of milling before evaluation. For a CEM III/A cement, for instance, the energy consumption is only 3720 MJ/T and the potential for global warming is 461 equivalent kilos of CO_2 per ton of product [ATI 11].

Thus, on average between 1990 and 2006, the quantity of CO_2 emitted per ton of cement produced the world over decreased from 890 kg [WBC 02] to 660 kg [WBC 09], essentially because of the commercial development of compound cements.

3.2. Lime

3.2.1. *General*

The use of lime in construction dates from the greatest ages of antiquity, from Asia Minor to Egypt and Greece. In actual fact, much like ceramics, the practice was established with the use of fire, in the Neolithic Age. However, it was the Romans who, by adding various pozzolanic materials to it, managed to turn it into a durable building material.

Lime is obtained by calcination and decarbonation of a limestone rock. The primary reaction is as follows:

$$CaCO_3 \longrightarrow CaO + CO_2$$

The lime thus obtained is called quicklime. In this state, it is unusable for construction (except for specific treatment of soils).

In order for this lime to be used, it must be slaked – i.e. combined with water, leading to the following reaction:

$$CaO + H_2O \longrightarrow Ca(OH)_2$$

This reaction is highly exothermic (1155 kJ/kg of CaO) and a significant increase in volume takes place. Depending on the amount of water, this slaked lime will be in the form of a powder, limewater or lime paste.

Depending on the nature of the limestone used, it is possible to create various different types of lime:

– calcic lime, made primarily of calcium oxides or hydroxides, with no added pozzolanic or hydraulic materials, creating by kiln-firing pure limestone;

– dolomitic lime, made primarily of oxides or hydroxides of calcium and magnesium, without added pozzolanic or hydraulic materials, created by kiln-firing a mixture of limestone (calcium carbonates) and dolomite (magnesium carbonates) or dolomite on its own;

– hydraulic lime, which may be simply hydraulic, natural hydraulic or natural hydraulic with added pozzolanic materials.

We can also differentiate the various types of lime on the basis of how they react in the presence of water:

– aerial lime, made primarily of calcium oxides or hydroxides, hardens slowly in air by carbonation under the effect of the carbon dioxide present in the air [LAW 06]. In general, this type of lime does not harden in reaction with water (no hydraulic properties);

– hydraulic lime, which besides $Ca(OH)_2$ also includes calcium silicate-type anhydrate compounds (C_2S and sometimes C_3S) at calcium aluminates (C_3A and

C_4AF), like in Portland cements, and presents a rapid hydraulic reaction followed by a slower carbonation reaction on contact with air.

3.2.2. *Aerial lime*

3.2.2.1. *Production*

The industrial manufacture of aerial lime is performed in several steps:

– extraction of lime or dolomite from a clearly-identified and relatively pure source (otherwise, the raw materials have to be washed);

– crushing, sieving and possibly washing of the raw material to make it compatible with the kiln (preferable use of the 30–150 mm fraction in vertical kilns, 10–60 mm in horizontal rotary kilns);

– calcination (decarbonation) between 800 and 1200°C in vertical furnaces (most with the raw material fed in from the top, remaining in the furnace for approximately a day and production of between 100 to over 300 tons per day) or horizontal furnaces (the type used in constantly-active cement plants, with the raw material remaining in the furnace for six–eight hours and a production which may be greater than 1000 tons per day);

– sieving and possibly crushing;

– hydration or slaking of the quicklime by dry means in a hydrator. The amount of water introduced into the hydrator is adjusted so that it is sufficient to ensure hydration and the evacuation of the heat given off by the reaction.

3.2.2.2. *Chemical and mineralogical composition*

From a chemical point of view, aerial lime essentially contains calcium- and magnesium oxides, water from the hydration of these compounds and CO_2 from the carbonate phases (the loss on ignition (LOI) is around 25% [ARA 05; LAN 05; LAW 07]).

From a mineralogical point of view, these substances are compounds of lime and magnesium hydrate ($Ca(OH)_2$ and $Mg(OH)_2$) and carbonates of calcium and/or magnesium ($CaCO_3$ and $MgCO_3$). There should be very little, or no, CaO or MgO.

3.2.2.3. *Properties*

3.2.2.3.1. Applicable standard

All forms of lime used in construction are defined by the norm NF EN 459-1 [NFE 02].

As regards aerial limes, this norm classifies them on the basis of their content in terms of calcium and magnesium oxides (CaO + MgO) with the denominations CL 90, 80 or 70 for calcic limes and DL 85 or 80 for dolomitic limes. The letter S is added to indicate that the lime is slaked (Q if it is quicklime).

From a chemical point of view, aside from the minimal content of CaO and MgO, it limits the free water, CO_2 and sulfate content. From a physical point of view, the main requirements relate to the fineness and volumic stability after slaking.

No mechanical requirement is prescribed.

3.2.2.3.2. Setting mechanisms

The process of setting of aerial lime results from the evaporation of excess free water and carbonation – i.e. it dries and then hardens by reacting with atmospheric CO_2. This very slow reaction (which takes several months) takes place in a humid environment and is essentially a classic acid/base reaction:

1) the carbonic gas first reacts with water to form carbonic acid, which is a weak acid:

$$CO_2 + H_2O \longrightarrow H_2CO_3$$

2) the carbonic acid then reacts with the hydrated lime (alkaline) to give a salt (calcium carbonate or calcite) and water:

$$H_2CO_3 + Ca(OH)_2 \longrightarrow CaCO_3 + 2H_2O$$

Finally, we have the following reaction:

$$Ca(OH)_2 + CO_2 \longrightarrow CaCO_3 + H_2O$$

The salt formed (calcium carbonate) is less soluble than the original slaked lime. Furthermore, carbonation takes place with a 12% increase in volume, without overall swelling of the paste but a change in the size of the pores. This leads to an increase in the mechanical strength of the carbonated paste.

The setting time for aerial lime, which results primarily from the evaporation of the initial free water, is over three hours, but it can take far longer than this if the W/L ratio increases (IZA 11).

3.2.2.3.3. Mechanical properties

The mechanical performances of aerial limes are very poor, and also, as the hardening reaction is caused by natural carbonation, the development of mechanical strength is very slow. The mechanical strengths obtained with mortar are around 3-5 MPa after six months of being kept in the presence of CO_2 [LAN 03; MOR 05; IZA 11]. However, the carbonation reaction may continue for several years, with the performances achieved depending, notably, on the conditions of conservation (ambient relative humidity, and presence of SO_2).

3.2.2.3.4. Physical and chemical properties

Aerial lime pastes are characterized by a high degree of highly-interconnected capillary porosity. According to various studies, the open porosity of aerial lime mortars generally lies between 25 and 40% [LAN 03; MOS 06; IZA 11; VEJ 12], but it may be greater than 50% in certain cases [ARA 05]. The poral distribution is nearly unimodal with an average diameter between 0.5 and 1 µm, increasing with the W/L ratio [IZA 11]. Small pores (0.1 µm) are also present in aerial lime and correspond to the pores of the crystals making up the paste. Hence, their size can change during the course of carbonation, owing to the formation of calcite at the expense of portlandite [LAW 07].

3.2.2.3.5. Thermal and hydric properties

As is the case for all mineral binders, the thermal conductivity of aerial lime-based plasters or coatings is relatively high. The values of this property vary depending on the density of the mortar in question, which notably is influenced by the W/L ratio and the dosage of aggregates. According to France's thermal regulation from 2005 (RT2005, [THU 05]), the thermal conductivity of a lime mortar varies between 0.3 and 1.8 W/m.K, but the values cited in the literature are between 0.65 and 0.84 W/m.K in the dry state [CER 06; VEJ 12a; VEJ 12b].

Aerial lime pastes exhibit a high coefficient of capillary absorption due to their highly-interconnected porosity. This coefficient varies greatly depending on the W/L ratio (between 0.05 and 0.09 $g/cm^2.s^{1/2}$ for W/L ratios between 0.8 and 1.3 [ARA 05]) but it can be very significantly reduced (–97%) by incorporation of a hydrophobing agent [IZA 10]. This characteristic may give rise to problems of durability [IZA 11], facilitating the diffusion of liquid water until it reaches the structural material, and favoring the penetration of aggressive agents.

As regards the transport of water vapor, aerial limes are very permeable (around 10 times more so than cement [STR 00]), which facilitates the diffusion of water vapor through the wall and its evacuation into the outside atmosphere, whilst limiting the danger of its condensing within the concrete [IZA 11].

3.2.2.3.6. Durability

Many physical and chemical phenomena may cause the degradation of aerial lime mortars. Primarily we can cite the cycles of wetting/drying, exposure to rain and the resulting leaching, the freeze/thaw cycles, exposure to pollutants and particularly SO_2, or indeed solar radiation.

The properties which influence the durability of these mortars are their chemical composition, their air content, the proportion and size of their pores, and their permeability to vapor and their capillary absorption. The incorporation of adjuvants such as hydrophobing agents, or indeed water reduction agents (derived from starch), can significantly improve the durability of aerial lime mortars [IZA 10].

3.2.3. *Natural hydraulic limes*

3.2.3.1. *Production*

The stages of formation of natural hydraulic limes are calcination, slaking and possibly grinding. The method and temperature of the firing of the limestone are the factors which will influence the amount of silica (SiO_2) that will combine with calcium oxide (CaO) to form calcium silicates (C_2S mainly, and possibly C_3S), which are responsible for the material's hydraulicity:

– the firing takes place in a vertical furnace at a maximum temperature of 1000°C (as opposed to 1450°C for the production of Portland cement) using very pure coal as the fuel (anthracite coal containing 92–95% carbon). The main reaction that takes place during calcination is the following:

$$CaCO_3 + x\,SiO_2 \longrightarrow (1\text{-}2x)\,CaO + x\,2CaO.SiO_2 + CO_2$$

Depending on the composition of the raw material (presence of aluminum and iron oxides), we can also obtain calcium aluminates (C_3A) and calcium aluminoferrites (C_4AF), which are also hydraulic[5];

5 The calcination temperature is generally considered to be insufficient to form C_3S.

– the next stage is slaking, which corresponds to the addition of water to the calcinated rock so as to slake the quicklime, and also render the material powdery (the slaking takes place with a massive release of heat and a 30% increase in volume). This part of the process is very carefully controlled so as to obtain a finished product that is dry and contains almost no quicklime (CaO content < 1%) by the reaction:

$$\begin{aligned}&(1-2x)\,\mathrm{CaO} + \mathrm{x}\,2\mathrm{CaO.SiO_2} + (1\text{-}2\mathrm{x})\,\mathrm{H_2O} \longrightarrow \\ &(1-2x)\,\mathrm{Ca(OH)_2} + \mathrm{x}\,2\mathrm{CaO.SiO_2}\end{aligned}$$

During this slaking process, a certain degree of carbonation between the newly-formed slaked lime and the atmospheric CO_2 may also take place, producing calcite;

– possibly, grinding may complete this phase of slaking.

3.2.3.2. *Chemical and mineralogical composition*

The chemical composition of natural hydraulic limes depends, of course, upon the source used. Primarily, we find calcium oxides (50–70%) and silicon oxides (10–20%), and to a lesser extent oxides of aluminum, iron, magnesium and sulfur. The loss on ignition, related to the water associated with the calcium oxide in the slaked lime or the CO_2 also associated with the calcium oxide in the calcite, may vary from 10 to 20% [LAN 05; MER 07; BRA 12].

From a mineralogical point of view, the main phases are hydrated lime ($Ca(OH)_2$ or CH: 15–30%), dicalcium silicate ($2CaO.SiO_2$ or C_2S: 20–40%) and calcite ($CaCO_3$ or $C\overline{C}$: 20–30%). In a minor proportion (less than 5%), we may find other hydraulic phases ($3CaO.SiO_2$ or C_3S; $3CaO.Al_2O_3$ or C_3A; $4CaO.Al_2O_3.Fe_2O_3$ or C_4AF; $2CaO.Al_2O_3.SiO_2$ or C_2AS) and even calcium sulfate ($CaSO_4$ or $C\overline{S}$).

3.2.3.3. *Properties*

3.2.3.3.1. Applicable standard

The norm NF EN 459-1 of October 2002 [NFE 02] is applicable here as well. Hydraulic limes are called HLs and natural hydraulic limes NHLs. Certain special products, which may contain added adapted hydraulic or pozzolanic materials, up to 20% mass, are denoted by the symbol *Z*.

Hydraulic limes are distinguished by their compressive strength at 28 days: HL 2 or NHL 2 for strengths between 2 and 7 MPa; HL 3.5 or NHL 3.5 for strengths between 3.5 and 10 MPa; and HL5 or NHL 5 for strengths between 5 and 15 MPa.

From a chemical point of view, hydraulic limes must have limited contents in terms of free water, free CaO and SO_3. From the physical point of view, the main requirements relate to the fineness, stability and setting time.

3.2.3.3.2. Setting mechanisms

The setting of hydraulic lime takes place in two stages:

– first, a rapid hydraulic setting, which results from the hydration of the C_2S (as well as the C_3S, the C_4AF, the C_2AS and the C_3A). This mechanism causes the formation of calcium silicate hydrates and portlandite;

– in a second phase, there is a slower hardening (which takes around a year) which results from the carbonation of the aerial lime and portlandite formed previously on contact with the atmospheric CO_2.

The principal reactions governing these two stages of setting are shown below:

$$2CaO.SiO_2 + H_2O \longrightarrow C\text{-}S\text{-}H + Ca(OH)_2$$

$$Ca(OH)_2 + CO_2 \longrightarrow CaCO_3 + H_2O$$

3.2.3.4. *Mechanical properties*

The mechanical strengths of mortars made from natural hydraulic limes change gradually over time, which may cause problems at an early age (very poor strength before seven days). Thus, for an NHL 3.5-type lime, [DOM 06] reports a strength of standardized mortar of 1.0 MPa at 7 days; 3.6 MPa at 28 days (with a Young's modulus of 5200 MPa) and 3.9 MPa at 60 days. After natural or accelerated carbonation, the strength may reach up to 7.1 MPa (with a Young's modulus of 7800 MPa). The values obtained by [LAN 04] are slightly higher, but evolve in a similar way.

From these examples, we can see that the role of the "aerial" strength is just as important as that of the "hydraulic" strength. Furthermore, the mechanical performances will be highly dependent upon the degree of hydraulicity of the lime, the W/L ratio and the curing conditions.

3.2.3.4.1. Physical and chemical properties

Because of their two-fold reaction of hydration and carbonation, the performances of hydraulic limes evolve more quickly than do those of aerial limes.

Hydraulic lime mortars exhibit a degree of porosity between that of cement mortars and that of aerial lime mortars (porosity of 20–25%), with most of this porosity being open. The distribution of the pores is unimodal, with a main family of pores in the vast majority, between 0.1 and 1 μm in diameter [MOS 06].

3.2.3.4.2. Thermal and hydric properties

The thermal conductivity of hydraulic lime-based mortars is fairly high, ranging from 0.3 to 1.0 W/m.K, but is similar to that measured for aerial limes.

The capillary absorption coefficients are less than those measured with aerial lime (around 0.02 $g/cm^2.s^{1/2}$ for mortars with a 26% open porosity, according to [MAR 05]). In terms of the transport of water vapor, hydraulic limes are also less permeable than aerial limes but more permeable than Portland cements.

3.2.3.5. *Environmental impacts*

The production of natural hydraulic lime, much like that of cement, causes a significant CO_2 emission, related on the one hand to the consumption of fossil fuels for the calcination of the raw material, and on the other hand to the decarbonation of the raw materials. However, these emissions are reduced in comparison to those for cement, because the calcination temperature is lower (around 1000°C as compared to 1450°C for cement).

The data provided by the company Saint Astier on Natural Hydraulic Lime [SAI 12] regarding CO_2 emission during the production of NHL 5 report 275 kg of CO_2 per ton of cement from the combustion of coal, and 360 kg of CO_2 per ton from the decarbonation of the primary materials, which gives a total of 635 kg of CO_2 per ton of NHL 5, and fuel consumption of 75 kg of coal per ton of NHL 5.

3.3. Lime-pozzolan mixtures

Pozzolan-lime binders, used since the time of the Romans (see the structure of the Pantheon in Rome, for instance) and the setting of which is based on a relatively slow pozzolanic reaction, have now been replaced in most constructions by Portland cements, with rapid setting and high mechanical performances.

This may not be so for bio-aggregate-based concretes, which – on the one hand – do not need *as* good a mechanical strength and for which – on the other hand– hygrothermal behavior, and particularly permeability to water vapor, must be favored. Indeed, the plant particles require the mix-water to be evacuated in order to present appreciable mechanical performances [ROW 05]. In addition, in the medium and long term, these pozzolanic materials limit the pH of the interstitial solution, thereby decreasing the alkaline attack on the plant particles. Thus, Toledo *et al.* [TOL 03], studying cement-limewater mixtures, observed an improvement in the durability of a concrete including sisal fibers that are vulnerable to alkaline attack. They attribute this improvement to a decrease in the pH, owing to the presence of the limewater.

Thus, most plant-based concretes developed to date were developed from mineral binders formulated from aerial and/or hydraulic lime, pozzolanic admixtures and, possibly, Portland cement to improve the mechanical performances in the short term [COL 04; CER 05; CEN 07; DEB 09; MAG 10; NGU 10].

These mineral binders formulated enable us to combine different properties necessary to the development of such composites and ensure the compatibility of the mineral binder paste and the plant particles [CER 06]:

– clear improvement of the mechanical strengths in the short or medium term;

– correct porosity even though the addition of cement to aerial-lime mortars leads to a progressive decrease in the porosity and the size of the pores. The vapor permeability properties may therefore decrease;

– maintained thermal performances;

– water stability.

Many admixtures may fall into the category of pozzolanic admixtures. They may be natural pozzolans, such as pumice [NOZ 12], or calcinated natural pozzolans such as metakaolin [MAG 10], or industrial by-products such as fly ash (a by-product of electricity production in coal-burning power stations) or silica fume (a by-product of the production of aluminum or magnesium). Blast furnace slags (a by-product of steel manufacture by smelting) may also fall into this category, although they are classified as latent hydraulic admixtures.

3.3.1. *Natural pozzolans*

According to the standard ASTM-C-618 [AST 08], a pozzolan is defined as being a siliceous or silico-aluminous material which does not, on its own, have any (or has very little) binding value, but in powdered form and in the presence of water, reacts chemically with solubilized calcium hydroxide $Ca(OH)_2$ at normal temperatures to form compounds with binding properties. This reaction is said to be "pozzolanic".

3.3.1.1. *General*

Natural pozzolans are defined in the norm NF EN 197-1 [AFN 01] as "*products composed in essence of silica, aluminum and iron oxide, which – either naturally (when they are of volcanic origin) or after thermal activation, exhibit pozzolanic properties*".

The natural pozzolans can be formed in two very different ways [PIC 94]:

– volcanic: scoria, pumices, basalt;

– sedimentary or metamorphic: zeolites, sedimentary silica, opals.

There is a very great diversity of the type of sources and their location. Thus, natural pozzolans have been used since ancient times in mixture with lime in many places the world over. Edifices made using concretes based on these binders can still be seen today – some notable examples are the Pont du Gard in France or the Coliseum in Rome, Italy – and stand in evidence of the durability of this type of material. The Roman architect Vitruvius was the first to compile empirical data on these concretes and their use in construction. Yet the Romans are not the only ones to have used pozzolanic materials. For over 2000 years, the inhabitants of Bali have been using a mixture of coral lime and ash from their sacred volcano, Mount Agung, to build walls [DEL 96].

Yet although there have been numerous studies performed on the making of binders, mortars and concretes based on a mixture between natural pozzolans and lime, precious little research exists on the use of these mixtures with lignocellular particles [NOZ 12].

3.3.1.2. *Transformation*

To be activated, most natural pozzolans require only grinding, which makes the amorphous phase available. Yet like metakaolins, certain sources – particularly metamorphic sources – require calcination between 500 and 800°C in order to exhibit sufficient reactivity [LIE 98].

3.3.1.3. *Chemical and mineralogical composition*

Within the category of natural pozzolans, there is a great variety of compositions, textures and origins. In general, their chemical composition is similar to that of fly ash. Their two main compounds are silica SiO_2 (40–90% mass) and aluminum Al_2O_3 (10–20%). Then come secondary compounds such as iron oxide, Fe_2O_3 (1–15%), magnesium oxide, MgO (0–12%), calcium oxide, CaO (0–12%), sodium oxide, Na_2O (0–7%) or potassium oxide, K_2O (0–6%).

In addition to a very diverse chemical composition, the mineralogy of these materials is, itself, peculiar to each source. In general, there is an amorphous part, mainly silico-aluminous, including alkaline fondants. This amorphous phase constitutes between 30 and 90% of the material, and represents the reactive part which participates in the pozzolanic reaction [PIC 94]. In addition to the amorphous phase, there are numerous crystalline categories represented: plagioclases, alkaline feldspars, sanidine, quartz, biotite, pyroxene, etc.

3.3.1.4. *Properties*

3.3.1.4.1. Applicable standard

Natural pozzolans form part of the components of type II Portland cements defined in the norm NF EN 197-1 [AFN 01]. They are divided into two distinct classes:

– natural pozzolans (P): these come mainly from volcanic sources, whose pozzolanic nature is naturally activated;

– calcined natural pozzolans (Q): these come mainly from sedimentary sources, which must be heat activated.

These admixtures give rise to four products: CEM II/A-P (6–20% mass), CEM II/B-P (21–35% mass), CEM II/A-Q (6–20% mass) and CEM II/B-Q (21–35% mass). Natural pozzolans may also be added into so-called pozzolanic cements (CEM IV/A or B) and compound cements (CEM V/A or B). In the area of road-building, gravel-pozzolan-lime mixtures are the subject of the norm NF P 98-117 [AFN 91].

3.3.1.4.2. Reactivity

Because of the chemical composition of natural pozzolans, the hydrate phases that can be formed in a mixture with lime or in contact with fresh-formed portlandite are defined in the ternary system SiO_2-Al_2O_3-CaO hydrate (see Figure 3.2). The main mechanisms of attack and the reactions involved were defined by Dron *et al.* [DRO 78]. Although a certain degree of reactivity of the crystals could be observed [PIC 94], the alkaline attack upon which the process is founded relates mainly to the amorphous phase. The main products of the pozzolanic reaction taking place are calcium silicates hydrates (C-S-H), gehlenite hydrate (C_2ASH_8) and tetracalcic aluminate hydrates (C_4AH_{13}).

In addition, there have been numerous studies done on the means of activating the pozzolanic reaction by finer grinding, introducing admixtures such as Na_2SO_4 or increasing the curing temperature. On this point, Shi *et al.* [SHI 01] concluded that it increased the efficiency of the chemical activation.

Finally, many tests have been devised to determine the reactivity of a pozzolan. The most widely-used of these remains the Chapelle test. The determination of such reactivity is an essential pre-requisite for its use in cements or in the making of pozzolan-lime binders.

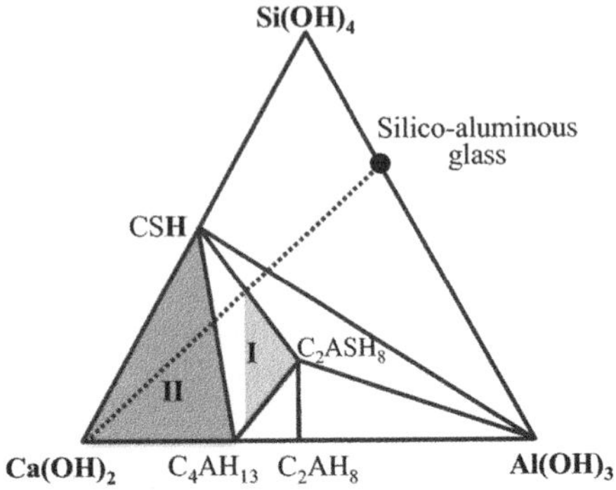

Figure 3.2. *Hydration diagram of the system Natural pozzolans – Lime – Water [DRO 78]*

3.3.1.4.3. Mechanical, physical, thermal and hydric properties

As regards the use of natural pozzolans in concretes, remarks identical to those formulated about metakaolins or fly ash can be made. Hence, the reader can refer to sections 3.3.2 and 3.3.3.

3.3.2. *Calcined natural pozzolans: metakaolin*

3.3.2.1. *General*

Metakaolin is a material made by firing milled kaolin at a temperature between 650 and 800°C.

The calcination entails the dehydroxylation and breakdown of the kaolinite's original crystalline structure in accordance with the following equation.

$$Al_2O_3.2SiO_2.2H_2O \longrightarrow Al_2O_3.2SiO_2 + 2H_2O$$

or $AS_2H_2 \longrightarrow AS_2 + 2H$ in cement-industry notation.

Metakaolin mainly contains amorphous aluminum (Al_2O_3) and silica (SiO_2) with an average mass ratio SiO_2/Al_2O_3 of 1.2. However, metakaolin may also contain impurities: primarily quartz.

Metakaolin is pozzolanic in nature – i.e. it will react with calcium hydroxide to form calcium silicate hydrates and calcium silico-aluminate hydrates.

3.3.2.2. *Production*

The kaolinite used for the production of metakaolin is a natural clay. It may come from primary or secondary sources:

– primary sources are the in-place metamorphosis of granite rocks. The resulting kaolinite is relatively pure, containing only small amounts of elements such as iron, titanium, etc., but it is associated with large amounts of quartz;

– secondary sources are obtained after the transport and depositing of kaolin, after the process of kaolinization. The kaolin is then richer in kaolinite (less quartz associated) and exhibits a finer particle size distribution. The kaolinite platelets are incompletely crystallized, and also contain structural iron.

There are two methods for industrial preparation:

– in the most commonly-used method, calcination is carried out in rotary kilns such as those found in cement works (60–90 m in length and 4–5 m in diameter). The starting material must be fairly pure, with a kaolinite content of over 75%. For a complete dehydroxylation of the kaolinite, this method requires the raw material to be fired between 600 and 800° for a sufficient length of time – around five hours. Following this calcination, it is crucial that the metakaolin be finely ground;

– a method that has developed more recently is based on "flash" calcination of powdery material [CAS 07]. In this process, the raw material is milled before calcination, which requires far less energy (the process merely involves crushing the sods of clay and possibly removing some of the quartz). Then, the procedure consists of a very rapid calcination, at approximately 750°C, of the particles injected into a stream of gas, with an exposure time of a few tenths of a second (hence the qualifier "flash"). This technique, which is less energy-hungry, also facilitates the calcination of less-pure kaolin (containing large amounts of quartz, up to 50%).

The properties of the metakaolins may differ depending on how they are produced. Thus, metakaolins obtained by flash calcination exhibit greater reactivity than those obtained by rotary kiln calcination [BIC 05]. This difference in reactivity is due to the difference in the morphology of the metakaolins obtained, which also alters the products' behavior in their raw state (water requirement, workability, etc.) [SAN 11], with flash calcination leading to the formation of reactive vitreous balls. This difference in morphology is illustrated in Figures 3.3 and 3.4.

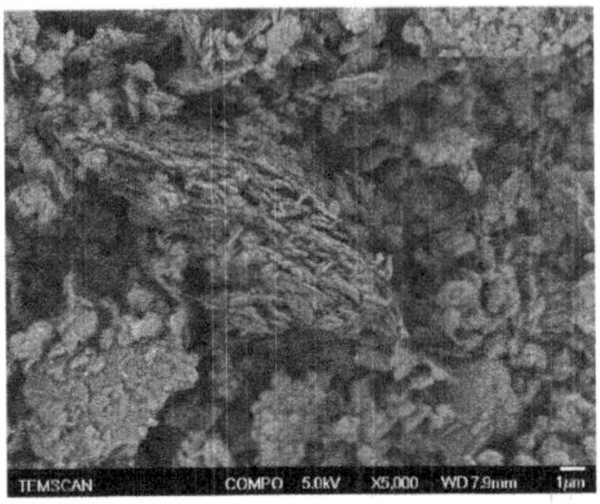

Figure 3.3. *SEM view of metakaolin obtained in a rotary kiln [SAN 11]*

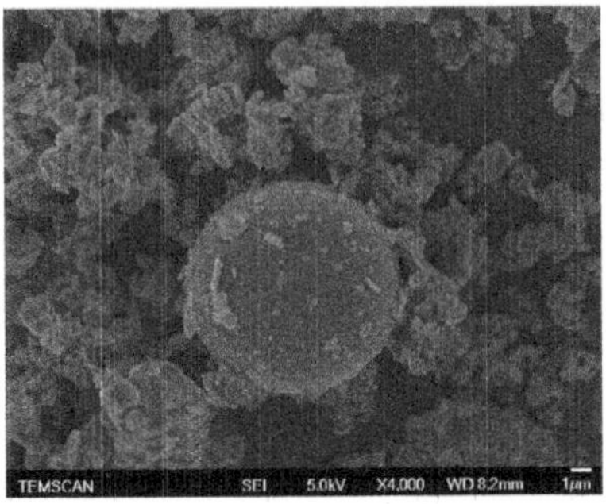

Figure 3.4. *SEM view of metakaolin obtained by flash calcination [SAN 11]*

3.3.2.3. *Chemical and mineralogical composition*

Metakaolins are essentially formed of silicon oxides (50–70%) and aluminum oxides (20–40%), with the whole being greater than 90%. The other elements which make up the remaining 10% are most often (in decreasing order) iron oxides, titanium oxides, potassium oxides, phosphorus oxides, sodium oxides, calcium oxides, magnesium oxides and so on. The nature of the source of kaolin determines the Al_2O_3-to-SiO_2 ratio, which is generally somewhere between 0.4 and 0.7.

From a mineralogical point of view, metakaolins are formed mainly of an amorphous part, characterized on an X-ray image by a halo centered on 3.8 Å, and a crystalline part composed primarily of quartz and, to a lesser extent, anatase. The presence of kaolin is a sign of its being under-fired, and the presence of mullite indicates that it has been over-fired: both are to be avoided.

3.3.2.4. *Properties*

3.3.2.4.1. Applicable standard

Many European countries use metakaolin, but only France has recently introduced a norm specific to this product. This norm, NF P 18-513, from July 2009 [NFP 09], gives the definitions, specifications and conformity criteria for this pozzolanic concrete admixture.

3.3.2.4.2. Reactivity

Metakaolin exhibits a high degree of pozzolanic activity – i.e. it reacts in the presence of water with the calcium hydroxide ($Ca(OH)_2$ or CH) originally present in the lime (hydraulic or aerial) or created by the hydration of calcium silicates. This pozzolanic reaction occurs in solution by dissolution/precipitation mechanisms between the calcium hydroxide and the silica and aluminum from the metakaolin. The balance equation for this reaction is:

$$AS_2 + 5CH + 5H \longrightarrow C_5AS_2H_{10}$$

$C_5AS_2H_{10}$ represents a typical composition of the products of this reaction, which are calcium silicate hydrates (C-S-H), calcium-aluminate silicate hydrates (C-A-S-H), a majority of gehlenite hydrate (C_2ASH_8) and possibly hydrogarnet (C_3ASH_6), and calcium aluminate hydrates (C_4AH_{13}) [FRI 01]. Also, in the presence of carbonates in the mix, the literature shows the formation of carboaluminates and hydroxycarboaluminates [FRI 03].

The reactivity of metakaolin is influenced by many factors: particularly its intrinsic properties (chemical composition, mineralogy, size and morphology of particles etc.) resulting from the nature of the kaolin, and the procedures used in its manufacture [CAS 07]. It also depends on the mixture realized (substitution of some of the cement or lime mixture, nature of that cement, etc.) and on the curing conditions, which may influence the kinetics of formation and the nature of the hydrated phases. The crucially important parameters seem to be the AS_2/CH ratio and the temperature [SAB 01].

3.3.2.4.3. Mechanical properties

The studies performed on the effect of metakaolin are based essentially on a partial substitution of the cement by metakaolin in concretes or mortars. Therefore, the properties are determined by comparison to the sample containing 100% cement.

Thus, in the fresh state, the introduction of metakaolin, because of its high water demand owing to its extensive specific surface, tends to decrease the workability of the mixture. However, by adjusting the W/L ratio or using a superplasticizer, it is possible to circumvent this problem.

In its hardened state, the incorporation of metakaolin into a concrete leads to an improvement in its compressive strength, particularly when recently poured. Thus, the activity index of the metakaolins, i.e. the ratio between the mechanical strength of the mortar with 15% substitution of a CEM I 52.5 N cement by metakaolin and the mechanical strength of the mortar with 100% of the same cement, is greater than or equal to 1.0 at 28 days. According to [KHA 05], this increase in strength is linked to:

– the densifying effect of the metakaolin, which increases the compactness of the mixture;

– the acceleration of the mechanisms of hydration of the cement by the phenomenon of heterogeneous nucleation;

– the pozzolanic reaction which decreases the portlandite content and leads to the formation of additional C-S-H, which improve strength.

With hydrated lime, Eires *et al.* [EIR 06] found that the best compressive strengths are obtained with a mixture comprising 25% hydrated lime and 75% metakaolin (8 MPa at 14 days and 10.5 MPa at 56 days).

Dimensional changes also appear to be limited in the presence of metakaolin. [BRO 01] notes the decrease in the total shrinkage and in particular the drying shrinkage in the presence of metakaolin, while [GLE 07] demonstrates a decrease in short- and long-term endogenous shrinkage because of the incorporation of metakaolin. However, this latter result appears to be dependent upon the content and the nature of metakaolin and on the cement used.

3.3.2.4.4. Physical and chemical properties

Metakaolins exhibit a high degree of fineness, which results in a flow-through of 0.063 mm greater than 70% and a very extensive specific surface, between 12.5 and 30 m^2/g. Their bulk density is approximately 700 kg/m^3 and the true density is 2600 kg/m^3.

3.3.2.4.5. Thermal and hydric properties

From the point of view of the thermal properties, replacing some of the cement with metakaolin has no effect on its measured characteristics (good inertia but poor thermal conductivity). In addition to aerial lime, the measured properties are similar to those obtained with natural hydraulic limes.

Yet the most significant improvement gained by substituting some of the cement with metakaolin relates to the durability of the material. Indeed, the metakaolin causes a reduction in the size of the pores of the paste and formation of denser and more stable hydrates such as C-A-S-H, which will greatly limit the penetration of aggressive agents [KHA 96]. Concretes in which some of the cement is replaced by metakaolin, therefore, exhibit better resistance to sulfates [KHA 98], to chlorides, whose bonding is also limited, to the freeze/thaw cycles [SAB 01] and to alkali-reaction [RAM 00].

3.3.2.5. *Environmental impacts*

Metakaolin presents advantages from an environmental point of view:

– unlike the production of cement or lime, which – because of the decarbonation of the raw material – causes the emission of a large amount of CO_2, the chemical mechanism of conversion of kaolin into metakaolin only gives off water vapor;

– the energy consumption related to the calcination of metakaolin (750°C) is far less than that necessary for the production of Portland cement or lime;

– the energy consumption related to milling is also minimized – particularly in the process of flash calcination.

Thus, for flash calcination, the energy consumption is 2210 MJ/T and the potential for global warming (greenhouse gases), expressed in equivalent kilos of CO_2 per ton of product, is 96 [LCP 10].

3.3.3. *Fly ash*

3.3.3.1. *General*

Fly ash is a by-product of the combustion of pulverized coal in conventional power plants. These silico-aluminous ashes are pozzolanic admixtures which can be used as a partial replacement for cement.

3.3.3.2. *Production*

Electricity production by combustion in pre-pulverized-coal-fired thermal plants yields two by-products: fly ash and bottom ash (grate ash). Fly ash is flying ashes, carried by the smoke (hence the name), recovered at the bottom of the chimney using an electrostatic precipitator, which are valuable in concretes. It takes the form, essentially, of smooth, vitrified balls, more or less hollow, and a few larger and porous elements, of indefinite shape. Their characteristics are directly related to the type of coal being burnt and the operation of the power plant (particularly the unburned carbon content). Only silico-aluminous ash, which contains a limited amount of CaO, can be used without the risk of later expansion.

The ashes may come from recent production (an active power plant) or from more or less old stores. In the former case, they are used dry; in the latter, they are used wet. The production of fresh fly ash is therefore directly dependent upon the production of electricity by coal-fired power plants, which is very limited in Western Europe (where these plants are used for extra power production), but is still very prevalent in Eastern Europe, China and the United States (around 40% of worldwide electricity production uses coal; in France this figure is less than 4%).

3.3.3.3. *Chemical and mineralogical composition*

The types of fly ash usable in concretes are essentially composed of silicon oxide (40–60%) and aluminum oxide (25–35%): hence the name, silico-aluminous ashes. They may also contain iron oxide (7–10%), calcium oxide (1–5%), potassium oxide

(5%) and magnesium oxide (2%). They contain practically no sulfates (less than 1%) or free lime (less than 0.5%).

From a mineralogical point of view, they are made of a silico-aluminous glass, which is reactive in an alkaline environment, and quartz, mullite and magnetite, which can be deemed inert.

3.3.3.4. *Properties*

3.3.3.4.1. Applicable standard

Fly ash for concrete must satisfy the requirements of the norm NF EN 450-1 of December 2007 [NFE 07].

3.3.3.4.2. Reactivity

The reactivity of the ash is related to the solubility of the vitreous part (silico-aluminous glass) in an alkaline medium: thus, it is a pozzolanic reaction.

However, the dissolution of silicon and aluminum is an extremely slow phenomenon, which accounts for the slow reaction kinetics of this type of compound in a cement medium.

3.3.3.4.3. Mechanical properties

Because of its pozzolanicity, fly ash improves the mechanical strength of concretes in the long term (typical dosage of 80–150 kg/m^3, replacing some of the cement). Conversely, in the short term (the first month), there is an appreciable loss of strength.

By decreasing the hydration heat, the ash also decreases the risk of the material's cracking.

3.3.3.4.4. Physical and chemical properties

Its true density ranges from 2.0–2.2 g/cm^3. In view of the spherical form of the particles, their size (0.5–100 μm in diameter, approximately) and their fineness (around 2000–3000 cm^2/g, slightly less than that of cements), they improve the behavior of the concretes in the fresh state (workability) and in the hardened state (compactness).

From a chemical point of view, the ash improves the behavior of concretes made with pure water or sulfate water by forming more stable hydrates (essentially C-S-H and C-A-H), instead of portlandite because of the pozzolanic reaction. It is also used to limit the alkali-silica reaction in the presence of reactive aggregates.

3.3.3.4.5. Thermal and hydric properties

The presence of fly ash in a concrete improves its resistance to fire and to thermal shocks. It does not have a significant impact on its thermal properties (e.g. thermal conductivity).

From the hydric point of view, and although the presence of fly ash refines the network of capillary pores, it has no impact on hydric exchanges.

3.3.3.5. *Environmental impacts*

As an industrial by-product, the environmental impact of fly ash may be taken to be null, both in terms of the energy requirement (fine-grained product obtained with no additional milling) and the release of CO_2 (this impact is entirely restricted to the production of electricity). Thus, only the transport of the ash to the place of use needs to be taken into account.

3.3.4. *Blast furnace slag*

3.3.4.1. *General*

Blast furnace slag is the by-product of industrial smelting. Its reactivity has been known since the beginning of the 19th Century and it has been used in combination with cement since the end of the 19th Century (e.g. in the building of the Paris metro).

The milled slag exhibits latent hydraulic properties – i.e. it needs to be activated (in an alkaline solution) in order to be able to form hydrates and develop mechanical strength. This activation can be done with lime or cement (thanks to the rapid formation of portlandite).

3.3.4.2. *Production*

In steel production, when smelting, the iron ore (essentially comprising oxides of silicon, iron and aluminum) and the fondant (lime and dolomite) are taken to a high temperature (fusion between 1400 and 1600°C) in a blast-furnace. Thereby, we obtain cast iron, on the one hand, which is denser, and on the other hand, the gangue from the ore, which is of lesser density and therefore floats on the surface of the molten iron, which enables them to be easily separated.

On coming out of the blast furnace, the liquefied gangue is subjected to very sudden cooling, generally performed with high-pressure water jets. Because of this soaking, the gangue solidifies and forms "granules" comparable in dimension to grains of sand (< 5 mm): this is known as granulated blast furnace slag, vitreous in structure with latent hydraulic properties.

This granulated slag (which can be stored outside and transported over long distances by boat) is dried and then milled into a powder, whose specific surface is generally around 4000–4500 cm^2/g in current production conditions.

3.3.4.3. *Chemical and mineralogical composition*

Blast furnace slags are made up of oxides of calcium (35–45%), silicon (30–35%), aluminum (10–20%) and magnesium (5–10%). It is also possible to find iron oxides in this material.

From a mineralogical point of view, these slags are entirely vitrified.

3.3.4.4. *Properties*

3.3.4.4.1. Applicable standard

The standard applicable to the blast furnace slag to be used as an admixture in concretes is the norm NF EN 15167-1 which defines its chemical, physico-chemical and physico-mechanical requirements. The means of monitoring, production and supply are also described in this norm [NFE 06].

The quality of the milled slag is monitored by determining, amongst other factors, its activity index (compressive strength in comparison to a standard), which must satisfy certain minimum requirements.

3.3.4.4.2. Reactivity

Blast furnace slag has a latent hydraulicity, i.e. under certain conditions it is likely to change into a stable crystallized form, developing mechanical strength (exactly as a cement would do). The vitrified slag, therefore, is a hydraulic binder whose reactivity depends on numerous factors such as its chemical composition, its fineness, its degree of vitrification, the degree of alkalinization of the interstitial solution, the temperature, etc.

Indeed, the vitrified slag is soluble only in alkalinized water and therefore need an additive agent, called an activator, in order to react. We essentially draw a distinction between alkaline activation (by sodium or calcium), sulfate activation (by gypsum) and sulfato-calcic activation (by a mixture of gypsum and lime) or sulfato-sodium activation (by a mixture of gypsum and caustic soda).

The main product of hydration of slag is C-S-H. Other products may also be formed, depending on the chemical composition of the slag, the duration of curing, the temperature and the activator used – e.g. hydrotalcite (M_5AH_{13}), a product rich in magnesium and aluminum, AFm substances (C_4AH_{13} and C_2ASH_8), ettringite and hydrogarnet ($C_6AFS_2H_8$) [CHE 07].

3.3.4.4.3. Mechanical properties

As a replacement for some of the cement (up to around 50%), the introduction of blast furnace slag helps maintain the mechanical strengths of concretes in the long term, but beyond 50% substitution, the mechanical strengths decrease. However, in recently-poured concrete, the mechanical performances are poorer than those of unmixed cements, because the dissolution of the slag requires prior hydration of some of the clinker.

When mixed with lime, the mechanical performances are poorer, and these binders are used primarily for the stabilization and treatment of fine clay-rich soils.

3.3.4.4.4. Physical and chemical properties

The introduction of slag into a cement mix has no significant bearing on its water requirement or its ease of implementation. However, the material is more susceptible to desiccation because of lesser reactivity at a young age. It therefore needs to be protected during the setting phase and at the start of the hardening phase (requires moist curing).

The introduction of slag enables the hydration heats to be reduced, which is favorable from the point of view of thermal shrinkage and secondary reactions (thus it limits the formation of differed ettringite). Conversely, the slag increases the endogenous shrinkage of its mixtures, and their fragility.

Binders including blast furnace slag are mainly used for their durability in harsh chemical environments. This good behavior relates to the decrease in the C/S ratio in the C-S-H and the consumption of portlandite by the slag.

3.3.4.4.5. Thermal and hydric properties

The presence of blast furnace slag does not have a significant impact on the thermal properties (e.g. thermal conductivity) of the materials in which it is included.

From the hydric point of view, although the presence of milled slag refines the network of capillary pores, it has no impact on hydric exchanges.

3.3.4.5. *Environmental impacts*

As an industrial by-product, the environmental impact of blast furnace slag may be taken to be very slight because, at present, the CO_2 release associated with their production is entirely restricted to the production of steel. Thus, only the milling and transport to its place of use need to be taken into account.

3.4. Plaster

3.4.1. *General*

Plaster is one of the oldest man-made construction materials (used since Antiquity). It is obtained by dehydration of gypsum (calcium sulfate hydrate or $CaSO_4.2H_2O$) into hemi-hydrate ($CaSO_4.0.5H_2O$) and anhydrite ($CaSO_4$). Once mixed with water, the mixture sets and hardens, forming gypsum once more (a reversible reaction). However it is not a hydraulic binder, and it remains sensitive to water and to humidity.

3.4.2. *Production*

The calcium sulfate required for the production of plaster may be of natural origin or be a by-product of certain industries:

– in its natural state, it is found in the form of gypsum ($CaSO_4.2H_2O$) or anhydrite ($CaSO_4$) and more infrequently bassanite ($CaSO_4.0.5H_2O$). These are sedimentary rocks which result from the evaporation of the water from oversaturated marine lagoons (evaporites) and are present in a relatively abundant amount in nature (e.g. the Paris Basin);

– as a by-product of the desulfuration of gases and smoke from power stations (desulfogypsum, resulting from the oxidation of sulfurous anhydride into sulfuric anhydride and then its reaction with lime) or the manufacture of acids (phosphoric, boric, hydrofluoric acids, etc.). However, this latter industrial source (phosphogypsum, borogypsum, fluorogypsum, etc.) is gradually being abandoned because it is technically difficult and economically not massively advantageous.

The preparation of the raw material is as follows:

– extraction (subterranean or open mining), crushing and sieving for natural rocks;

– drying and accumulation of gypsums from desulfuration.

Then comes the phase of dehydration of that gypsum, which – depending on the temperature and the procedure employed – yields a series of products which are more or less hydrated and more or less soluble:

– around 100°C, formation of hemi-hydrates α or β (depending on whether steam pressure or free air is used), which are soluble, by the reaction:

$$CaSO_4.2H_2O \longrightarrow CaSO_4.0.5H_2O + 1.5H_2O$$

– around 200°C, formation of anhydrite III, which is soluble, by the reaction:

$$CaSO_4.0.5H_2O \longrightarrow CaSO_{4_{III}} + 0.5H_2O$$

– between 220 and 350°C, formation of anhydrite II, which is insoluble, by the reaction:

$$CaSO_{4III} \longrightarrow CaSO_{4_{II}}$$

Firing (at around 150°C) can be done in a dry or humid atmosphere:

– in a dry atmosphere, with contact between the gypsum to be processed and the combustion gases. Such chambers are essentially rotary kilns (1–3 m in diameter and 10–30 m in length) which are used with a constant treatment of 120–700 tons of gypsum per day;

– in a humid atmosphere with indirect firing of the gypsum. Autoclave kilns (to obtain hemi-hydrate α), or fixed or rotary shells or closed rotary kilns (BEAU kilns – to obtain hemi-hydrate β) are used, and yield very similar products but with a lesser rate of production (maximum 100 tons per day).

Following this phase of firing, a final phase of milling and possibly mixing takes place:

– the fineness of plaster is a factor of primary importance, and therefore a well-monitored and efficient phase of milling/separation is crucial;

– the final mixing stage is necessary to obtain different categories of plaster by mixing hemi-hydrates α and β and anhydrites II and III and adjuvants in appropriate proportions.

Thus, coating plasters contain, in varying proportions, semi-hydrate β with soluble anhydrite III and insoluble anhydrite II, whereas plasters for prefabrication (which account for over 80% of plaster production) are composed entirely of semi-hydrate β.

3.4.3. *Chemical and mineralogical composition*

From the chemical point of view, plaster is composed of calcium oxide (CaO), sulfate oxide (SO_3) and water (LOI of 6.8% at 400°C [VAL 05]).

From the mineralogical point of view, plaster is made up of the different products of dehydration of gypsum – i.e. semi-hydrates α and β ($CaSO_4.0.5H_2O$), anhydrite III (soluble $CaSO_4$) and anhydrite II (insoluble or over-fired $CaSO_4$). Gypsum is also to be found ($CaSO_4.2H_2O$).

3.4.4. *Properties*

3.4.4.1. *Referential standard*

Plaster-binders and coating plasters are subject to the norm NF EN 13279-1 from November 2008 [NFE 08].

The norm defines plaster-binders (notated A), coating plasters (B) and coating plasters for special applications (C). The requirements relate to the end use conditions (reaction to fire, fire resistance, soundproofing performance, heat resistance, release of dangerous substances, etc.) but also to the properties of the materials (plaster content, fineness, set start time, flexural and compressive strengths, adherence, etc.).

3.4.4.2. *Setting mechanisms*

On contact with water, products made by thermal dehydration of gypsum will recover their initial degree of hydration and form gypsum once again. This is the setting of the plaster, which takes place, schematically, in three success stages:

– partial dissolution of the compounds, creating an over-saturated solution;

– crystallization of the gypsum, which is the setting in the true sense of the word (crystallization due to the difference in solubility between the anhydrite, the hemi-hydrates and the gypsum, which is the least soluble compound);

– hardening, with increase in the internal cohesion of the crystals.

These phenomena of setting and hardening are accompanied by variations in volume of the solid phases (+60% for the anhydrite and +40% for the hemi-hydrate) whilst remaining smaller than the sum of the absolute volumes of the constituent parts. This is part of the reason for the intrinsic porosity of hardened plaster.

3.4.4.3. *Mechanical properties*

From the point of view of standardization, coating plasters must exhibit flexural strength greater than or equal to 1 (B1 to B6) or 2 (B7) MPa, and compressive strength greater than or equal to 2 (B1 to B6) or 6 (B7) MPa. Their adherence must be greater than 0.1 MPa.

Yet the mechanical performances of plaster may greatly surpass these values: they depend, in particular, on the degree of tempering [HER 92], and can be further improved by application of a pre-stress during the setting stage or indeed by introduction of grains of gypsum at the time of tempering [GMO 03].

The fragile mechanical property of this material may be compensated by adding synthetic or plant fibers as reinforcement, for instance [HER 92; EVE 02].

3.4.4.4. *Physical and chemical properties*

The set start times are over 20 minutes for a plaster coating applied manually, and over 50 minutes for "shot" plaster. These may be altered by the addition of certain mineral or organic products which act on the solubility of the initial components (accelerators, retarders, thickeners and water retainers).

Hardened plaster exhibits extensive porosity, which is at the root of its various properties. This results, essentially, from the large amount of tempering water needed for the implementation of the material (two–five times the quantity of stoichiometric water).

However, this causes a period of extensive natural drying, which is necessary for the attainment of these properties. It is generally considered that plaster loses all of its uncombined water – i.e. around 50% of its mass – in less than 30 days (ambient temperature of 20°C and less than 95% relative humidity).

One of the main characteristics of plaster is its resistance to fire. It is incombustible, and a poor heat conductor, but above all, it experiences an endothermic reaction when exposed to an increased temperature, absorbing heat and releasing water vapor. Indeed, the approx. 20% of water that constitutes hardened plaster will absorb 1255 KJ per kilogram of plaster, and therefore maintain the temperature of the plaster below 140°C for longer.

From the chemical point of view and in normal use conditions, plaster is sensitive only to water – particularly if it is not hydrofuged. Therefore, prolonged contact of the hardened plaster with water is to be avoided, and hence it should not be used outdoors.

3.4.4.5. *Thermal and hydric properties*

Hardened plaster exhibits heat conductivity which varies from 0.18 to 0.56 W/(m.K) depending on its density (respectively 600–1500 kg/m^3). These values, obtained on dry material used indoors (23°C and 50% RH), must be adjusted if the material is damper.

From the hydric point of view, plaster is permeable to water vapor and can exchange water with the atmosphere (between 0.1 and 0.2% of the mass of the plaster). Therefore, it is capable of absorbing surface condensation of water and then expelling it once more when the hygrometric conditions become favorable.

From an acoustic point of view, hardened plaster as such, presents no particular advantage. Rather, the form or the assembly of different elements containing plaster may provide additional acoustic comfort.

3.4.5. ***Environmental impacts***

The plaster industry has only a very slight impact on the emission of CO_2, which is due to the low firing temperature (less than 180°C) and the lack of any chemical reaction of decarbonation.

The environmental impact comes mainly at the end of its lifecycle, because since 2002, France's system for categorizing waste has classified plaster-based waste as non-dangerous and non-inert waste products. Therefore, since 2006, plaster-based waste products have had to be stored in purpose-specific chambers in *Installations de Stockage de Déchets Non Dangereux* (ISDND – Non-Dangerous Waste Storage Installations).

3.5. Summary

The advantages and disadvantages to all the different binders can be summed up as follows:

– Portland cements: although they are mechanically very strong, they have low vapor permeability which limits any properties of perspiration or regulation of the internal humidity which might result from the use of highly hygroscopic plant-based aggregates. They also have a significant impact in terms of CO_2 emissions;

– aerial limes: they take a very long time to set, and initially have poor mechanical strength which will develop very gradually. The main advantage to these materials stems from their high permeability to water vapor;

– hydraulic limes: in comparison to aerial lime, the main advantage is their increased short-term mechanical strength;

– lime/pozzolan mixtures: the pozzolanic reaction leads to a significant increase in mechanical strength without having too great an impact on the thermal properties or the permeability to water vapor. This proves an excellent compromise for making plant-based concretes;

– plaster: its major disadvantage is its sensitivity to water, which precludes any application outdoors or in humid rooms (except if a hydrofuging agent is included). Its recycling at the end of its lifecycle also poses a problem.

3.6. Bibliography

[AFN 01] AFNOR, Norme NF EN 197-1 - Ciment - Partie 1 : composition, spécifications et critères de conformité des ciments courants, AFNOR, 2001.

[ARA 05] ARANDIGOYEN M., PÉREZ BERNAL J.L., BELLO LÓPEZ M.A., ALVAREZ J.I., "Lime-pastes with different kneading water: Pore structure and capillary porosity", *Applied Surface Science*, 252, p. 1449–1459, 2005.

[AST 08] ASTM, C618 - 08a, Standard Specification for Coal Fly Ash and Raw or Calcined Natural Pozzolan for Use in Concrete, ASTM, 2008.

[ATI 11] ATILH, Module d'informations environnementales de la production de ciments courants en France, June 2011.

[BAR 07] BAROGHEL BOUNY V., "Water vapour sorption experiments on hardened cementitious materials - Part I: Essential tool for analysis of hygral behaviour and its relation to pore structure", *Cement and Concrete Research*, 37, p. 414–437, 2007.

[BIC 05] BICH C., Contribution à l'étude de l'activité thermique du kaolin : évolution de la structure cristallographique et activité pouzzolanique, Doctoral thesis, Institut National des Sciences Appliquées de Lyon, 2005.

[BRA 12] BRAS A., HENRIQUES F.M.A., "Natural hydraulic lime based grouts – The selection of grout injection parameters for masonry consolidation", *Construction and Building Materials*, 26, p. 135–144, 2012.

[BRO 01] BROOKS J.J., MEGAT JOHARI M.A., "Effect of metakaolin on creep and shrinkage of concrete", *Cement and Concrete Composites*, 23–6, p. 495–502, 2001

[BRU 09] BRUIJN P.B., JEPPSSON K.H., SANDIN K., NILSSON C., "Mechanical properties of lime – hemp concrete containing shives and fibres", *Biosystems Engineering*, 103, p. 474–479, 2009.

[CAS 07] CASSAGNABÈRE F., Produits préfabriqués en béton filé : vers l'amélioration des performances du matériau pour mieux gérer le procédé de production, Doctoral thesis, University of Toulouse, 2007

[CEN 07] CONSTRUIRE EN CHANVRE, Règles Professionnelles d'exécution, April 2007.

[CÉR 05] CÉRÉZO V., Propriétés mécaniques, thermiques et acoustiques d'un matériau à base de particules végétales : approche expérimentale et modélisation théorique, Doctoral thesis, ENTPE Lyon, 2005

[ČER 06] ČERNÝ R., KUNCA A., TYDLITÁT V., DRCHALOVÁ J., ROVNANÍKOVÁ P., "Effect of pozzolanic admixture on mechanical, thermal and hygroc properties of lime plasters", *Construction and Building Materials*, 20, p. 849–857, 2006.

[CHE 07] CHEN W., Hydration of slag cement: Theory, Modeling and Application, PhD Thesis, University of Twente, 2007.

[COL 04] COLLET F., Caractéristique hydrique et thermique de matériaux de génie civil à faibles impacts environnementaux, Doctoral thesis, INSA de Rennes, 2004.

[DEL 96] DELOYE F-X., "La chaux à travers les âges", *Bulletin du Laboratoire des Ponts et Chaussées*, vol. 201, January-February 1996, p. 94–98.

[DOM 06] DOMÈDE N., Méthode de requalification des ponts en maçonnerie, Doctoral thesis, Institut National des Sciences Appliquées de Toulouse, 2006.

[DRO 78] DRON R., "L'activité pouzzolanique", *Bulletin de liaison Laboratoire des Ponts et Chaussées*, vol. 93, January-February 1978, pp. 66–69.

[EIR 06] EIRES R., NUNES J. P., FANGUEIRO R., JALALI S., CAMÕES A., "New eco-friendly hybrid composite materials for civil construction", *European Conference on Composite Materials*, Biarritz, France, 2006.

[EVE 02] EVE S., GOMINA M., GMOUH A., SAMDI A., MOUSSA R., ORANGE G., "Microstructural and mechanical behaviour of polyamide fibre-reinforced plaster composites", *Journal of the European Ceramic Society*, 22, p. 2269–2275, 2002.

[FRI 01] FRIAS M., CABRERA J., "Influence of MK on the reaction kinetic in MK-lime and MK-blended cement systems at 20°C", *Cement and Concrete Research*, Vol. 31, p. 519–527, 2001.

[FRI 03] FRÍAS ROJAS M., SÁNCHEZ DE ROJAS M.I., "The effect of high curing temperature on the reaction kinetics in MK/lime and MK-blended cement matrices at 60°C", *Cement and Concrete Research*, Vol. 33, p. 643–649, 2003.

[GLE 07] GLEIZE P.J.P., CYR M., ESCADEILLAS G., "Effect of metakaolin on autogenous shrinkage of cement paste", *Cement and Concrete Composites*, Volume 29, p. 80–87, 2007.

[GMO 03] GMOUH A., EVE S., SAMDI A., MOUSSA R., HAMEL J., GOMINA M., "Changes in plaster microstructure by pre-stressing or by adding gypsum grains: microstructural and mechanical investigations", *Materials Science and Engineering*, A352, p. 325–332, 2003.

[HER 92] HERNANDEZ-OLIVARES F., OTEIZA I., DE VILLANUEVA L., "Experimental analysis of toughness and modulus of rupture increase of sisal short fiber reinforced hemihydrated gypsum", *Composites Structures*, 22, p. 123–137, 1992.

[IZA 10] IZAGUIRRE A., LANAS J., ÁLVAREZ J.I., "Ageing of lime mortars with admixtures: Durability and strength assessment", *Cement and Concrete Research*, 40, p. 1081–1095, 2010.

[IZA 11] IZAGUIRRE A., LANAS J., ALVAREZ J.I., "Effect of a polypropylene fibre on the behaviour of aerial lime-based mortars", *Construction and Building Materials*, 25, p. 992–1000, 2011.

[KHA 96] KHATIB J.M., WILD S., "Pore size distribution of metakaolin paste", *Cement and Concrete Research*, 26, p. 1545–1553, 1996.

[KHA 98] KHATIB J.M., WILD S., "Sulphate resistance of metakaolin mortar", *Cement and Concrete Research*, 28, p. 83–92, 1998.

[KHA 05] KHATIB J.M., HIBBERT J.J., "Selected engineering properties of concrete incorporating slag and metakaolin", *Construction and Building Materials*, 19, p. 460–472, 2005.

[LAN 03] LANAS J., ALVAREZ J.I., "Masonry repair lime-based mortars: Factors affecting the mechanical behaviour", *Cement and Concrete Research*, 33, p. 1867–1876, 2003.

[LAN 04] LANAS J., PEREZ BERNAL J.L., BELLO M.A., ALVAREZ GALINDO J.I., "Mechanical properties of natural hydraulic lime-based mortar", *Cement and Concrete Research*, 34, p. 2191–2201, 2004.

[LAN 05] LANAS J., SIRERA R., ALVAREZ J.I., "Compositional changes in lime-based mortars exposed to different environments", *Thermochimica Acta*, 429, p. 219–226, 2005.

[LAW 06] LAWRENCE R.M.H., MAYS T.J., WALKER P., D'AYALA D., "Determination of carbonation profiles in non–hydraulic lime mortars using thermogravimetric analysis", *Thermochimica acta*, 444, p. 179–189, 2006.

[LAW 07] LAWRENCE R. M., MAYS T. J, RIGBY S.P., WALEKER P., D'AYALA D., "Effects of carbonation on the pore structure of non-hydraulic lime mortars", *Cement and Concrete Research*, 37, p. 1059–1069, 2007.

[LCP 10] LCPC, Evaluation environnementale d'un procédé de fabrication de métakaolin (Calcination flash), 2010

[LIE 98] LIEBIG E., ALTHAUS E., "Pozzolanic activity of volcanic tuff and suevite: effects of calcination", *Cement and Concrete Research*, vol. 28, p. 567–575, 1998.

[MAG 10] MAGNIONT C., Contribution à la formulation et à la caractérisation d'un écomatériau de construction à base d'agroressources, Doctoral thesis, University of Toulouse, 2010.

[MAR 05] MARAVELAKI-KALAITZAKI P., BAKOLAS A., KARATASIOS I., KILIKOGLOU V., "Hydraulic lime mortars for the restoration of historic masonry in Crete", *Cement and Concrete Research*, 35, p. 1577–1586, 2005.

[MER 07] MERTENS G., MADAU P., DURINCK D., BLANPAIN B., ELSEN J., "Quantitative mineralogical analysis of hydraulic limes by X-ray diffraction", *Cement and Concrete Research*, 37, p. 1524–1530, 2007.

[MOR 05] MOROPOULOU A., BAKOLAS A., MOUNDOULAS P., AGGELAKOPOULOU E., ANAGNOSTOPOULOU S., "Strength development and lime reaction in mortars for repairing historic masonries", *Cement and Concrete Composites*, 27, p. 289–294, 2005.

[MOS 06] MOSQUERA M.J., SILVA B., PRIETO B., RUIZ-HERRERA E., "Addition of cement to lime-based mortars: Effect on pore structure and vapor transport", *Cement and Concrete Research*, 36, p. 1635–1642, 2006.

[NFE 08] NF EN 13279-1, Liants-plâtres et enduits à base de plâtre pour le bâtiment. Partie 1 : Définitions et exigences, 2008.

[NFE 06b] NF EN 15167-1, Laitier granulé de haut-fourneau moulu pour utilisation dans le béton, mortier et coulis – Partie 1 définitions, exigences et critères de conformité, 2006.

[NFE 01] NF EN 197-1, Ciment. Partie 1 : composition, spécifications et critères de conformité des ciments courants, 2001.

[NFE 02] NF EN 459-1, Chaux de construction. Partie 1 : définitions, spécifications et critères de conformité, 2002.

[NFE 06a] NF EN 196-1, Méthodes d'essai des ciments. Partie 1 : détermination des résistances mécaniques, 2006.

[NFE 07] NF EN 450-1, Cendres volantes pour béton. Partie 1: définition, spécifications et critères de conformité, 2007.

[NFP 09] NF P 18-513, Addition pouzzolanique pour bétons – Métakaolin – Définitions, spécifications et critères de conformité, 2009.

[NGU 10] NGUYEN T.T., Contribution à l'étude de la formulation et du procédé de fabrication d'éléments de construction en béton de chanvre, Doctoral thesis, University of Brittany, 2010.

[NOZ 12] NOZAHIC V., AMZIANE S., TORRENT G., SAÏDI K., DE BAYNAST H., "Design of green concrete made of plant-derived aggregates and a pumice-lime binder", *Cement and Concrete Composites,* vol. 34, p. 231–241, 2012.

[PIC 94] PICHON H., Le système « Pouzzolanes naturelles – Chaux – Eau » à 28 et 100°C – Relations entre la réactivité chimique, les phases néoformées et les conséquences physico-mécaniques (application aux matériaux volcaniques du Massif Central Français), Doctoral thesis, University Joseph Fourier – Grenoble I, 1994.

[RAM 00] RAMLOCHAN T., THOMAS M., GRUBER K.A., "The effect of metakaolin on alkali-silica reaction in concrete", *Cement and Concrete Research*, vol. 30, p. 339–344, 2000.

[ROW 05] ROWELL R., "Moisture properties", *Handbook of Wood Chemistry and Wood Composites*, CRC Press, 21p, 2005.

[SAB 01] SABIR B.B., WILD S., BAI J., "Metakaolin and calcined clays as pozzolans for concrete: a review", *Cement and Concrete Composites*, 23, p. 441–454, 2001.

[SAI 12] SAINT ASTIER NATURAL HYDRAULIC LIME, http://www.limes.us/index.php

[SAM 08] SAMRI D., Analyse physique et caractérisation hygrothermique des matériaux de construction : approche expérimentale et modélisation numérique, Doctoral thesis, Institut National des Sciences Appliquées de Lyon, 2008.

[SAN 11] SAN NICOLAS R., Approche performantielle des bétons avec métakaolins obtenus par calcination flash, Doctoral thesis, University of Toulouse, 2011.

[SHI 01] SHI C., DAY R.L., "Comparison of different methods for enhancing reactivity of pozzolans", *Cement and Concrete Research*, vol. 31, p. 813–818, 2001.

[STR 00] STRAUBE J., "Moisture Properties of Plaster and Stucco for Strawbale Buildings", *Report for Canada Mortgage and Housing Corporation*, June 2000.

[THU 05] Réglementation Thermique 2005, Règles TH-U 2005.

[TOL 03] TOLEDO FILHO R.D., GHAVAMI K., ENGLAND G.L., SCRIVENER K., "Development of vegetable fibre-mortar composites of improved durability", *Cement and Concrete Composites*, vol. 25, p. 185–196, 2003.

[VAL 05] VALANCIUS Z; NIZEVICIENE D; LESKEVICIENE V, KYBARTIENE N., "Influence of the technological parameters on the structure and properties of hemi-hydrate phosphogypsum", *Ceramics Silikaty*, 49–2, p. 120–125, 2005.

[VEJ 12a] VEJMELKOVA E., KEPPERT M., KERŠNER Z., ROVNANIKOVA P., CERNY R., "Mechanical, fracture-mechanical, hydric, thermal, and durability properties of lime-metakaolin plasters for renovation of historical buildings", *Construction and Building Materials*, 31, p. 22–28, 2012.

[VEJ 12b] VEJMELKOVA E., KEPPERT M., ROVNANIKOVA P., KERSNER Z., CERNY R., "Application of burnt clay shale as pozzolan addition to lime mortar", *Cement and Concrete Composites*, Volume 34, Issue 4, p. 486–492, 2012.

[WBC 02] WBCSD, Toward a sustainable cement industry, Publication of World Business Council for Sustainable Development, 2002.

[WBC 09] WBCSD, Cement Industry Energy and CO_2 performance, Getting the numbers right, Publication of World Business Council for Sustainable Development, 2009.

Chapter 4

Formulation and Implementation

4.1. Objectives

4.1.1. *Preamble*

In a manner of speaking, constructing the walls of buildings by coupling plant-based straw with a mineral matrix is not a new phenomenon. This is done with the earth/straw mixtures, earth/reeds or indeed cob (coiled straw or hay mixed with mud) which are systematically to be found in the act of building throughout human history. This is certainly attributable to the fact that these construction techniques are simple and exploit local resources usually available in the vicinity of the construction site.

In the industrial societies of the 19th and 20th Centuries, the advent of construction techniques based on transformed materials (steel, Portland cement), led to a major overhaul of construction traditions. The erection of complex or extremely tall buildings became possible, because of an optimization of the performances of the materials, coupled with increased mastery of design. These changes led to the coming into force of formulation rules (e.g. for concrete) and design (design of reinforced concrete or steel construction in a European normative framework: Eurocodes).

In parallel to these technical evolutions, the artisanal know-how related to building with earth was gradually lost. It was only from the 1980s onwards that

---

Chapter written by Christophe LANOS, Florence COLLET, Gérard LENAIN and Yves HUSTACHE.

builders began using various additional tissues with a view to updating the historical methods of building (adobe, rammed earth, etc.) or adapt them by using "modern" components (mechanically stabilized earth (MSE), use of pozzolanic binders, and so on).

In the same vein, the coupling of hemp straw (or hemp shiv) with a mineral binder to create a modern variant of cob was envisaged by Charles Rasetti (a master-builder close to *La chanvrière de l'aube* – a well-known hemp production cooperative and pressure group for the use of hemp) around 1985 [EKO 11].

We can also cite a second artisanal approach to hemp concrete (or hempcrete), attributable to Yves Kühn, who set out to design a procedure for constructing with raw agricultural materials (the Canosmose procedure), to build houses entirely from hemp.

Later on, a number of different "experimental" builds were performed between 1990 and 1995.

From the different phases of this learning curve in construction, it emerged that hemp concrete made by combining hemp shiv and a mineral matrix can be used to build a wall whose thermo-hydric performances appear advantageous. However, on a structural level, it is useful to couple hempcrete filling with a structure that is able to support vertical load transfer and the bracing of the building. Most often a wooden structure is chosen. A number of developments based on the use of a reinforced concrete structure (concrete walls or beam-columns) are currently under study.

The techniques for implementing hempcrete used today correspond to traditional techniques (casting in a formwork, as is done with cast concrete or traditional rammed earth) or techniques based on the adaptation of more modern procedures (vibro-compacting to make a building block, in-situ spraying, etc.). Depending on the type of application of hempcrete – in a wall (vertical filling), on a roof or in flooring – professionals in this area offer different types of hempcrete.

In any case, it is useful to optimize the formulation of the mixture in view of the manufacturing method, and of the objectives in terms of the performances of the finished product (thermal, hydric and mechanical performances, or even acoustic performances, etc.).

Ultimately mastering the performances of hempcrete requires the enforcement of relevant formulation rules. The aim of this chapter is to propose various tools to guide the user in the formulation of hempcrete. The criteria for formulation are discussed in view of the type of application being envisaged (wall, roof, floor). The discussion is supplemented with various examples of formulations.

The second part of this chapter is devoted to the presentation of the various manufacturing methods and the professional rules associated with its use. Thus, the applier will find all the elements necessary for the mastery of execution in conformity with the requirements and standards in a quality framework which contributes to the development of the whole sector of "hemp construction".

4.1.2. *Traditional applications*

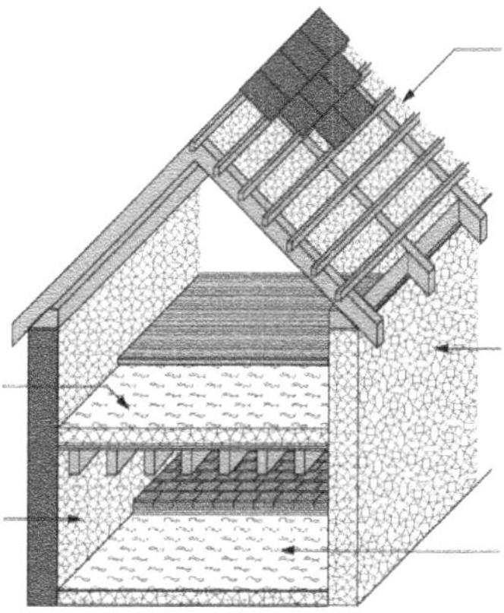

Figure 4.1. *Traditional applications of hempcrete [CON 12]*

Hempcrete can be used to make various surfaces in a construction. Thus, we distinguish applications in walls, roof and floor (Figure 4.1). For each type of application, the following criteria are generally adhered to [ASS 07]:

– the hempcretes commonly used to build walls have apparent densities of around 400 kg/m^3 and a heat conductivity of around 0.1 W/(m.K) (at 0% RH). Sometimes, the walls are built with a partial formwork, in which case a less dense hempcrete, such as that used for roofs, can be used to fill the walls.

– the hempcrete used as a roofing insulation material has a very low binder content (typically between 100 and 120 kg of binder per m^3) and exhibits an apparent density of around 200–250 kg/m^3 and a heat conductivity of around 0.06 W/(m.K) (at 0% RH).

– the hempcretes used to make ground insulation (floor application) have an apparent density of around 500 kg/m^3 and a heat conductivity of around 0.12 W/(m.K) (at 0% RH). These criteria specific to the floor application are indicative, and are no longer covered by the professional regulations promulgated by the association *Construire en Chanvre* (review of the insurance situation).

These values are indicative, but they are able to ensure an acceptable energy balance for the project and thermo-hydric transfer quality of the surface without adversely affecting the thickness of the structure.

In addition to these thermo-physical criteria, there are minimum mechanical performances which must be respected so as to limit any structural problems. The minimum compressive strengths and rigidity are proposed on the basis of the use to be made of the material [ASS 07] (see Tables 4.1 and 4.12). For memory, in the case of a floor application, we can use thresholds of 15 MPa for the elasticity modulus and 0.3 MPa for the compressive strength (for details see Table 4.12).

WALL	Elasticity Modulus (MPa)	Compressive strength (MPa)
Threshold value (minimum) *in standard condition*	> 15 MPa	> 0.2 MPa

ROOF	Elasticity Modulus (MPa)	Compressive strength (MPa)
Threshold value (minimum) *in standard condition*	> 3 MPa	> 0.05 MPa

RENDERING	Elasticity Modulus (MPa)	Compressive strength (MPa)
Threshold value (minimum) *in standard condition*	> 20 MPa	

Table 4.1. *Minimum mechanical performances depending on the application required for samples stabilized at 20°C and 50% RH [ASS 07]*

4.1.3. *Constituents and mixture*

Like a traditional cement concrete, hemp concrete exhibits a fresh state characterized by sufficient fluidity that ensures easy manufacturing (casting or spraying), and a hardened state due to the eventual setting of the mineral binder used, its hardening and drying. Thus, hempcrete is obtained by mixing a mineral binder, hemp shiv and water. The mineral binder may be an aerated lime or, more usually, a formulated binder combining hydraulic compounds (hydraulic limes, cements, etc.), and pozzolanic compounds (pozzolans, metakaolin) with lime.

The hemp shiv plays the simple role of an aggregate – i.e. an "inert" component. However, its capacity to absorb water and the rate at which it does so are elements which will affect the fluidity of the mixture and the stability of that fluidity.

One of the additional phases in hempcrete is air, which may be trapped in the porosity of hemp shiv or in the form of pores in the mineral paste. The complexity of the formulation of hempcrete therefore lies in getting the right ratio between the fluid phases (air and water) and the solid phases (binder and hemp shiv).

The mixing can be done in the cement mixer, on site. In this case, it is preferable to form a paste with the binder and all the water and then gradually introduce the hemp shiv. It is generally noted that a too long duration of mixing can result in the formation of lumps of hemp shiv, characteristic of de-mixing (a phenomenon which is more common with fibered hemp shiv) or a too high air content. The hempcrete then acquires a remarkable fluidity, caused by the effect of foaming. However, such mixtures may exhibit very significant settlement under their own weight when they are molded.

The use of a concrete mixer, developing greater mixing energy, enables us to mix the hemp shiv and binder in the dry state, and then wet the mixture by sprinkling. With this type of mixing, we can obtain uniform concretes containing relatively little water. The molding of such concretes is often coupled with compacting in order to ensure good cohesion of the material. Water absorption by the hemp shiv can rapidly affect the rheological properties of these mixtures. Therefore, it is helpful to have a short processing time, which is easier to ensure in a prefabrication workshop.

In the case of sprayed or shot concrete (shotcrete), the mixture is obtained either by injecting water into a stream of a mixture of hemp shiv and binder from a hose, using a technique similar to that used for spraying of hydraulic concrete (dry-mixing of the hemp shiv and binder, done with a mixer), or by pulverization of a paste (binder + water) onto a stream of hemp shiv granulate. The paste can be produced by a spray machine for instance (continuously mixing or not). The water content in such mixtures tends to be low. This type of ratio leads to good cohesion of the sprayed mixture, which reduces its aptitude for uses with gravitational molding.

4.1.4. *Methods of implementation*

Hempcretes may be made in a workshop for the prefabrication of elements of walls (building blocks or prefabricated elements) or on building sites for manual or mechanized application.

Hempcrete can be put in place using a traditional framing technique (either on-site or by prefabrication). The concrete is simply poured between two formwork plates or into a dismountable mold. The process may be performed with successive layers, slightly compacted or not. Such an implementation technique requires that the hempcrete be sufficiently fluid. The use of a cement mixer or concrete mixer is common, with the time taken to use the concrete often being over half an hour. One advantage to these molding techniques, similar to those used for hydraulic concrete, is that professionals in construction are already very familiar with their use.

The spraying method is able to limit the molding stages. The hempcrete is sprayed onto a provisional or lost cast (left embedded in the structure), or onto an existing wall (for the purposes of renovation). The non-cast surface is adjusted and worked by hand. This highly-mechanized technique is fairly rapid (around 6 m^3 per hour for a hempcrete wall). However, it does require technical personnel trained in its use.

The final mode of implementation of hempcrete is to produce pre-cast blocks which are assembled on site, much like brick-laying. These products are made in a prefabrication workshop by molding or vibro-compacting. In this latter case, the procedure is fairly similar to that used by a concrete block machine. A fairly dry mixture is poured into a dosing hopper placed above a mold. The material is then run into the mold and fairly highly compacted, possibly with vibration of the mold. The block is removed almost immediately. The use of mold inserts enables us to obtain blocks with a rather complex cross-section. For such shaping, the mixture is obtained either with a slusher (a Bivis mixer able to guarantee continuous supply) or with a concrete mixer.

4.2. Rules of formulation

4.2.1. *Basis of usual formulations*

The formulation criteria that are most commonly used in the profession relate to the ratio between the mass of hemp shiv and the mass of binder:

– a ratio near to one is considered for hempcrete intended for a roof application.

– a ratio of around half is considered for hempcrete intended for a wall application.

The amount of water is generally adjusted to ensure a satisfactory rheology of the fresh product in connection with the mixing and manufacturing methods chosen. Thus, the ratio between the mass of water and the mass of binder can vary between 0.5 and 1.5.

Yet it should be borne in mind that some of the water in the mixture may quickly be trapped in the hemp shiv by absorption (between 300 and 400% of the mass of the hemp shiv [CHA 11], measured by applying the test protocol from [NFV 93]).

Figure 4.2. *Fabrication and implementation of hempcrete: (a) mixture made with a cement mixer [CON 12], application for a wall (by casting) [TER 11] and flooring [CON 12]; (b) dry mix obtained with a mixer [GLO 11] and then applied by spraying (water added at the nozzle) [CON 12]; (c) prefabrication of blocks [EAS 11] with a block press [GLO 11]*

4.2.2. *Influence of the proportion of paste in the mixture*

The physical characteristics of the hemp shiv in relation to the objectives in terms of density of the hardened product may give rise to a number of reflections.

If we consider that the average characteristics of hemp shiv are [NGU 10]:

– specific density: 1450 kg/m^3;

– apparent density: 110 kg/m^3 to 60 kg/m^3 depending on the fibrousness;

– particle density: 260 kg/m^3;

we can calculate the following data:

– the particles of (non-fibered) hemp shiv occupy a volume of 423 l/m^3;

– the internal porosity of hemp shiv is 82%;

– the mass of water required to saturate hemp shiv represents 316% of the mass of hemp shiv. This theoretical value, estimated without taking account of potential swelling of the hemp shiv, appears to be in agreement with the measurements of absorption [CHA 11].

Considering that the stacking of the hemp shiv is not altered by the presence of the paste, the maximum volume that can be occupied by the hardened binder would be 577 l/m^3 with non-fibered hemp shiv and 769 l/m^3 for fibered hemp shiv. The extensive volume available does not, in itself, constitute a criterion that is exploitable for the formulation. The usual paste contents are a great deal lower than this saturation limit.

It should be noted that, depending on the mode of implementation chosen, the hempcrete is more or less compacted: compacting during the building of a banked wall, compacting in the press for the manufacture of blocks, compacting caused by spraying. It is therefore helpful to introduce a compacting coefficient which expresses the reduction in volume accessible to the paste. The compacting coefficient c_c can be expressed as:

$$c_c = \rho_{ch} / \rho_h \qquad [4.1]$$

where ρ_h is the apparent density of the hemp shiv and ρ_{ch} the density of the hemp shiv in its compacted state due to manufacturing method. [NGU 10] shows that attaining a compacting coefficient of 1.5 requires a fairly small amount of effort, but that a compacting coefficient of 2 (for non-fibered hemp shiv; possibly 3.5 for fibered hemp shiv) is unattainable with traditional techniques of implementation.

Introducing the target density of hardened hempcrete on the basis of the type of use enables us to go further in the analysis of its composition.

Let us use ρ to denote the target density. The mass m of hardened binder for a cubic meter of hardened hempcrete is given by:

$$m = \rho - \rho_{ch} \tag{4.2}$$

Knowing the necessary concentration of hardened binder, it is possible to calculate the amount of binder B of the freshly-applied mixture, in the knowledge of the mass degree of hydration of the binder chosen – t:

$$B = m/t \tag{4.3}$$

knowing that t ranges from 1 for an aerated lime to 1.25 for a Portland cement or indeed 1.2 for a hemi-hydrate calcium sulfate (plaster).

The value of t can be estimated simply by taking samples of pure paste (binder plus water). The mass of binder used to make the sample is known (M_1). The mass of the sample after hardening and complete drying is measured (M_2). The mass degree of hydration is given by:

$$t = M_2/M_1 \tag{4.4}$$

Note that the estimation of B takes account of the effects of hydration related to the hydraulic nature of the binder. However, the effects of carbonation are not taken into account, with the objective in terms of the target density of the mixture being conceived of after a reasonable time-period of 60-90 days after the hempcrete's installation. Over such a period, the effects of carbonation are limited.

The amount of binder per cubic meter of fresh-applied hempcrete can therefore be estimated on the basis of the target density ρ. The expression can include the amount of hemp shiv aggregates G per cubic meter of fresh-applied hempcrete:

$$\begin{aligned} B &= (\rho - G)/t \\ G &= c_c.\rho_h \end{aligned} \tag{4.5}$$

Table 4.2 presents the results obtained for a hemp shiv whose apparent density is 110 kg/m^3 (a value which is characteristic of non-fibered hemp shiv) and t ranging from 1 (aerated lime) to 1.3 (highly hydraulic binder) for a compacting coefficient of 1. Table 4.3 presents the same type of results for a highly fibered hemp shiv whose apparent density is 60 kg/m^3.

ρ (kg/m^3) t	1	1.05	1.1	1.15	1.2	1.25	1.3
110	0	0	0	0	0	0	0
130	20	19	18	17	17	16	15
150	40	38	36	35	33	32	31
170	60	57	55	52	50	48	46
190	80	76	73	70	67	64	62
210	100	95	91	87	83	80	77
230	120	114	109	104	100	96	92
250	140	133	127	122	117	112	108
270	160	152	145	139	133	128	123
290	180	171	164	157	150	144	138
310	200	190	182	174	167	160	154
330	220	210	200	191	183	176	169
350	240	229	218	209	200	192	185
370	260	248	236	226	217	208	200
390	280	267	255	243	233	224	215
410	300	286	273	261	250	240	231
430	320	305	291	278	267	256	246
450	340	324	309	296	283	272	262
470	360	343	327	313	300	288	277
490	380	362	345	330	317	304	292
510	400	381	364	348	333	320	308
530	420	400	382	365	350	336	323
550	440	419	400	383	367	352	338
570	460	438	418	400	383	368	354
590	480	457	436	417	400	384	369
610	500	476	455	435	417	400	385
630	520	495	473	452	433	416	400
650	540	514	491	470	450	432	415
670	560	533	509	487	467	448	431
690	580	552	527	504	483	464	446
710	600	571	545	522	500	480	462
730	620	590	564	539	517	496	477
750	640	610	582	557	533	512	492
770	660	629	600	574	550	528	508
790	680	648	618	591	567	544	523
810	700	667	636	609	583	560	538

Table 4.2. *Amount of binder (kg) to be added to 1 m^3 of non-fibered hemp shiv (ρ_c = 110 kg/m^3) to obtain 1 m^3 of hempcrete whose density in the hardened state is ρ, depending on the mass degree of hydration of the binder t*

ρ (kg/m^3) t	1	1.05	1.1	1.15	1.2	1.25	1.3
110	50	48	45	43	42	40	38
130	70	67	64	61	58	56	54
150	90	86	82	78	75	72	69
170	110	105	100	96	92	88	85
190	130	124	118	113	108	104	100
210	150	143	136	130	125	120	115
230	170	162	155	148	142	136	131
250	190	181	173	165	158	152	146
270	210	200	191	183	175	168	162
290	230	219	209	200	192	184	177
310	250	238	227	217	208	200	192
330	270	257	245	235	225	216	208
350	290	276	264	252	242	232	223
370	310	295	282	270	258	248	238
390	330	314	300	287	275	264	254
410	350	333	318	304	292	280	269
430	370	352	336	322	308	296	285
450	390	371	355	339	325	312	300
470	410	390	373	357	342	328	315
490	430	410	391	374	358	344	331
510	450	429	409	391	375	360	346
530	470	448	427	409	392	376	362
550	490	467	445	426	408	392	377
570	510	486	464	443	425	408	392
590	530	505	482	461	442	424	408
610	550	524	500	478	458	440	423
630	570	543	518	496	475	456	438
650	590	562	536	513	492	472	454
670	610	581	555	530	508	488	469
690	630	600	573	548	525	504	485
710	650	619	591	565	542	520	500
730	670	638	609	583	558	536	515
750	690	657	627	600	575	552	531
770	710	676	645	617	592	568	546
790	730	695	664	635	608	584	562
810	750	714	682	652	625	600	577

Table 4.3. *Amount of binder (kg) to be added to 1 m^3 of highly fibered hemp shiv (ρ_c = 60 kg/m^3) to obtain 1 m^3 of hempcrete whose density in the hardened state is ρ, depending on the mass degree of hydration of the binder*

We note that, no matter what the type of binder (value of t) and the value of the target density for the hempcrete, the amounts of binder estimated are far from sufficient to saturate the inter-particle vacuums left by the hemp shiv, which contributes to the formation of a high degree of porosity, favorable for thermo-hydric performances but unfavorable for mechanical strengths.

4.2.3. *Quality of the paste and water content*

As yet, it is difficult to estimate the amount of water required for the right reaction of the binder and an optimal manufacturing of hempcrete. However, the water content will have a significant impact on the hempcrete's performances.

With exactly the same amount of binder, an excess of water will dilute the paste, which will negatively impact the mechanical performances. Too little water, coupled with too-rapid drying of the hempcrete, may result in poor hydration of the binder, which, again, is unfavorable in terms of mechanical strength.

Yet in order to identify the appropriate water content, we can build upon the following physical observation. The porosity of the hemp shiv is characterized by a set of capillaries whose minimum diameters are generally between 10 µm and 50 µm.

In addition, the granulometry of the binders generally used to formulate hempcretes is characterized by particle diameters between 5 µm and 30 µm.

When we bring the concentrated suspension corresponding to the water/binder mix into contact with the capillaries of the hemp shiv, capillary suction effects will lead to the absorption of the water and therefore an increase in the solid volume fraction of the paste. This phenomenon is possible only because the capillary forces due to the size of the pores in the hemp shiv remain greater than those developed in the inter-particle channels in the concentrated suspension.

Consequently, water absorption by the capillaries in the hemp shiv will be prevented if the distance between the clusters of particles in the concentrated suspension is around the same size as the capillaries.

As the size of the capillaries in the hemp shiv (average diameter D) is greater than or similar to the size of the clusters of particles of binder, it is possible to estimate a volume of water to be associated with the volume of binder to block the absorption. This volume of water is calculated by estimating the volume of water necessary to obtain a pattern similar to that shown in Figure 4.3.

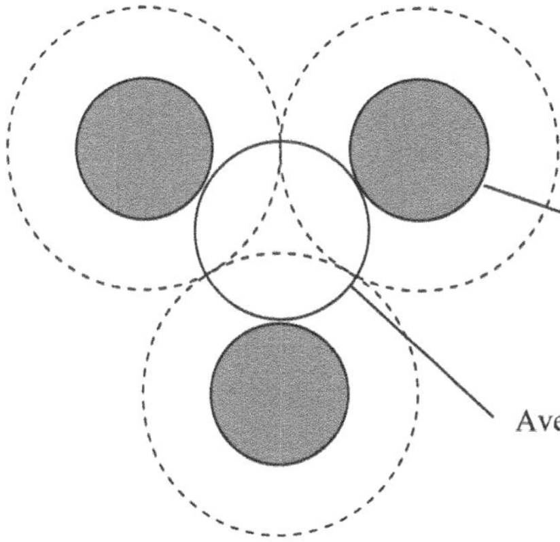

Figure 4.3. *Principle for calculating the maximum amount of water in the concentrated suspension so that the capillary effects between particles will be comparable to the capillary effects of the hemp shiv*

The distance between the centers of the particles to achieve such a state is given by:

$$X = \sqrt{3}\left(R + \frac{D}{2}\right) \tag{4.6}$$

D: diameter of a typical capillary of the hemp shiv;

R: radius of the particle of binder.

The volume of water V_w to be combined with a particle whose volume is V_p can be approximated by calculating the volume of water necessary to surround each particle in order to obtain the arrangement shown in Figure 4.3:

$$V_e = \frac{4}{3}\pi\left(\frac{X}{2}\right)^3 - \frac{4}{3}\pi R^3 \tag{4.7}$$

If we consider that R is approximately $D/2$, we get:

$$V_e = \frac{4}{3}\pi(\sqrt{3}^3 - 1)R^3 = 4.2.V_p \tag{4.8}$$

This case would correspond to a suspension of grains 5 μm in diameter very well dispersed (where D = 10 μm). The mass ratio of water to binder W/B can then be calculated on the basis of the specific density of the binder ρ_B:

$$\frac{W}{B} = \frac{4200}{\rho_B} \quad [4.9]$$

For a typical binder (with ρ_B of around 2500 kg/m^3), the water content calculated in this way is prohibitive. It should be noted, however, that this solution corresponds to a suspension of particles of binder which are perfectly dispersed.

The method of calculation may be revised by considering clusters of particles rather than isolated particles. Typically, the clusters of particles have sizes of around 30 μm (i.e. five-six times the size of the particles). These clusters present a solid volume fraction of around $\phi = 0.64$ (disorganized cluster of spheres).

The volume of water V_w to be added to a cluster of particles of solid volume $\phi . Vp$ can be corrected from equation [4.7]:

$$V_e = \frac{4}{3}\pi\left(\frac{X^*}{2}\right)^3 - \phi\frac{4}{3}\pi R^{*3} \quad [4.10]$$

where X^* is the distance between clusters of radius R^*.

If we consider that R^* is around 15 μm and $D = 10$ μm, we obtain a mass W/B ratio of:

$$\frac{E}{B} = \frac{\sqrt{3}^3}{8}\left(1 + \frac{D}{2R^*}\right)^3 \frac{1}{\varphi . \rho_B} - \frac{1000}{\rho_B} = \frac{1540}{\varphi . \rho_B} - \frac{1000}{\rho_B} \quad [4.11]$$

For a normal binder (with ρ_B near to 2500 kg/m^3) and $\phi = 0.64$ (spherical particles), the amount of water becomes such that W/B = 0.56 – a value which proves to be acceptable for formulations intended for spraying or vibro-compaction.

It is interesting to note that the amounts of water estimated by this method remain far greater than the requirements linked to the hydration of the binder (value of t). Thus, hempcretes will always be characterized by a significant loss of mass on drying.

Equation [4.11] is obtained making different hypotheses about the distribution of the material (shape and constitution of the clusters of binder particles, porosity of the hemp shiv, etc.). However, the results that can be achieved with this equation are consistent. If the binder is formulated so as best to disperse its particles (e.g. by adding a deflocculating agent), a large amount of water (see equation [4.6]) can be used without being adversely affected by the effects of absorption. However, the dilution of the paste may soon become harmful to the mechanical properties of the hardened material. If, on the other hand, the binder is formulated to favor the formation of clusters of binder particles, the water content will be reduced. The result of this will be a loss of fluidity, but possibly better mechanical characteristics of the hardened mixture.

In physical terms, the formulation model may be understood thus:

– if the water content is lower than that estimated by equation [4.11], capillary absorption in the hemp shiv is practically non-existent. The rheology of the mixture is essentially dependent upon the properties of flow of the paste. Any change in the water content of the paste will have a noticeable bearing on the product's rheological behavior. The mechanical properties of the mixture will essentially be dictated by the volume of paste in the mixture and its quality [ELF 08]. This type of formulation corresponds more to methods of manufacture by spraying or vibro-compaction.

– if the water content is greater than that estimated by equation [4.11], but lower than that estimated by equation [4.11] plus the content required because of the absorbency of the hemp shiv, the rheology of the mixture will change over time (by absorption of the water into the hemp shiv) and then stabilize around a common value. This type of formulation, therefore, is fairly robust, because the variation in water content is counterbalanced by the effects of absorption. The mechanical behavior of such mixtures should be independent of the amount of water in this range of dosage, because after the hemp shiv has absorbed the excess water, the concentration of the paste will converge toward an identical value. This type of formulation corresponds more closely to a method of manufacture by a mixer.

– if the water content is greater than that estimated by equation [4.11] plus that required by the absorbency of the hemp shiv, significant fluidification is to be expected in the paste. Such content ratios are not robust, and any variation in the water content will greatly harm the mechanical characteristics of the mixture.

On these theoretical bases, therefore, it is possible to propose a range of water contents that should be prioritized when making a hempcrete. The low value for the water content corresponds to the value identified with equation [4.11]. The high value corresponds to the content estimated by equation [4.11] plus the content required because of the absorbency of the hemp shiv. The results are presented in Tables 4.4 to 4.6 for a compacting coefficient equal to one.

ρ_B (kg/m^3)	2100	2300	2500	2700	2900	3100	3300
D/(2R*)	-	-	-	-	-	-	-
0.05	0.08	0.08	0.07	0.06	0.06	0.06	0.05
0.1	0.17	0.15	0.14	0.13	0.12	0.11	0.11
0.15	0.26	0.24	0.22	0.20	0.19	0.18	0.16
0.2	0.36	0.33	0.30	0.28	0.26	0.24	0.23
0.25	0.47	0.43	0.39	0.36	0.34	0.32	0.30
0.3	0.59	0.53	0.49	0.46	0.42	0.40	0.37
0.35	0.71	0.65	0.60	0.55	0.52	0.48	0.45
0.4	0.85	0.78	0.71	0.66	0.62	0.58	0.54
0.45	1.00	0.91	0.84	0.78	0.72	0.68	0.63
0.5	1.15	1.05	0.97	0.90	0.84	0.78	0.73
0.55	1.32	1.21	1.11	1.03	0.96	0.90	0.84
0.6	1.50	1.37	1.26	1.17	1.09	1.02	0.96
0.65	1.69	1.55	1.42	1.32	1.23	1.15	1.08
0.7	1.90	1.73	1.59	1.48	1.37	1.29	1.21
0.75	2.11	1.93	1.78	1.64	1.53	1.43	1.35
0.8	2.34	2.14	1.97	1.82	1.70	1.59	1.49
0.85	2.58	2.36	2.17	2.01	1.87	1.75	1.64
0.9	2.84	2.59	2.38	2.21	2.06	1.92	1.81
0.95	3.11	2.84	2.61	2.42	2.25	2.10	1.98
1	3.39	3.10	2.85	2.64	2.45	2.30	2.16
1.05	3.69	3.37	3.10	2.87	2.67	2.50	2.35
1.1	4.00	3.65	3.36	3.11	2.90	2.71	2.55
1.15	4.33	3.95	3.63	3.37	3.13	2.93	2.75
1.2	4.67	4.26	3.92	3.63	3.38	3.16	2.97
1.25	5.03	4.59	4.22	3.91	3.64	3.41	3.20
1.3	5.40	4.93	4.54	4.20	3.91	3.66	3.44
1.35	5.80	5.29	4.87	4.51	4.20	3.93	3.69
1.4	6.20	5.67	5.21	4.83	4.49	4.20	3.95
1.45	6.63	6.05	5.57	5.16	4.80	4.49	4.22
1.5	7.07	6.46	5.94	5.50	5.12	4.79	4.50
1.55	7.54	6.88	6.33	5.86	5.46	5.11	4.80
1.6	8.02	7.32	6.73	6.24	5.81	5.43	5.10
1.65	8.52	7.78	7.15	6.62	6.17	5.77	5.42
1.7	9.04	8.25	7.59	7.03	6.54	6.12	5.75
1.75	9.57	8.74	8.04	7.45	6.93	6.49	6.09
1.8	10.13	9.25	8.51	7.88	7.34	6.86	6.45

Table 4.4. *Mass ratio W/B necessary to avoid any phenomenon of water absorption by the hemp shiv depending on the specific density of the binder used and the D/(2R*) ratio ($\phi = 0.64$)*

ρ (kg/m^3) t	1	1.05	1.1	1.15	1.2	1.25	1.3
110	-	-	-	-	-	-	-
130	18.03	18.84	19.66	20.49	21.32	22.15	23.00
150	9.35	9.73	10.11	10.50	10.90	11.30	11.71
170	6.46	6.69	6.93	7.18	7.43	7.69	7.95
190	5.01	5.17	5.34	5.51	5.69	5.88	6.07
210	4.14	4.26	4.38	4.51	4.65	4.79	4.94
230	3.56	3.65	3.75	3.85	3.96	4.07	4.19
250	3.15	3.22	3.29	3.37	3.46	3.55	3.65
270	2.84	2.89	2.95	3.02	3.09	3.17	3.25
290	2.60	2.64	2.68	2.74	2.80	2.86	2.93
310	2.41	2.43	2.47	2.52	2.57	2.62	2.68
330	2.25	2.27	2.30	2.34	2.38	2.43	2.48
350	2.12	2.13	2.15	2.18	2.22	2.26	2.31
370	2.01	2.01	2.03	2.06	2.09	2.12	2.16
390	1.91	1.91	1.93	1.95	1.97	2.00	2.04
410	1.83	1.83	1.84	1.85	1.87	1.90	1.93
430	1.75	1.75	1.76	1.77	1.79	1.81	1.84
450	1.69	1.68	1.69	1.70	1.71	1.73	1.75
470	1.63	1.62	1.62	1.63	1.64	1.66	1.68
490	1.58	1.57	1.57	1.57	1.58	1.60	1.61
510	1.54	1.52	1.52	1.52	1.53	1.54	1.55
530	1.50	1.48	1.47	1.47	1.48	1.49	1.50
550	1.46	1.44	1.43	1.43	1.43	1.44	1.45
570	1.42	1.40	1.39	1.39	1.39	1.40	1.41
590	1.39	1.37	1.36	1.35	1.35	1.36	1.37
610	1.36	1.34	1.33	1.32	1.32	1.32	1.33
630	1.34	1.31	1.30	1.29	1.29	1.29	1.29
650	1.31	1.29	1.27	1.26	1.26	1.26	1.26
670	1.29	1.26	1.24	1.23	1.23	1.23	1.23
690	1.27	1.24	1.22	1.21	1.20	1.20	1.20
710	1.25	1.22	1.20	1.19	1.18	1.18	1.18
730	1.23	1.20	1.18	1.16	1.16	1.15	1.15
750	1.21	1.18	1.16	1.14	1.14	1.13	1.13
770	1.20	1.16	1.14	1.13	1.12	1.11	1.11
790	1.18	1.15	1.12	1.11	1.10	1.09	1.09
810	1.17	1.13	1.11	1.09	1.08	1.07	1.07

Table 4.5. *Maximum mass ratio W/B to be used in the presence of non-fibered hemp shiv (ρ_h = 110 kg/m^3) to obtain 1 m^3 of hempcrete whose density in the hardened state is ρ, depending on the mass degree of hydration of the binder t (the specific density of the binder ρ_B = 2500 kg/m^3, D = 10 µm, R* = 15 µm, ϕ = 0.64)*

ρ (kg/m^3) t	1	1.05	1.1	1.15	1.2	1.25	1.3
110	3.77	3.86	3.97	4.08	4.20	4.33	4.45
130	2.88	2.94	3.00	3.07	3.14	3.22	3.30
150	2.39	2.42	2.46	2.50	2.55	2.61	2.66
170	2.08	2.09	2.11	2.14	2.17	2.21	2.26
190	1.86	1.86	1.87	1.89	1.91	1.94	1.98
210	1.70	1.70	1.70	1.71	1.72	1.74	1.77
230	1.58	1.57	1.56	1.57	1.58	1.59	1.61
250	1.48	1.47	1.46	1.46	1.46	1.47	1.49
270	1.41	1.39	1.37	1.37	1.37	1.38	1.39
290	1.34	1.32	1.30	1.30	1.29	1.30	1.30
310	1.29	1.26	1.24	1.23	1.23	1.23	1.23
330	1.24	1.21	1.19	1.18	1.17	1.17	1.17
350	1.20	1.17	1.15	1.14	1.13	1.12	1.12
370	1.17	1.14	1.11	1.10	1.08	1.08	1.08
390	1.14	1.10	1.08	1.06	1.05	1.04	1.04
410	1.11	1.08	1.05	1.03	1.02	1.01	1.00
430	1.09	1.05	1.02	1.00	0.99	0.98	0.97
450	1.07	1.03	1.00	0.98	0.96	0.95	0.94
470	1.05	1.01	0.98	0.96	0.94	0.93	0.92
490	1.03	0.99	0.96	0.94	0.92	0.90	0.89
510	1.01	0.97	0.94	0.92	0.90	0.88	0.87
530	1.00	0.96	0.93	0.90	0.88	0.87	0.85
550	0.99	0.94	0.91	0.88	0.86	0.85	0.84
570	0.97	0.93	0.90	0.87	0.85	0.83	0.82
590	0.96	0.92	0.88	0.86	0.84	0.82	0.81
610	0.95	0.91	0.87	0.84	0.82	0.81	0.79
630	0.94	0.90	0.86	0.83	0.81	0.79	0.78
650	0.93	0.89	0.85	0.82	0.80	0.78	0.77
670	0.92	0.88	0.84	0.81	0.79	0.77	0.76
690	0.92	0.87	0.83	0.80	0.78	0.76	0.75
710	0.91	0.86	0.82	0.79	0.77	0.75	0.74
730	0.90	0.85	0.82	0.79	0.76	0.74	0.73
750	0.89	0.85	0.81	0.78	0.75	0.73	0.72
770	0.89	0.84	0.80	0.77	0.75	0.73	0.71
790	0.88	0.83	0.80	0.76	0.74	0.72	0.70
810	0.88	0.83	0.79	0.76	0.73	0.71	0.69

Table 4.6. *Maximum mass ratio W/B to be used in the presence of highly fibered hemp shiv (ρ_h = 60 kg/m^3) to obtain 1 m^3 of hempcrete whose density in the hardened state is ρ, depending on the mass degree of hydration of the binder t (the specific density of the binder ρ_B = 2500 kg/m^3, D = 10 µm, R* = 15 µm, ϕ = 0.64)*

We note that the range of W/B ratios acceptable is fairly large for a particular target density of hempcrete. Thus, for instance, in the presence of non-fibered hemp shiv, with $\rho_B = 2500$ kg/m^3, $t = 1$, $D = 10$ µm, $R^* = 15$ µm and $\phi = 0.64$, we identify the amount of binder B to be added to 1 m^3 of hemp shiv and the mass ratio W/B:

– for a roofing application (250 kg/m^3): B = 100 kg; $0.56 < W/B < 3.15$

– for a walling application (410 kg/m^3): B = 340 kg; $0.56 < W/B < 1.83$

– for a flooring application (650 kg/m^3): B = 540 kg; $0.56 < W/B < 1.31$

These limits correspond to a hempcrete that has not been compacted during its application.

In the same conditions but with fibered hemp shiv, we get:

– for a roofing application (250 kg/m^3): B = 190 kg; $0.56 < W/B < 1.48$

– for a walling application (410 kg/m^3): B = 350 kg; $0.56 < W/B < 1.11$

– for a flooring application (650 kg/m^3): B = 590 kg; $0.56 < W/B < 0.96$

We also note that switching from a non-fibered hemp shiv to a fibered hemp shiv necessitates a higher binder content, whilst reducing the range of W/B ratios that are acceptable.

4.2.4. *Homogeneity of the paste*

When identifying the ranges of W/B ratios acceptable for the various applications, account has been taken of the aptitude of the hemp shiv to absorb some of the water from the mixture. The theoretical calculation is performed considering that the amount of water in the paste will evolve as the hemp shiv absorbs the water. The homogeneity of the paste, though, may locally be compromised – particularly on contact with the hemp shiv. Indeed, the phenomenon of absorption is driven by two events:

– on the one hand, the pure speed of water absorption by capillary action by the hemp shiv. Immersion tests [TER 11] show that this phenomenon is very quick (only taking a few minutes).

– on the other hand, the water absorbed by the hemp shiv has to percolate through the concentrated suspension (formed by the diluted binder in the water). This phenomenon is based on the suspension's aptitude for drainage, with the water percolating through a network of particles. This drainage will lead to an increase in the concentration of particles in the paste. The increase in the concentration of particles will be greater in the vicinity of the hemp shiv than further away from it. A coating layer of binder with low W/B may therefore form around the hemp shiv and lead to the stoppage of water absorption by the hemp shiv if the local W/B ratio falls below the limit value given by equation [4.11].

The consequences of stoppage of absorption by the formation of a coating layer of dense binder around the hemp shiv are multiple:

– the rheology of the mixture is fixed, as the hemp shiv is surrounded by a coating of "impermeable" paste. This effect is beneficial in relation to the time required for the stabilization of the rheology, which is directly linked to the mode of implementation chosen. Remember that a 30-minute period of stable rheology is desirable for production of hempcrete in a concrete mixer and implementation by formwork.

– the final properties of the paste exhibit a gradient as a function of the distance from the hemp shiv. The binder located close to the hemp shiv will exhibit high compactness and strength characterized by a W/B ratio similar to that given by equation [4.11]. The binder located far away from the hemp shiv will present low compactness and poor strengths, characteristic of a high W/B ratio (which may attain the values cited in Tables 4.5 and 4.6).

Thus, we can distinguish two philosophies on formulation with regard to the rheological properties of fresh hempcrete:

– either the formation of an "impermeable" coating is desired so as to guarantee the long-term stability of the material's rheology – in which case, a binder with quick setting kinetics is preferred. In this case, binders exhibiting limited colloidal effects are prioritized (e.g. Portland cement, natural cement, etc.).

– or the gradient of the W/B ratio is limited. In this case, the kinetics of change of the W/B ratio is essentially related to the permeability of the paste. The stability of the rheology is less satisfactory than in the previous case, but the homogeneity of the paste is assured. This type of paste is obtained by the use of a binder in which the colloidal effects are privileged (e.g. aerated lime or metakaolin).

In rheological terms, binders such as hydraulic limes and calcium sulfates will exhibit intermediary properties.

An alternative is to alter the conditions of water absorption by the hemp shiv by treating it. Immersion in a solution of alkaline silicate (sodium or potassium) constitutes a frequently-cited solution. Just like the hydrophobing solutions based on fluoride or silicone, these solutions prove too costly. In any case, although they affect the conditions of wetting of the hemp shiv, they are not able to prevent the capillary effects directly linked to the morphology of the hemp shiv.

The W/B gradient in the paste can, after hardening and drying of the hempcrete, lead to crumbling of the binder. In areas where the binder has a high water-to-binder ratio, the hardening/drying of the binder causes the formation of insolated particles that are too poorly coherent. These areas exhibit poor mechanical strength which is characterized by powdering that is evident when the concrete is analyzed. Professionals note that these powdering effects occur more frequently with mixtures made with a high W/B ratio.

The type of binder, which may be formulated, will have a controlling impact of the stability of the paste's homogeneity, and therefore the occurrence of the risk of crumbling.

Currently, no laboratory test is used to assess the aptitude for drainage of a hempcrete binder. However, we can cite the protocols developed by [PIC 11] about pastes of Portland cement to identify parameters characteristic of the phenomena of drainage and estimate the permeability of such media subjected to great mechanical stresses.

It is useful to adjust these methods by adjusting the level of mechanical stress in order to characterize hempcrete binders, which are far more dilute.

4.2.5. *The relationship between formulation and strength*

The strength of hardened hempcrete is essentially caused by the properties of the binder chosen for its formulation. The mechanical properties of mineral binders are generally expressed as a function of the amount of mixing water/binder ration (W/B tempering ratio). For illustrative purposes, Figure 4.4 shows the change in compressive strength measured on samples of hardened paste by [NGU 10] on a pre-formulated binder (Tradical 70) depending on the amount of tempering. A trend similar to a power law can be derived from these results. This type of result can be generalized to all types of binder.

In order to get around any variation of the intrinsic characteristics of the binder, these results can be modeled by a power law based on the volume fraction v occupied by the hardened binder in the sample of hardened paste:

$$\sigma = \sigma_0 \, v^a = \sigma_0 \, (\rho/\rho_0)^a \qquad [4.12]$$

where σ_0, obtained for $v = 1$, is the intrinsic strength of the (non-porous) binder. The parameter a usually takes a value of around two. The values of σ_0 and a can be obtained simply by adjusting the results of compression tests obtained for different degrees of tempering (see the example in Figure 4.4).

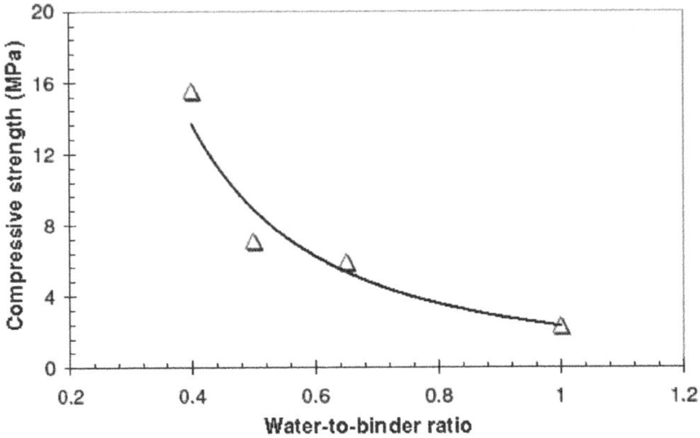

Figure 4.4. *Change in compressive strength measured on samples of hardened paste by [NGU 10] on a pre-formulated binder (Tradical 70) depending on the amount of tempering:* $\sigma = 126.2.(v)^{2.98}$

For the hardened hempcrete, this law of variation of the binder's strength can be adjusted, if we consider that the inert load corresponding to the hemp shiv plays no part in the strength of the mixture, other than by the effect of dilution.

This being the case, the strength of the mixture is given by:

$$\sigma = \sigma_0 \, v^a \, (V)^a \qquad [4.13]$$

where V is the volume proportion of hardened paste (of volume fraction v) in a cubic meter of hempcrete. This type of law is in accordance with the trends indicated by [GLO 11], [ELF 08] and [ARN 11] concerning the change in strength of different hempcretes as a function of their density.

In order to draw the connection between the mechanical properties of the hardened hempcrete and its formulation, we need to evaluate the parameters v and V. The former corresponds to the quality of the paste (in other words the W/B ratio) and the latter to the quantity of paste in the mixture (in other words the amount of binder, m).

In the case of a pre-formulated hempcrete with just enough water not to cause water absorption by the hemp shiv, the value of v is directly related to the W/B ratio used.

If water absorption phenomena leads to a change in the concentration of the paste, the value of v that needs to be taken into account in the calculation cannot simply be linked to the W/B ratio used.

If some drainage effects in the paste are associated with water absorption by the hemp shiv, the value of v needing to be taken into account may correspond to the lower bound value of W/B (equation [4.11]), but in this case, the volume proportion of paste V to consider in equation [4.13] would correspond only to the volume of drained paste forming a hardened hull around the hemp shiv. As the situation currently stands, it is still difficult to estimate this volume.

Yet it is clear that the strengths of such hempcretes remain less than those of concretes formulated with the minimum W/B ratio.

For illustrative purposes, the strengths expected in the case of hempcrete formulated with a water content corresponding to equation [4.11] (with no water absorption by the hemp shiv) and exhibiting properties of paste in line with those presented in Figure 4.4 are calculated with $\rho_B = 2500$ kg/m^3, $t = 1$, $D = 10$ µm, $R^* = 15$ µm and $\phi = 064$. The results are presented in Table 4.7. In this case, the W/B ratio in the calculation is 0.56. The amounts of binder and water are also indicated. The results correspond to a compacting coefficient of one.

From these estimations of compressive strength (Cs), we note that the values required for the different applications of hempcrete are satisfied for:

- roofing application: non-fibered hemp shiv: Cs > 50 kPa if $\rho > 290$ kg/m^3
  highly fibered hemp shiv: Cs > 50 kPa if $\rho > 240$ kg/m^3
- walling application: non-fibered hemp shiv: Cs > 200 kPa if $\rho > 400$ kg/m^3
  highly fibered hemp shiv: Cs > 200 kPa if $\rho > 350$ kg/m^3
- flooring application: non-fibered hemp shiv: Cs > 300 kPa if $\rho > 440$ kg/m^3
  highly fibered hemp shiv: Cs > 300 kPa if $\rho > 390$ kg/m^3

Thus, we note that such a binder can be used to formulate hempcretes that are usable for walling and flooring applications, but to satisfy the criteria related to the roofing application, the densities required become too great.

Once again, we note that the low apparent density of fibered hemp shiv is more readily able to satisfy the criteria relating to the resistance.

Non-fibered hemp shiv				Highly fibered hemp shiv			
ρ (kg/m³)	**B (kg/m³)**	**W (kg/m³)**	**Cs (kPa)**	**ρ (kg/m³)**	**B (kg/m³)**	**W (kg/m³)**	**Cs (kPa)**
110	0	0.0	0	**110**	50	28.0	1
130	20	11.2	0	**130**	70	39.2	3
150	40	22.4	1	**150**	90	50.4	6
170	60	33.6	2	**170**	110	61.6	11
190	80	44.8	4	**190**	130	72.8	19
210	100	56.0	9	**210**	150	84.0	29
230	120	67.2	15	**230**	170	95.2	42
250	140	78.4	24	**250**	190	106.4	58
270	160	89.6	35	**270**	210	117.6	79
290	180	100.8	50	**290**	230	128.8	103
310	200	112.0	68	**310**	250	140.0	132
330	220	123.2	90	**330**	270	151.2	166
350	240	134.4	117	**350**	290	162.4	206
370	260	145.6	149	**370**	310	173.6	251
390	280	156.8	185	**390**	330	184.8	303
410	300	168.0	228	**410**	350	196.0	360
430	320	179.2	276	**430**	370	207.2	425
450	340	190.4	331	**450**	390	218.4	498
470	360	201.6	392	**470**	410	229.6	578
490	380	212.8	461	**490**	430	240.8	666
510	400	224.0	537	**510**	450	252.0	762
530	420	235.2	621	**530**	470	263.2	868
550	440	246.4	713	**550**	490	274.4	982
570	460	257.6	814	**570**	510	285.6	1107
590	480	268.8	924	**590**	530	296.8	1241
610	500	280.0	1043	**610**	550	308.0	1386
630	520	291.2	1173	**630**	570	319.2	1542
650	540	302.4	1312	**650**	590	330.4	1708
670	560	313.6	1462	**670**	610	341.6	1887
690	580	324.8	1624	**690**	630	352.8	2077
710	600	336.0	1796	**710**	650	364.0	2280
730	620	347.2	1981	**730**	670	375.2	2495
750	640	358.4	2177	**750**	690	386.4	2724
770	660	369.6	2386	**770**	710	397.6	2966
790	680	380.8	2608	**790**	730	408.8	3222
810	700	392.0	2843	**810**	750	420.0	3492

Table 4.7. *Compressive strength Cs expected for a hempcrete whose density in the hardened state is* ρ, *with* $W/B = 0.56$, $t = 1$, $\rho_B = 2500\ kg/m^3$, $D = 10\ \mu m$, $R^* = 15\ \mu m$, $\phi = 0.64$, $\sigma_0 = 126.2$, $a = 2.98$

4.2.6. *The relationship between formulation and thermo-hydric properties*

The proportions of binder and hemp shiv, the type of hemp shiv, and so on, will condition the thermo-hydric properties of the hardened hempcrete. At present, it is not possible to establish rules of formulation on the basis of these criteria. However, we can point out that the apparent density of the hempcrete constitutes an effective parameter to gage the variation of parameters such as heat conductivity, for instance. A proper understanding of such tendencies therefore enables us to estimate the properties to be expected for a particular formulation. Various elements relating to the study of such changes are largely presented in Chapter 5, devoted to the study of the thermo-hydric properties of hempcrete.

4.3. Examples of formulations

4.3.1. *Origin of the data*

Often, professionals have used experimentation (trial and error) to define the proportions of the various components needed to ensure the criteria of fluidity necessary for the manufacturing, depending on the application for which the hempcrete is intended. This section brings together various formulation data put forward by the suppliers of binder or communicated by the construction experts who apply the products. As we saw previously, the formulations proposed are given for a particular type of binder – usually pre-formulated.

Finally, these typical formulations are not an absolute guarantee that the mechanical criteria of the hardened hempcrete will be fulfilled.

4.3.2. *Walling application*

Type of binder (manufacturer)	Hemp shiv (liters)	Binder content (kg)	Water content (kg)	Comments
Tradical PF 70 (CBC)	100	22	30–35	
Prompt natural cement (Vicat)	100	25	24–30	Addition of a retarder
Socli or Rabot lime (Socli)	100	35	40–44	
Renocal lime (Socli)	100	35	36	
Bâtichanvre [Buildhemp] (St. Astier)	100	25–35	35–40	
Hydrated lime Plaster	100	7.5 11.5		Addition of chamotte (grog): 1.5 kg

Table 4.8. *Examples of content of hempcretes: walling application*

4.3.3. *Flooring application*

Type of binder (manufacturer)	Hemp shiv (liters)	Binder content (kg)	Water content (kg)	Comments
Prompt natural cement (Vicat)	100	30	25–27	Addition of a retarder
hydrated lime CL90 (St. Astier) Plâtre gros	200	30–40 15–40	30–40	Pozzolanic aggregate: 10–60 liters Sand: 0–20 liters
Socli or Rabot lime (Socli)	100	35	40	
Bâtichanvre (St. Astier)	100	25-35	35-40	

Table 4.9. *Examples of content of hempcretes: flooring application*

4.3.4. *Roofing application*

Type of binder (manufacturer)	Hemp shiv (liters)	Binder content (kg)	Water content (kg)	Comments
Tradical PF 70 (CBC)	100	11	20-25	
Prompt natural cement (Vicat)	100	12.5	22	Addition of a retarder
Socli or Rabot lime or Renocal (Socli)	100	17.5	17.5-20	
Bâtichanvre (St. Astier)	100	12.5-17.5	17.5-20	

Table 4.10. *Examples of content of hempcretes: roofing application*

4.3.5. *Other applications*

Other applications of hempcrete may rely on the use of formulations derived from those presented above: e.g. resurfacing of a wall, external insulation of an existing building, etc.

4.4. Installation techniques

4.4.1. *Building a wall using formwork*

This technique is reminiscent of the bases for formwork used for traditional concrete. The form panels or molds may be plywood boards 16 mm thick. These will be sufficiently rigid so as not to be too greatly deformed under the influence of the pressure when the fresh hempcrete is poured, which remains light. These panels can be kept in place simply by affixing them to the load-bearing structure of the building (the wooden skeleton): four screws per m^2 are sufficient to ensure the boards hold their position. The form panels, which are 60 cm in height, are put in place as and when the hempcrete is ready to be poured into them.

The filling is done in carefully-equalized layers, around twenty centimeters in thickness. With each layer, a slight compression is performed, only at the edges of the surfaces and at the points of contact with the wooden skeleton. This compression is necessary in order to obtain good homogeneity of the surface and to ensure that the mixture is properly compacted. Excessive compression of the material would slow down the drying and hardening within the surface. It is preferable to remove the formwork soon after pouring – ideally within a day – to facilitate the start of carbonation.

In order to avoid a construction joint, as soon as the setting of the binder is sufficiently advanced, it is necessary to daub the joint surfaces with a mixture of binder and water. It is wise to prioritize horizontal joints, and if the pouring finishes on a vertical line it is helpful to reshape the material to increase the contact surface by an oblique cut.

The filling of the coffer can be done manually, but also by pouring of the hempcrete using a pipe or a spraying nozzle. These latter techniques are continuous. They help ensure a good homogeneity of the product. However, the operation of tamping is still necessary.

4.4.2. *Application by spraying*

For application by spraying, a single formwork surface is sufficient. In order to avoid the fresh hempcrete being pulled off when the formwork is removed, a polythane film is laid on the boards before spraying.

The hempcrete, compelled by a stream of compressed air, is sprayed against the boards to create the wall. In order to ensure good holding of the fresh hempcrete while the wall is being built, it is preferable to lay it down in horizontal layers in the form of linear rows rather than overlap vertical layers in the way of yarrow. Hence, the application is performed in a series of horizontal layers, less than 20 cm in height, applied throughout the section of wall to be created. The spraying is continuous, and the whole wall is created in a single pass, from bottom to top.

4.4.3. *Laying of a floor*

Whether with a manual or mechanized process, the procedure for putting the hempcrete in place for a flooring application will be the same. The material is poured onto surfaces prepared for this purpose in accordance with professional rules. The spread is equalized using a rake, moderately compressed on the whole surface of the floor and leveled. A powered float ensures a flat, properly compacted surface. If the hempcrete is thicker than 15 cm, it is recommended to lay it in two separate layers, put in place two weeks apart in order to facilitate the hardening of the material throughout its thickness.

4.4.4. *Creating a roof*

Hempcrete with a low binder content for roof insulation is usually put in when the roof covering is being totally redone, in cases of restoration or renovation of old buildings. In all cases, in a new construction or an existing building, the filling of the roof cavity is done before the cover is laid. Indeed, in the case of intervention after the cover has been put in place, the operation is difficult to perform under the roof and the results of perfect insulation will not be guaranteed.

To begin with, a formwork is laid under the roof framework, chambers or other structures to be over-filled. Pouring the hempcrete from outside enables all the gaps to be filled and thus ensures the most complete filling possible. The hempcrete can be put in place by spraying or pouring using a raising device – a conveyor belt, a bucket, a hydraulic lift cab, a crane or other tools adapted for tall buildings. After filling, very slight compacting is performed with a float. The material is then serried at the surface to obtain a stable whole without settling. In accordance with the regulations in force, a rain guard must be installed to ensure air-tightness and protection against any possible dripping from the roof.

For this type of application, particular care and attention is given to the mixtures and the appropriate water content so as to avoid any run-off and exposure of the structural pieces to excessive damp.

4.4.5. *Other uses*

In the case of an existing wall requiring additional insulation, hempcrete can be used in a layer of greater or lesser thickness, applied to the face of the wall. We then speak of rendering. This solution is considered more as a thermal repair – particularly if applied manually. High measures of the binder are necessary for the mixture to stick to the wall when troweled on. Consequently, the thermal properties of these rendering mixtures are less advantageous than those of hempcretes intended for walling and roofing applications. Conversely, mechanized application by projection offers these rendering mixtures qualities in terms of hydric behavior which are very similar to those for the walling application.

Finally, these mixtures, which are sometimes applied to a pellicular layer to correct surfaces that are not quite flat, fulfill a more aesthetic rather than a practical performing role.

4.5. Professional rules for buildings using hempcrete and hemp mortars

4.5.1. *History*

The actors in the hemp construction sector in France, grouped together in the association *Construire en Chanvre*, have been working since 1998 to promote the development of the use of hemp in building by awareness-raising, training, R&D and standardization.

In order to ensure the continuation of this development, the need soon became evident to register hemp-based construction products in the regulatory and normative framework governing the construction sector.

Furthermore, in the face of an increasing demand for renovations, refurbishing and new constructions integrating hemp material, it was necessary to provide construction companies with a framework to help them cater well for their customers, and provide them all with the guarantees that are so crucial in terms of insurance.

In 2004, therefore, professionals in hemp construction decided to draft professional rules for the use of hemp-based concretes and mortars. Initially devoted solely to the realization, this document soon became a wide-reaching quality framework, dealing not only with the realization but also the quality of the raw materials, the design and training of companies.

An early version of the *Règles Professionnelles* (Professional Rules) was established as part of a project led by the main actors in the fields of agriculture and construction in France (*Ministère de l'Agriculture*, *Ministère de l'Ecologie*, *Fédération Nationale des producteurs de chanvre*, *Fédération Française du Bâtiment* et *Construire en Chanvre*). A commission of experts including construction professionals (artisans, architects, study bureaus), scientists (ENTPE, CEBTP), industrial actors and a supervising office (APAVE) was put in charge of drafting four reference documents to describe the correct practices for use of hemp-based concretes and mortars in four different applications:

– building of walls from hempcrete;

– rendering;

– hempcrete flooring insulation;

– hempcrete roofing insulation.

Presented to the C2P (*Commission Prévention Produits* – Commission for Product Prevention) of the *Agence Qualité Construction*, these rules were accepted as experimental professional regulations between 2007 and 2009.

In 2010, the association *Construire en Chanvre* founded a revision committee to establish a second version of these professional regulations so that this version would be put definitively on the C2P's green list.

Presented to the C2P at the end of 2011, version 2 of the professional regulations for use of hemp-based concrete and mortar in construction was accepted in January 2012. Thus, today these regulations are the reference document upon which any company wishing to use hemp-based concretes and mortars can base its practices [ASS 07].

4.5.2. *Principles and content of the professional regulations*

Besides simple documents on realization, the professional regulations for building using hemp-based concretes and mortars now define an overarching framework, the aim of which is to ensure the quality and durability of constructions using hemp-based concretes and mortars.

Thus, these professional Regulations manifest themselves in different areas, and include the following key points:

– four documents corresponding to the four applications for hemp-based concretes and mortars;

– a book of construction details;

– a document describing the testing procedures;

– the exclusive use of a hemp granulate bearing the label "*chanvre bâtiment*" (building hemp);

– the training of companies.

4.5.2.1. *Four documents corresponding to the four applications of hemp-based concretes and mortars: building of walls, rendering, roofing insulation and flooring insulation*

These documents, on the one hand, describe the key points of the use of hempcretes in different applications and, on the other hand, specify the requirements in terms of performances of the concretes and mortars, and more particularly the mechanical performances (compressive strength and elasticity modulus). This requirement in terms of mechanical performances enables us to guarantee the "proper functioning" of a binder/hemp granulate pair used to manufacture hempcrete. In addition, defining a requirement in terms of performance is not a hindrance to innovation and does not preclude any solution or future development.

The four documents were written in accordance with the same template of a "DTU (*Document Technique Unifié* – Unified Technical Document)".

The chapters of these documents are as follows:

– an *"Avant-Propos" (preface)*, defining the major principles governing the professional regulations, particularly the notion of compatibility of the "binder-granulate" pair, the notion of performances expected and requirement in terms of skills of the applier.

– a chapter on the *"domaine d'application" (domain of use)*, particularly specifying that these regulations apply to the following buildings: dwellings and Category 5 ERPs (public buildings).

– a chapter on *"références normatives" (normative references)*, recalling other normative documents (standards and DTUs) to which the readers can refer. Indeed, the professional regulations do not redefine the entirety of the normative framework, but build on the existing regulations and provide the particularities of hempcretes.

– a chapter on *"termes et définitions" (terms and definitions)*, defining around ten terms associated with the use of hemp-based concretes and mortars.

– a chapter on the material, *"Les bétons et mortiers de chanvre" (hemp-based concretes and mortars)* ", specifying:

- the nature of the components of the concretes and mortars: the binder (based on lime, regular cement or prompt cement, and with possible adjuvants), the hemp granulate (with exclusive use of a labeled granulate – see the paragraph about the label) and water,

- the ranges of composition specified in Table 4.11.

	Binder mass content % in relation to total dry weight	"Hemp aggregate" mass content % in relation to total dry weight	Dry density (indicative)
Maximum (rich mixture)	90%	10%	1000 kg/m^3
Minimum (lean mixture)	50%	50%	200 kg/m^3

Table 4.11. *Examples of content of hempcretes: roofing application*

- the minimum performances expected of hemp-based concretes and mortars in terms of compressive strength and elasticity modulus. These characteristics are given in standard conditions (20°C and 50% RH) and at 60 and 90 days. Table 4.12, which supplements the indicative Table 4.1, for each application, recalls the threshold performances expected of hemp-based concretes and mortars.

WALLING	Elasticity modulus (MPa)	Compressive strength (MPa)
Threshold value (minimum) *in standard condition*	>15 MPa	>0.2 MPa

FLOORING	Elasticity modulus (MPa)	Compressive strength (MPa)
Threshold value (minimum) *in standard condition*	>15 MPa	>0.3 MPa

ROOFING	Elasticity modulus (MPa)	Compressive strength (MPa)
Threshold value (minimum) *in standard condition*	>3 MPa	>0.05 MPa

FLAGGING	Elasticity modulus (MPa)	Compressive strength (MPa)
Threshold value (minimum) *in standard condition*	>20 MPa	>0.3 MPa

Table 4.12. *Range of composition of hempcretes*

- the manufacture of a hempcrete. This section recaps, on the one hand, the way of manufacturing of a hemp concrete or mortar with a concrete mixer or a cement mixer – and on the other hand the importance of the binder and aggregate content.

– a chapter relating to the *execution of the construction* in hemp concrete and mortar and dealing with the following points:

- the precise definition of the application in question.

Thus, for "*wall*" use, three arrangements are discussed:

– the case of a hempcrete used as wall filling, with a completely flooded skeleton;

– the case of a hempcrete used as wall filling, with a bare skeleton on one surface;

– the case of reinforcement of a vertical wall – load-bearing or otherwise; mechanically independent support.

For the application "*floor insulation*", the document details the implementation of hempcrete to create an insulating form on an upstairs landing or a crawl-space;

For the application "*roof insulation*", the document deals with the use of hempcretes to create roofing insulation in the rempits;

and for the application "*render*", the document specifies that "hemp render" is a render product, used indoors or outdoors, comprising: a gobetis (substrate preparation) if need be, a render substance made of hemp mortar and, inevitably in an outdoor environment, a finishing layer.

– the general preparations and prescriptions such as, e.g., the preparation of the supports in the case of an application for renders, the interfaces and networks in the case of a wall application, and the limitations of use in terms of temperature (between 8°C and 25°C) for all applications.

– the installation of hemp-based concretes and mortars:

- by deposition (manual or mechanical) of horizontal layers between two formworks or by mechanical spraying against a support (or a provisional or permanent formwork) in the case of a wall application;

- by pouring onto a support and smoothing in the case of applications for flooring insulation or roofing insulation;

- by manual or mechanical spraying in the case of a render application.

– the coatings, protections and finishings needing to be applied to hemp-based concretes and mortars, with an important point shared by all applications: the use exclusively of coatings and finishings that are permeable to water vapor.

4.5.2.2. *A set of construction details*

This set of details presents around forty details on construction in accordance with the regulations in force, relating to five topics:

– positioning of the skeleton, with different types of skeleton: buried skeleton, external half-timber structure, external skeleton with cladding, internal skeleton, skeleton with extended return, skeleton with a re-entrant angle.

– basements and ground floors;

– external woodwork, with different types of frame (tunnel type installation, outside frame, inside frame) and different solutions for fixation of the shutters.

– skeleton and roofing with different types of structural work: cross joists, traditional joists.

It is important to note that this is not an exhaustive list of construction details, but rather of putting forward constructive solutions which have been tested on site and which are able to deal with many arrangements that are to be found on building sites.

4.5.2.3. *A document describing the testing procedures*

A document describing the testing procedures to be used to measure the performances of hemp-based concretes and mortars – both mechanical and thermal performances – and presenting a list of independent laboratories able to realize these kinds of tests. These procedures are based on standardized protocols, adapted for hemp-based concretes and mortars. They were established by the *Collège Développement Scientifique et Technique de Construire en Chanvre* (College for Scientific and Technical Development on Hemp Construction) which encompasses around ten French laboratories specialized in construction and plant matter.

4.5.2.4. *Sole use of a hemp aggregate with the label "chanvre bâtiment"*

This label is a label of quality, which guarantees the user (the construction company) the stability of the "hemp aggregate" raw materials, and hence the stability of the performances of the construction. This label is subject to a referential framework which defines the characteristics which have to be adhered to by the supplier of the aggregate, the methods for measuring these characteristics, the frequency of the measurements and the requirements regarding these characteristics.

Within this referential framework, the characteristics which are required to be monitored are:

– the amount of material from a hemp crop. So as not to deceive the consumer, it is important that the label "*chanvre bâtiment*" should not be attributed to products obtained by mixing hemp granulate and other kinds of granulates;

– the density;

– the particle size distribution. To evaluate this PSD, a simple method by image analysis has been developed by ENTPE;

– the amount of dust obtained by sieving;

– the degree of humidity of the straw, generally measured upon entering the factory before the hemp straw is transformed;

– the color. Indeed, the parameter of color may express the degradation of the granulate by micro-organisms;

In the interests of full disclosure to the consumer, this label is placed on sacks of material with a "*chanvre bâtiment*" type of logo.

4.5.2.5. *Training of companies*

In view of the specificity and newness of the material, it is important for companies to be trained by people with good knowledge of the characteristics of the material and to be able to prove their skills to their insurer.

Consequently, the association *Construire en Chanvre* has put in place a procedure for accreditation of trainers able to offer training courses to companies in accordance with the professional regulations and provide them with certification of completion of these courses to show to their insurers.

This accreditation is obtained on the basis of the presentation of a file and participation in a two-day training/evaluation session during which the candidates for accreditation are evaluated on their skills in the area of building, hemp construction and teaching ability. The trainers are accredited for a period of one year, with the accreditation being renewed yearly after participation in a refresher session.

All the elements presented above have been put in place in the interests of safety in the use of hemp-based concretes and mortars, and thus in order to be able to guarantee, to the end client, the quality and performance of the constructions. They are based on scientific, technical and practical expertise accrued over more than twenty years.

The approach of the professional regulations thus put in place is not immobile: it is intended to be open-ended so as not to exclude any practice, so long as it is able to provide guarantees of quality. Furthermore, these professional regulations will doubtless evolve on the basis of new knowledge about the material and practices on site. In any case, today they represent a very solid basis for the design of future DTUs on hemp-based concretes and mortars.

4.6. Bibliography

[ARN 11] ARNAUD L., GOURLAY E., "Experimental study of parameters influencing mechanical properties of hemp concretes", *Construction and Building Materials,* 28 50–56 (2011).

[ASS 07] Règles Professionnelles d'Exécution d'Ouvrage en Bétons de Chanvre. Association Construire en Chanvre, 2007.

[CHA 11] CHAMOIN J., COLLET F., PRETOT S., LANOS C., "Réduction du pouvoir absorbant de chènevottes par traitement imperméabilisant", *Matériaux & Techniques,* EDP Sciences, Volume 99, N° 6, 633–641 (2011).

[CON 12] http://www.construction-chanvre.asso.fr.

[EAS 11] photo: www.easychanvre.fr.

[EKO 11] http://fr.ekopedia.org/Construction_en_chanvre.

[ELF 08] ELFORDY S., LUCAS F., TANCRET F., SCUDELLER Y., GOUDET L., "Mechanical and thermal properties of lime and hemp concrete ("hempcrete") manufactured by a projection process", *Construction and Building Materials,* 22, 2116–2123 (2008).

[GLO 11] GLOUANNEC P., COLLET F., LANOS C., MOUNANGA P., PIERRE T., POULLAIN P., PRETOT S., CHAMOIN J. AND ZAKNOUNE A., "Propriétés physiques de bétons de chanvre", *Matériaux & Techniques,* EDP Sciences, Volume 99, N° 6, 657–665 (2011).

[NFV 93] NF V19-002, Norme, Détermination du pouvoir absorbant, 1993.

[NGU 10] NGUYEN T. T., Contribution à l'étude de la formulation et du procédé de fabrication d'éléments de construction en béton de chanvre, Thesis, University Bretagne Sud, 2010.

[PIC 11] PICANDET V., RANGEARD D., PERROT A., LECOMPTE T., "Permeability measurement of fresh cement paste", *Cement and Concrete Research,* Vol 41, Issue 3, 330–338 (2011).

[TER 11] photo: terrevivante.org.

Chapter 5

Mechanical Behavior

5.1. Composite material

Hempcrete is a mixture, in very variable proportions, of two very different components: a plant-based aggregate and a hydraulic and aerated setting binder. It exhibits mechanical behavior which is unusual in the domain of construction materials. Indeed, the particles of hemp shiv are characterized by a high degree of porosity which results in a high capacity to deform: this is one of the essential characteristics which set hempcretes apart from tradition mineral-based concretes for which the aggregates are considered non-deformable. This difference has major effects on the performances of these concretes in all areas – acoustic, thermal, etc. – and of course, it has a direct impact on the mechanical properties of these innovative concretes. The aim of this chapter is to show why and how these concretes are distinguished from conventional concretes, and in particular to establish the properties of use and the characteristic values necessary for their use in construction.

The high porosity of the particles of hemp, seen in Chapter 2, has many important consequences for hempcretes: lightness, deformability and absorbent capacity are directly linked to the microscopic and macroscopic porosity of the concretes.

These particles are mixed with one or more binders, whose mechanical strength – because of the flexibility of the hemp shiv – will play a crucial role in the behavior of the concretes. Today, many different binders are used: natural binders based on

---

Chapter written by Laurent ARNAUD, Sofiane AMZIANE, Vincent NOZAHIC and Etienne GOURLAY.

lime – hydraulic or aerated – or cement, or indeed pre-formulated binders. More recently, experimental results have become available concerning the use of new binders such as metakaolin or even clay. Unfortunately, such results cannot be presented in detail here owing to a lack of references.

Given the high degree of porosity of the hempcrete analyzed in the previous chapters, and the resulting high compressibility of the concrete, the mechanical limitations logically relate to the mechanical strength. This is why, at present, it is not used as a load-bearing material. However, there is a current trend toward the development of formulations and methods of implementation which will enable it to be used as such [BUT 04; NGU 09; MAG 10].

In the context of this overview, we have chosen mainly to outline the results in terms of compressive strength σ_C. Indeed, the studies performed hitherto show close correlation between that value and the flexural strength σ_F of the material. However, parallels could be drawn with Young's quasi-elasticity modulus and the Poisson coefficient.

5.1.1. *Making of the test tubes*

Cerezo [CER 05] put forward a process for making conventional cylindrical test tubes 16 cm in diameter and 32 cm in height. This approach was then integrated into the *Règles professionnelles d'exécution d'ouvrages en béton de chanvre* (Professional Regulations on the Edification of Buildings Using Hempcrete) [RP2C 07].

This process consists of filling the cylindrical molds with successive layers, 5 cm in height, and compacting each of them under a quasi-static stress of 0.05 MPa. The samples are then kept in chambers with controlled temperature and relative humidity in stable conditions equal to 20 ± 1°C and 50 ± 5% RH. This *modus operandi* enables us, for the characterization tests, to obtain excellent reproducibility of the results.

5.1.2. *Mechanical behavior*

The behavior of hempcrete can be considered to be elasto-plastic with a very extensive range of possible strains [ARN 01].

More specifically, [CER 05] shows that it is quasi-elastic during the first part of compression (I: Elastic Zone – see Figure 5.1, the window showing the close-up of the first part, showing the linear change in the stress/strain curves). Taking the small deformation theory to be true, it defines – in a conventional manner in accordance with Hooke's law – the elastic modulus E as being the slope at the origin of the stress/strain curve $\sigma_C = f(\varepsilon)$ and the Poisson coefficient ν as the ratio between the radial strain ε_{radial} and the axial strain ε_{axial} (see Figure 5.2).

The same author describes the behavior of BCC (compacted hempcrete) after the elastic part as non-linear. She observes a progressive inflection of the stress/strain curve, which she explains as a progressive cracking of the binder matrix to the point of rupture (II: Pre-peak Zone). These results are borne out by loading/unloading tests which clearly show the appearance of permanent plastic strains: thus, the material's behavior has become elasto-plastic. Finally, after the peak of stress (III: Post-peak Zone), the binder would be totally degraded and the strain essentially passed on to the plant particles. This can be observed in concretes at least six months old.

The behavior is a little different in recently-laid concrete (see Figure 5.1). The material is then far more deformable and it is not possible to distinguish a stress peak, with the stress being absorbed in the ductile phase. Cerezo attributes this difference to the network of hydrates which, at an early age, has not yet fully formed. Therefore, the behavior is influenced mainly by the particles of hemp shiv.

Overall, the high compressibility of plant aggregates means that hempcretes exhibit unusual behavior in comparison to the behavior of traditional, mineral aggregate-based concretes. The main characteristics of hempcretes are (see Figure 5.1):

– a small, initial quasi-elastic domain, and at the same time, we observe quasi-linear variations in the radial strains in relation to the axial ones. Thus, it is possible to calculate the value of the Poisson coefficient;

– then, a elasto-plastic behavior, with more marked permanent deformations. It should be highlighted that the modulus during loading/unloading cycles differs noticeably from that of the monotonous curve;

– and finally post-peak behavior (when there *is* a peak), with a high degree of ductility.

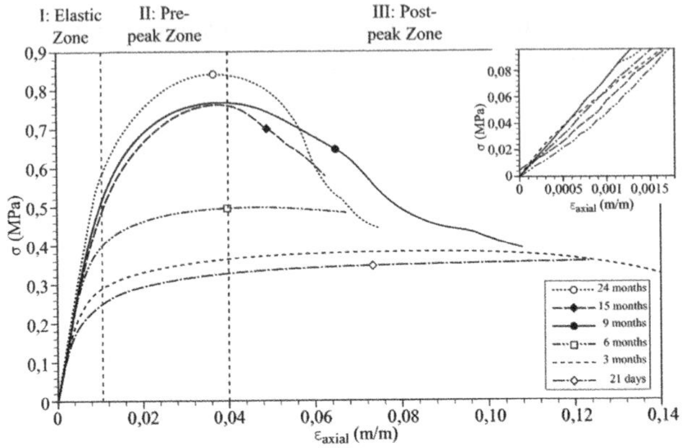

Figure 5.1. *Change over time in the compressive mechanical behavior of BCC A4-1.5 [CER 05]*

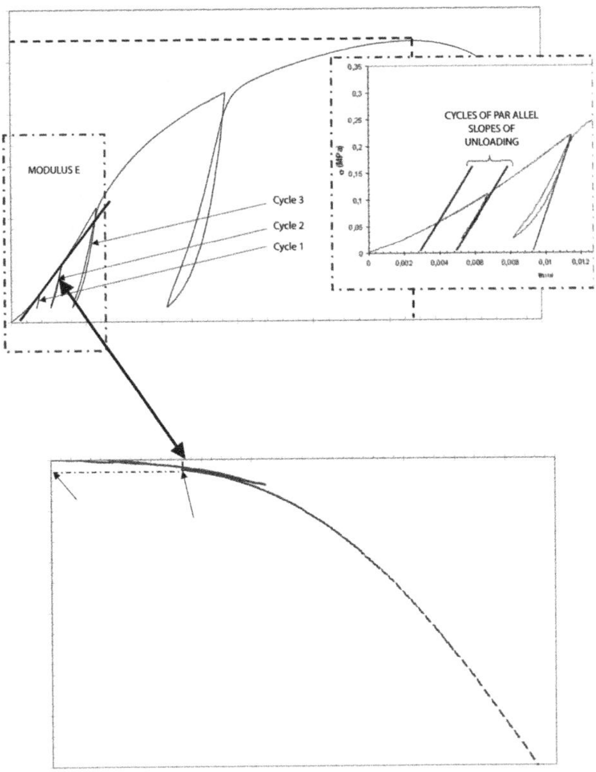

Figure 5.2. *Mechanical behavior with cyclic compression and measurement of the Poisson coefficient of BCC A4-1.5 [CER 05]*

This general behavior is modulated by a great many parameters of formulation, such as:

– the binder and water content;

– the nature of the binder and interactions between the binder and aggregate during the phases of hardening and setting;

– the size of the particles of hemp;

– the installation and particularly the compactness level of the mixture;

– the conditions of conservation;

– the age of the concrete;

– and so on.

The following sections offer an analysis of the influence of these parameters on the mechanical behavior of hempcretes.

5.1.3. *Effect of initial compression*

[NGU 09a] performs a study of mechanical behavior at an early age (28 days) for concretes which have been greatly compacted when the samples were taken. It consists of a mold made of a PVC tube, reinforced with a clamping ring. A piston moving down applies the compaction force. The initial apparent volume of the mixture poured into the cylinder is reduced by a factor of up to three.

After drying, the specimens are tested in terms of compressive strength. Nguyen describes their behavior according to 4 distinct phases (see Figure 5.3) after having studied the correlation between the stress/strain curve and the field of strain obtained by image analysis:

Zone I: homogeneous elastic behavior.

Zone II: elasto-plastic zone: development of damage to the matrix binder and/or the interface between the particles of hemp shiv and the matrix.

Zone III: distribution of the stress into the particles of hemp shiv and beginning of their compaction (closure of the capillaries in the particles). Very great deformation.

Zone IV: possible rupture of the sample.

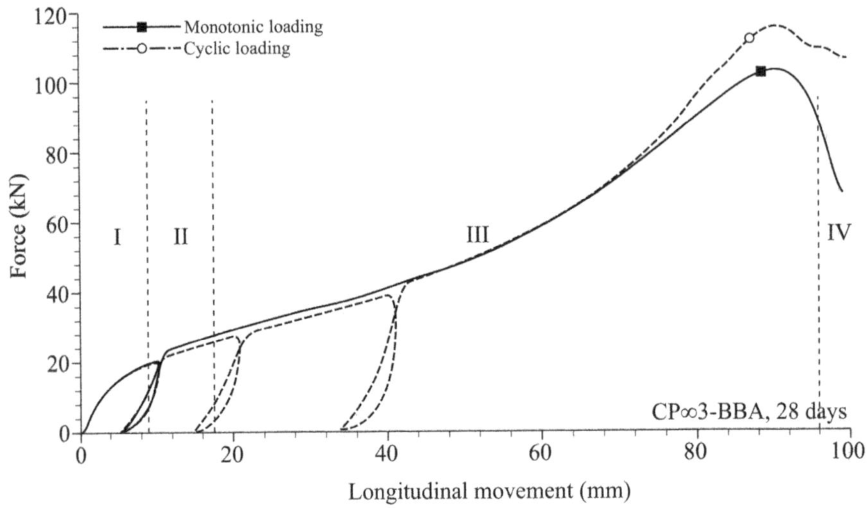

Figure 5.3. *Characteristic curves of loading / unloading of samples of hempcrete under simple compression*

The compaction applied leads to an alteration of the microstructure and results in a decrease of the porosity of the material. This porous network plays a crucial role in the phases of drying and setting, and of course, in the final mechanical performances.

As regards the setting of the binder, e.g. for aerated binders, the presence of air facilitates a carbonation reaction between the $Ca(OH)_2$ and the CO_2. This aerated setting is therefore favored when the material presents a network of many, interconnected pores (and thus a high degree of open porosity, leading to good permeability to air).

Similarly, the presence of an open porous network favors the evacuation of water to the outside environment by transfer in liquid and gaseous phases: that is, it facilitates the drying.

Finally, the porosity has a bearing on the level of mechanical performance of a material. The air bubbles represent weak points in a material. The compressive strength and the elastic modulus are inversely proportional to the porosity.

5.1.4. *Effect of the nature of the binder*

Today, various binders are validated and commercialized for use with hemp shiv. Tests were performed with three binders based on hydraulic limes belonging to different classes of standardization. The results were then compared to the performances of pre-prepared concretes based on Tradical PF70 from BCB. Thus, we tested:

– NHL2: White Line from Pavier,

– NHL3,5: Pure White Line from Saint Astier LC EN 459-1,

– NHL3,5 Z: White Line from Lafarge,

– Pre-formulated binder Tradical® PF70 based on aerated lime.

So as to ensure confidentiality of the results, the graphs and references made to these binders display only the letters A to D to distinguish between them on the basis of the primary materials. As a control sample for comparison, a unique formulation [RP2C 07] was used. In order to preserve the same initial rheology and not penalize binders with the water content, this content was adapted depending on the different binders. Thus, for binder B, the water content was reduced by 3.5 liters in relation to the normal mix – i.e. a reduction of 14.9% – and for binder C, the water content was reduced by 4.88 liters in relation to the normal mix – a 20.4% reduction.

The samples were taken as prescribed by the professional regulations, and then conditioned at a certain temperature and relative humidity for 28 days until the testing date. Finally, a last measurement was taken at 90 days – i.e. around 3 months – in order to judge the evolution of the properties in the longer term.

To begin with, let us analyze the influence of the binder on the mixture. Clearly, significant differences appear both in terms of the levels of stress attained and of the behavior of the concrete. The important differences, which can be seen on the curves, in fact reveal more or less advanced states of setting depending on the binder in question.

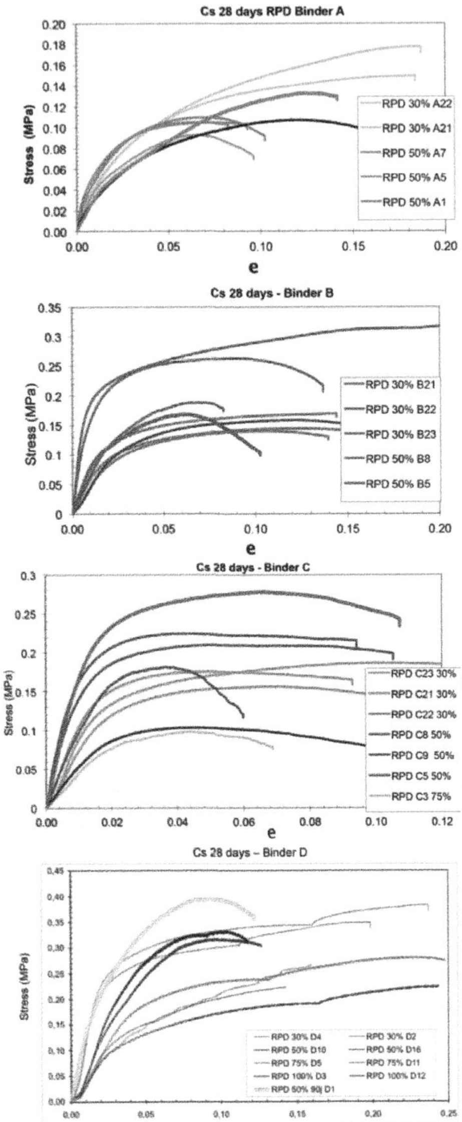

Figure 5.4. *Comparison of the mechanical performances with simple compression for hempcretes based on various binders and having been kept in different aging conditions (variable relative humidity), for the same formulation and the same length of time [ARN 07]*

Table 5.1 summarizes the average values of maximum stress reached and the Young's modulus for these concretes at 28 days.

Binder used	*Curing conditions*	$\sigma_{C\text{-}MAX}$ *(MPa)*	*E (MPa)*
A	*30%*	*0.11*	*5.0*
	50%	*0.16*	*3.8*
	75%	*0.09*	*3.5*
	100%	*0.11*	*2.5*
B	*30%*	*0.16*	*6.0*
	50%	*0.29*	*25.0*
	75%	*0.19*	*6.6*
	100%	*0.16*	*5.3*
C	*30%*	*0.17*	*10.8*
	50%	*0.22*	*20.0*
	75%	*0.10*	*4.6*
	100%	*0.10*	*5.3*
D	*30%*	*0.32*	*14.5*
	50%	*0.32*	*10.8*
	75%	*0.23*	*5.5*
	100%	*0.19*	*7.4*

Table 5.1. *Maximum stress* $\sigma_{C\text{-}MAX}$ *and average Young's modulus E at 28 days for hempcretes formulated using different types of binders*

5.1.5. *Influence of the binder content*

Depending on the binder content, Cerezo defines three kinds of mechanical behavior (see Figure 5.5):

– low binder content: the material behaves like a collection of particles linked by rigid "bridges" of binder;

– intermediary binder content: the material is composed of particles surrounded by layers of hydrates, of varying thickness;

– high binder content: the particles are buried in the binder matrix, which is continuous and determines the mechanical properties of the composite.

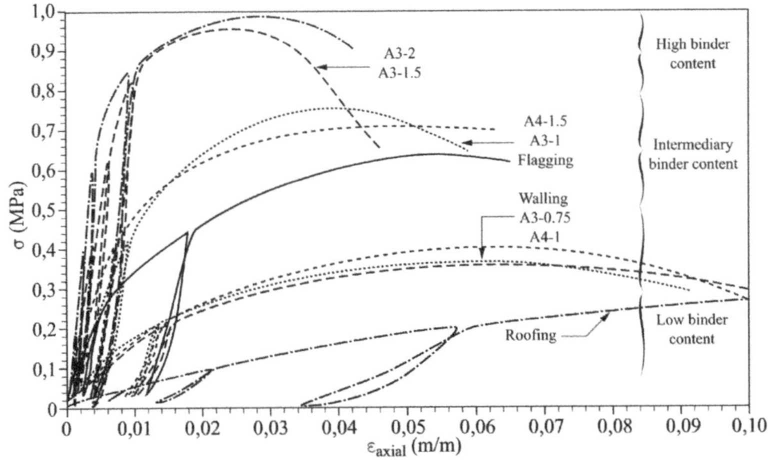

Figure 5.5. *Influence of the binder content on the compressive strength of the hempcrete shaped by compaction (0.05MPa) after 1 year of setting [CER 05]*

It is commonly accepted that the higher the binder content, the more similar the final mechanical properties will be to those of the binder, and therefore important (see Figure 5.5) in view of those of the plant particles. Similarly, it appears that the higher the binder content, the more closely the mechanical behavior of the material mimics the fragile mechanical behavior of the binder. Conversely, for a formulation with a high proportion of plant materials, as is used for roofing, the deformability of the material is extremely high – hence the presence of a ductile phase, observed by [CER 05] and [ARN 12], among others.

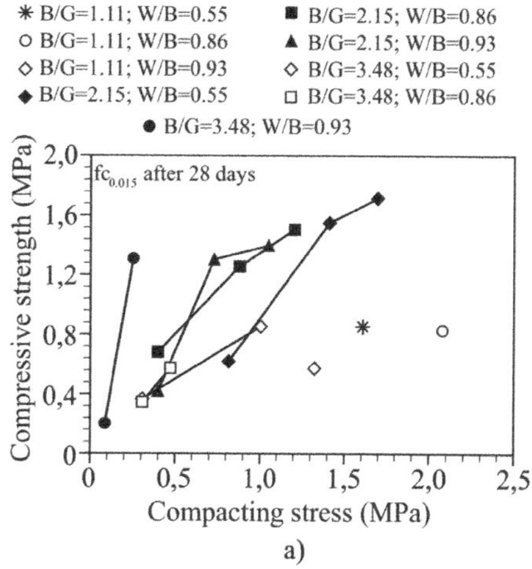

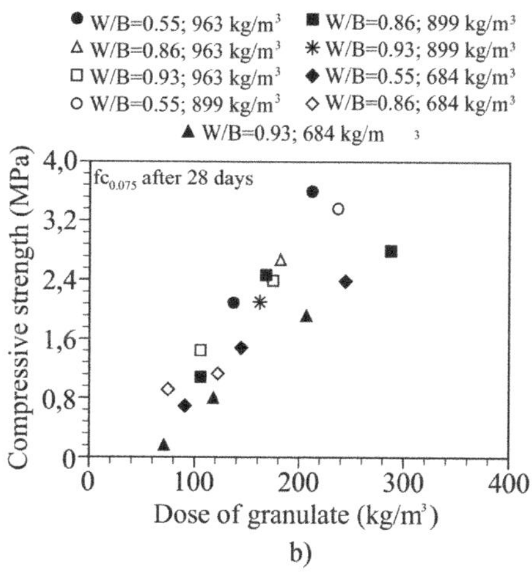

Figure 5.6. *Change in the compressive strength at 28 days when measured at 1.5% and 7.5% strain respectively, depending on the compacting stress (a) and depending on the aggregate content (b) [NGU 09a]*

However, the observation made by Nguyen for highly compacted hempcretes shows the opposite (see Figure 5.6b). Irrespectively of the density of the fresh concrete and the W/B ratio, he observes an increase in the compressive strength at 7.5% strain when the amount of aggregate increases. Yet this trend is less pronounced for the elastic modulus [NGU 09a]. However, we must take account of the fact that the higher the quantity of aggregate, the greater the compacting force that needs to be applied. Nguyen also observes an increase in compressive strength with the compacting stress (see Figure 5.6.a), as does Cerezo in a previous study

[CER 05]. In Nguyen's study, though, he theoretically determined that the levels of compaction inevitably caused squashing of the particles of hemp shiv. In other words, the increase in strength is due not only to the reorganization of the particles, but also to the decrease in intra-particle porosity. If the amount of aggregate per mass increases, there is no dramatic corresponding increase in the content per volume.

Depending on the proportions, the parameters of installation and the length of drying time, the behaviors of compacted hempcretes will be very different. Currently, however, there is no consensus as to the measurement of $\sigma_{C\text{-}MAX}$ or any other parameter that is characteristic of the compression. In addition, the maximum stress is often attained for values of stress that are incompatible with usage in construction. Thus, Elfordy *et al.* [ELF 08] take account of the threshold elastic stress $\sigma_{S\text{-}EL}$ rather than the maximum stress. In the same way, Nguyen chooses to take account of σ_C for 1.5 and 7.5% strain. Cerezo for her part chooses to use $\sigma_{C\text{-}MAX}$ as the basis for her measurements, without taking account of the high deformability of the material.

5.1.6. *Influence of the particle size*

Hempcretes have been installed using a pre-formulated binder based on aerated lime and three PSDs of the same shiv [GOU 09; ARN 12]: PSD 1 is slightly coarser than 2 and PSD 3 is significantly finer than both the others. The three concretes made from these mixtures were tested at 28 days, 4 months and 14 months (see Figure 5.7).

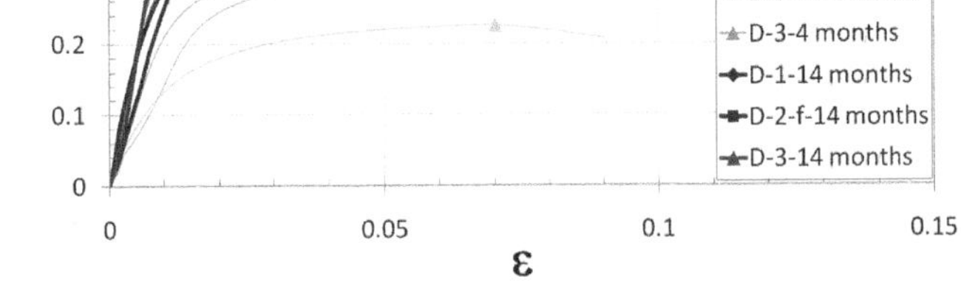

Figure 5.7. *Stress/strain curves at different times for hempcretes made using hemp shiv samples with different PSDs [GOU 09; ARN 12]*

At 28 days, all the hempcretes have a very ductile behavior, characterized by the presence of a long plateau on the curve. This is due to the fact that the hydrates of binder have not yet formed an interconnected network: the behavior of the concretes is therefore similar to that of the highly deformable hemp particles. Over time, the hydrates gradually connect and progressively create a continuous network in which the forces are transferred. The characteristics of the binder gradually become predominant in the mixture: the concretes then become able to deal with more and more significant strains and their behavior becomes less and less ductile.

The mechanical properties of the concretes formulated using PSDs 1 and 2 barely change any further after 4 months of drying. In addition, because the hemp particles of PSD 1 are slightly longer than those of PSD 2, we note that the hempcretes made with PSD 1 are less dense and therefore have poorer mechanical properties than those formulated using PSD 2.

The concretes made using PSD 3, which is much finer than the other two, exhibit mechanical properties which at 28 days are poorer than the others, but which beyond 4 months are better than those of the concretes formulated using PSDs 1 and 2. Also, the compressive strength of the concretes made from PSD 3 changes even more significantly after 4 months of drying, in contrast to that of the other two concretes.

These observations can be explained by the fact that the concretes formulated using PSD 3 are less porous than the others, because the PSD is finer: the diffusion of CO_2 in these concretes therefore takes place less quickly, which slows down the kinetics of their setting. This slow kinetics results in a slower evolution of the mechanical properties. Furthermore, given that PSD 3 is finer, the particles of hemp are better covered by the binder during the making of the concrete, which explains the better mechanical properties in the long term with concretes made using PSD 3. Note, however, that the binder used to formulate the concretes is an aerated lime-based binder: the carbonation reaction between the $Ca(OH)_2$ and the CO_2 is therefore greatly subordinate to the diffusion of CO_2 through the material.

5.1.7. *Influence of the curing conditions*

Whether the binder used to formulate a hempcrete is based on cement, hydraulic lime or aerated lime, it is vital that the water be able to be mobilized by the binder in order to set mechanically. Various curing conditions have been studied with a view to optimizing the setting and hardening of the material [ARN 07; ARN 12].

The samples of hempcretes made from binders based on hydraulic or aerated limes were kept, between the day of their making and the day of the mechanical test, at 20°C and at four distinct environments in terms of relative humidity regulated using saturated saline solutions: 30%, 50%, 75% and 98% RH.

The results for each binder are presented in Figure 5.4. For nearly all cases, the extreme conditions tested at low (30%) and high relative humidity (75% and 98%) create environments which are not very advantageous for the setting and hardening of the binders. The stress/strain curves show poorer mechanical behaviors, both in terms of the modulus and the maximum stress attained. However, a very great increase in cohesion in conditions of low relative humidity has been observed for the hempcrete created using the aerated lime-based binder.

It appears clearly for all the mixtures that a humidity-saturated environment greatly slows down the setting. This result is well known for the air setting of a binder: the diffusion of CO_2 from the air through the pores of a lime mortar is indeed greatly slowed when the material's internal relative humidity is very high, because when the pores of the mortar are entirely saturated, the transfers take place in the liquid phase, and therefore more slowly than if they were to take place in the gaseous phase [VAN 94]. However, the poor compressive strengths measured for the hempcretes prepared using hydraulic lime-based binders are more unexpected: it appears that in conditions of high relative humidity, a physico-chemical interaction occurs between the binder and the hemp shiv, which interferes with the setting and hardening of the material.

Finally, it seems that curing conditions at 20°C and 50% RH are the most favorable to the setting and hardening of binders because these are the conditions which yielded the highest compressive strengths for all the hempcretes tested.

5.1.8. *Evolution over time*

In the case of mixtures made with an aerated lime-based binder, the key mechanical values – the maximum stress $\sigma_{C\text{-}MAX}$ and Young's modulus E – increase as the degree of setting advances. Three phases can be observed on the curves. Between 0 and 3 months, the evolution is very significant. Between 3 months and a year, the values increase moderately; this evolution represents between 10 and 50%, depending on the formulation. After a year, a quasi-stabilized state is observed. This temporal breakdown differs somewhat from that observed with the pure binder. Thus, the presence of plant particles appears to slow down the setting of the material. This could be explained by an interaction between the plant particles and the water supposed to hydrate the binder.

In addition, we note a clear dependency of the level of performance on the volume concentration of binder. The higher this concentration, the greater will be the compressive strength. The volume concentrations of binder can also be used to explain the clusters of values observed for $\sigma_{C\text{-}MAX}$. We again see the three domains: low, intermediate and high content.

This result is consistent with the fact that increasing the amount of binder helps to connect the hydrates and create a more-or-less continuous network, responsible for the mechanical strength of the system. However, the levels of performances remain modest in comparison to other construction materials. With low binder content, the compressive strength is around 0.2 MPa. For intermediate proportions, it varies between 0.4 and 0.8 MPa, and for high content, it is 1.2 MPa. Hence, this material has to be used with a load-bearing structure in order to fulfill the structural requirements.

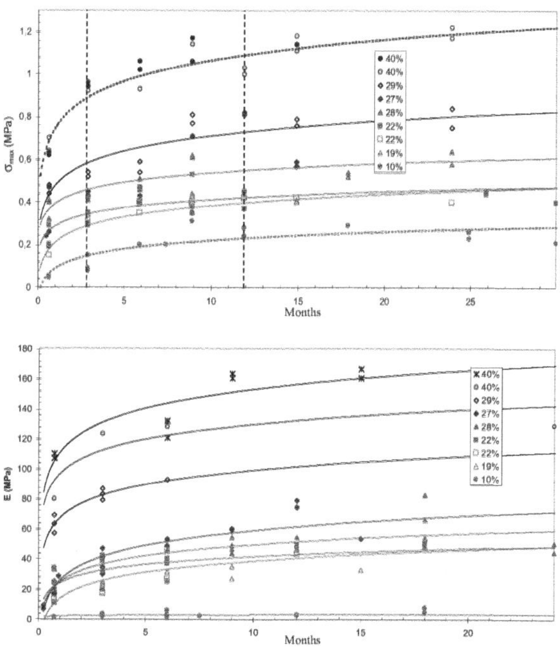

Figure 5.8. *Change over time in the mechanical characteristics of the hempcretes (simple compressive strength and Young's modulus), observed for aerated lime-based concretes for different binder content values per volume [CER 05]*

5.1.9. *Interaction between particles and binder*

Work done recently in France as part of a program called Prebat 2C2E in collaboration between the INRA (National Agricultural Research Institute) in Reims and the ENTPE[1] revealed and explained significant interactions between the particles of hemp and the setting reactions of the binders [DIQ 12]. The study showed that mixing hemp shiv into the water leads to the dissolution of chemical compounds, whose action on a clinker-based binder accounts for the false setting observed above [JOR 04].

For this reason, particles of hemp shiv of very different origins have been analyzed. There are numerous parameters which set them apart, such as their biological "family", the farming methods used to cultivate them, the agronomic conditions, the harvesting and the tools used to transform the product. The results show significant differences from the point of view of the chemical composition (Table 5.2).

Compounds (%)	CA	CB	CC
Cellulose	47.3	45.6	49.2
Hemicellulose	18.3	17.8	21
Klason lignin	21.8	23.3	21.9
Soxhlet extraction	6	5.1	6.2
Other	6.6	8.1	1.6
Ash	3.7	2.6	3.5

Table 5.2. *Chemical composition of three types of hemp shiv used for construction [DIQ 12]*

These lixiviates of hemp shiv were then used as tempering water for the cement pastes for a binder with a high clinker content. The results are crystal clear because the measurements obtained by a Vicat needle test show delays in setting ranging from 100 to 300 minutes depending on the hemp shiv in question (see Figure 5.9). Furthermore, the measurements show that then, the amounts of CSH in these pastes are greatly reduced.

In addition, the measurements taken on hempcrete demonstrate that after these delays, the concretes obtained then present poorer characteristics at 7, 28 and 90 days: the setting has been noticeably modified, so the term "false setting" is justified.

---

1 ENTPE: *Ecole Nationale des Travaux Publics d'Etat* – (French National School for Public Works).

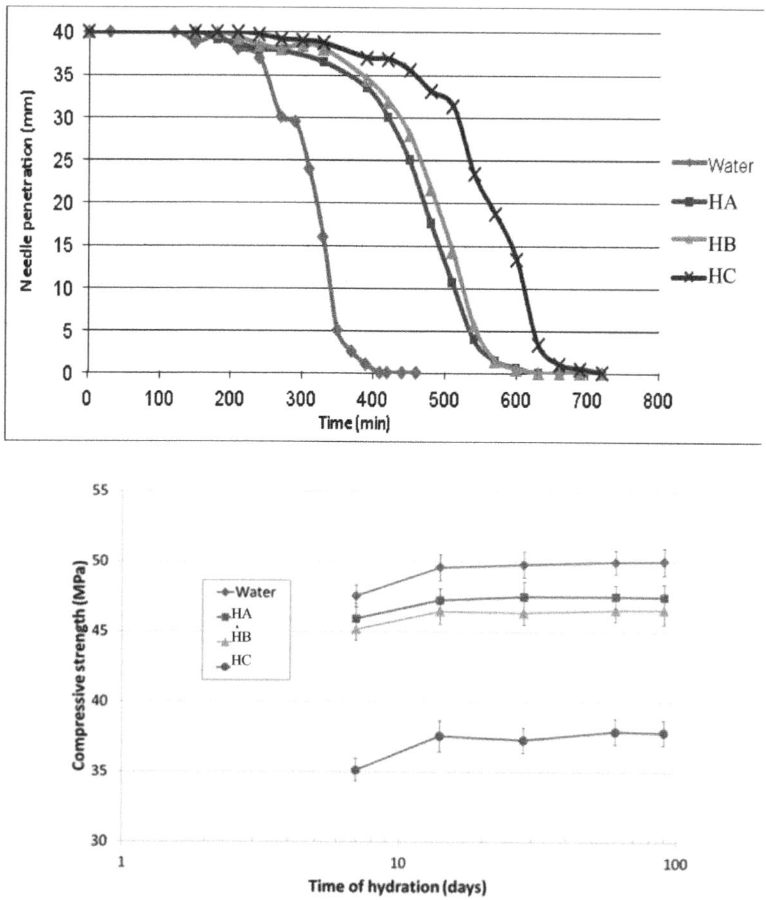

Figure 5.9. *Mechanical behavior of cement paste (evolution over time of the Vicat penetration, top) and hempcrete (simple compressive strength, bottom) for materials tempered using the lixiviate from 3 kinds of hemp shiv: CA, CB and CC [DIQ 12]*

Finally, a test is proposed in order to verify and quantify these phenomena. A particle is fully immersed in a cement matrix, and the evolutions at the interface are monitored. At the end of the test, a transition halo is easily detected by a difference in coloration (see Figure 5.10). Chemical analysis of the materials on both sides of this boundary confirms the lack of setting of the binder around the particles. The study now relates to the chemical analysis and the kinetics of these interactions.

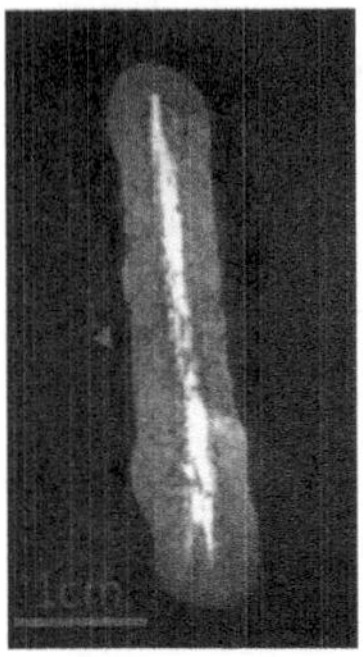

Figure 5.10. *Particle of hemp shiv in a cement matrix. Visualization of the transition halo showing the area where the binder undergoes false setting (zone 2) and where it exhibits normal setting (zone 1) [DIQ 12]*

5.1.10. *Anisotropic behavior*

Most research carried out on the compressive strength of compressed hempcretes – be they laid by projection [ELF 08], compacting [CER 05; NGU 09] or banking [DEB 09] – place strain on the samples in the direction of their formation. To date, the only research focusing on compression in a direction perpendicular to the force of compaction has been carried out by Mounanga *et al.* [MOU 09] and the CSTB [HUS 08]. However, the issue is discussed for projected or vibrated materials whose anisotropy is less pronounced than in the case of a highly compacted material. The study [MOU 09] reports values for $\sigma_{C\text{-}MAX}$ and Young's modulus E without specifying the strains for which this stress is achieved or the method used to measure the modulus. Yet the results demonstrate a significant increase in the mechanical strengths when strain is applied to the material perpendicularly to the direction of setting (see Figure 5.11). While this effect is less dramatic for the measurements of E, they also note an improvement in the values measured in this direction of strain. We must make a caveat as regards the comparability of results obtained with different formulations and implementations.

The results of the study by the CSTB [HUS 08] report a strength $\sigma_{C\text{-}MAX}$ equivalent in both directions of strain applied to the compressed hempcretes. Conversely, the behavior in the orthogonal direction proves to be much more fragile and rigid, with a definite rupture occurring just after the stress peak. This time, the samples are identical, both in terms of the material and the installation method.

These observations show that installation by projection, banking or compaction cause transverse isotropy of the material, with the particles forming strata. The mechanical behavior of the material is then affected and requires a more in-depth examination.

The behavior of the compressed hempcrete in the direction orthogonal to that of projection can be described in 3 stages, explaining the material's fragile behavior:

Zone I: homogeneous elastic behavior. The plant particles, the binder and the interface are subject to strain;

Zone II: zone of elastic limit and stress peak. The hemp shiv/matrix interface progressively gives way;

Zone III: rupture of the sample. The hemp shiv/matrix interface is destroyed and the material loses all cohesion.

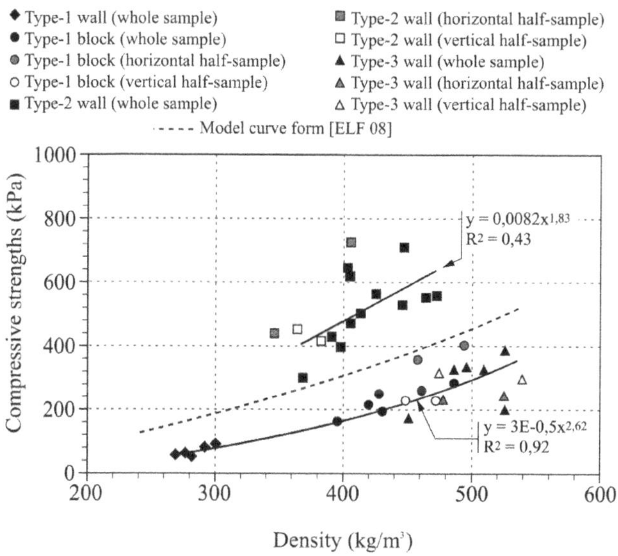

Figure 5.11. *Change in the rupture compressive strength depending on the density and type of installation [MOU 09]*

5.2. Modeling of the mechanical behavior

5.2.1. *Empirical approach*

In general, all writers agree on the fact that there is a linear relation between the compressive strength σ_C and the dry density ρ_{40} [CER 05; NGU 09a; MOU 09] (see Figure 5.11), as well as the heat conductivity. This relation is due to the greater compactness of the mixtures obtained by a larger amount of binder or a more optimal granular arrangement. However, this consideration is only valid if we are dealing with an identical binder [DEB 09].

Elfordy proposes an empirical analytical model established by linear regression and linking ρ_{A0} to the mechanical properties of the material by a power function [ELF 08]. The model is based on previous studies of porous materials. The general form proposed for the equation is as follows:

$$X = Y_0 . \rho_{A0}{}^{m} \quad \text{where } Y_0 = \frac{X_0}{\rho_S{}^{m}} \qquad [5.1]$$

X: mechanical value needing to be estimated;

X_0: value of the parameter to be estimated when ρ_{A0} tends toward ρ_S, i.e. when the porosity of the material tends toward 0;

m: a coefficient set at 1.8 by the author.

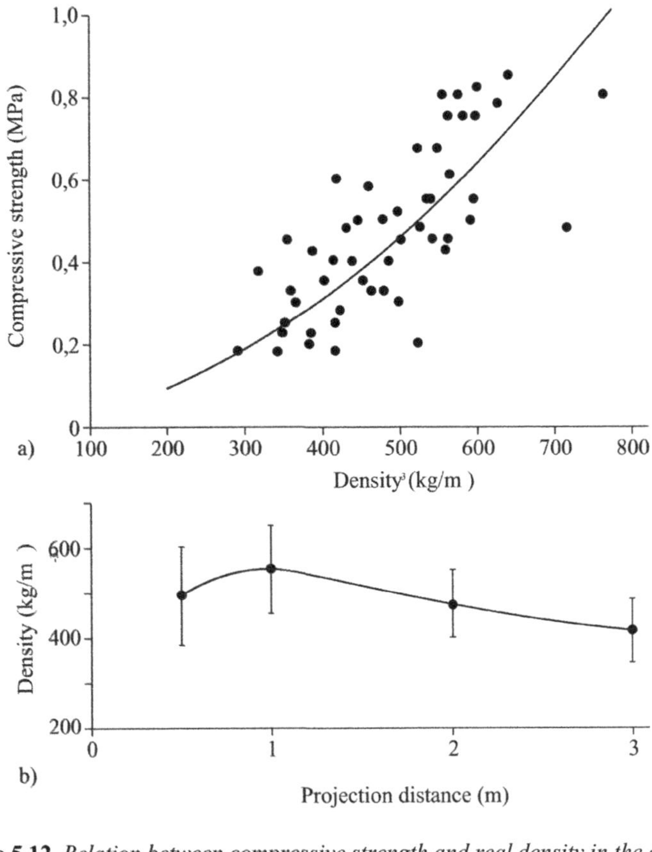

Figure 5.12. *Relation between compressive strength and real density in the case of casting by projection and influence of the projection distance [ELF 08]*

The results obtained with manual casting methods exhibit high variability, although overall they obey the linear relation σ_C/ρ_{A0}. The same is true for dry-mix projection of compressed hempcrete, with a greater degree of dispersion (see Figure 5.12). Elfordy [ELF 08] accounts for this variation by stressing the influence of the projection distance and the area from which the projected samples are taken.

5.2.2. *Self-consistent homogenization approach*

Experimental measuring campaigns have been able to determine three types of behavior depending on the concentration of binder, which in fact stem from the variable structures of the material. Self-consistent homogenization is a useful method for modeling these materials; it lends itself well to material presenting a "regular" distribution of heterogeneities. It offers direct access to the mechanical values of the material (modulus E), macroscopically equivalent.

The approach consists of identifying the material finely heterogeneous to a material which is equivalent on the macroscopic scale, by examining the influence of an addition to the continuous matrix made of equivalent material. Various models with different single, bi- or tri-compound additives have been studied by [CER 05], [SAM 08] and [DAR 09], notably to take account of the effects of anisotropy of the particles.

The material is considered as linear elastic and isotropic. Solving the numerical problem enables us to determine $K = \frac{E}{2(1+\upsilon)}$, the compressibility modulus and $\mu = \frac{E}{3(1-2\upsilon)}$, the shear modulus. We then work back to E, the elasticity modulus and ν, the Poisson coefficient. The application of a few classic results for self-consistent homogenization confirms the modification of the microstructure of the material depending on the binder content. The hempcrete with a low concentration of binder has a microstructure which is very far removed from that of the others, which explains why it is not possible to model its behavior using the same REV (Representative Elementary Volume). This is why there is a divergence between the model values and the experimental measurements. Conversely, for the other hempcretes (with a binder content per volume greater than 10%), the model with spherical tri-composite additives yields very favorable results which give the correct order of magnitude for the elasticity modulus E.

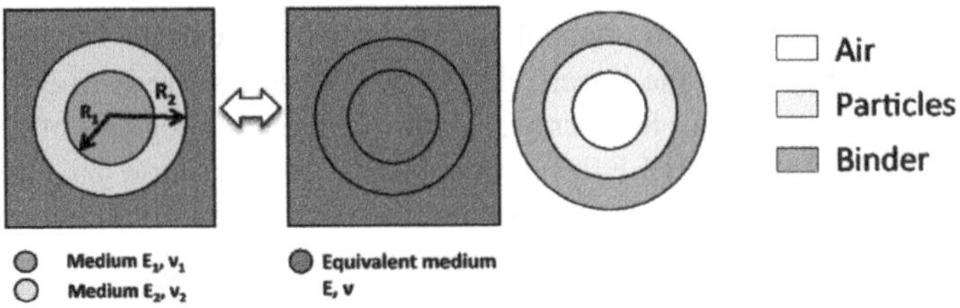

Formulation	Emeasured	Esimulated
ROOFING	3.4	21.4
A4-1	32	40.7
A3-0.75	45	49.5
WALLING	46	52
A3-1	64	72
FLOORING	54	72
A4-1.5	93	85
A3-1.5	130	136
A3-2	163	144

Figure 5.13. *Self-consistent homogenization: principle and model of inclusions chosen, tables of basic data and results obtained for the quasi-elastic modulus [CER 05]*

5.3. Toward the study of a stratified composite

Nguyen's study [NGU 09a] on compacted compressed hempcretes demonstrates several important points regarding the compacting of compressed hempcretes. To begin with, unlike the studies cited previously, the higher the amount of plant material, the better the mechanical strength. Also, this observation is globally verified no matter what the parameter used to quantify it – the compressive strength $\sigma_{C\text{-}MAX}$, the flexural strength σ_F or the elastic modulus E. Observation of the microstructure of these compacted compressed hempcretes shows that there is a very thin joint of binder between the plant particles (see Figure 5.14). The binder is distributed homogeneously and plays the role more of a glue than a bridge between the particles, as happens in "roofing"-type formulations rich in plant material, with the fundamental differences being the lack of inter-particle voids and the partial crushing of the particles.

In addition, Nguyen's experiments highlight an improvement of the mechanical performances when the size of the aggregate increases. He attributes this increase to a smaller interfacial surface between the aggregate and the binder, and therefore decreased problems of compatibility. This observation was also made in a study performed by ENTPE [HUS 08] which specifies that we also need to examine the factor of the shape of the particles. Indeed, if we wish to rearrange the granular stack in order to optimize the interfaces, the elongation of the particles could play an important role by increasing the overlap and the friction surfaces like fibers (see Figure 5.14). This is what was seen in [ARN 12]: a finer hemp shiv gives rise to a granular arrangement containing fewer inter-particle voids. Therefore, compressed hempcrete containing fine-grained hemp shiv will have better performances in the long term, because the binder covers fine hemp shiv particles better when mixed.

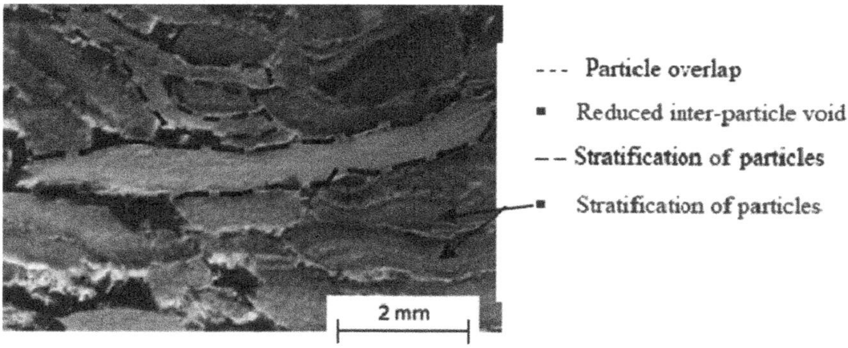

Figure 5.14. *Macroscopic structure of hempcrete with a high content of plant matter and compacted [NGU 09a]*

If we expand the scope of our investigation to other industrial products based on lignocellular materials, we discover the group of panels of particles. These products, which are mixtures of lignocellular particles and glue, are generally shaped by thermoforming or compacting. In comparison with compressed hempcrete, the mechanical strength of these materials is provided by the plant matter, with the binder's only role being to keep the particles together. Thus, it is toward this type of material that we begin to turn, increasing the amount of plant particles in lignocellular concretes. However, unlike panels of particles, the thickness involved in agro-concretes is far lesser. However, the existing body of literature [CER 05; NGU 09a] tells of problems of homogeneity when the compacted layers surpass 8 cm in thickness.

5.4. Conclusion

As a mixture of plant particles and binder, hempcrete is an unusual construction material: the high flexibility of the aggregates in conjunction with the rigidity of the cement matrix leads to a non-fragile elasto-plastic behavior. Thus, it is distinguished from other construction materials by a high deformability under stress, lack of fracturing and marked ductility with absorbance of the strains ever after having reached the maximum mechanical strength.

It is also helpful to highlight another peculiarity: the variability of the behavior depending on the formulation enables us to adjust and optimize the performances of this material for diverse applications as a roof filling material, in walling or as flagging. Depending on the concentration of binder, three types of behaviors of hempcrete are observed. For small proportions of binder, we have a material with poor mechanical strength and an elasticity modulus of less than 5 MPa. The levels of deformation are very high (> 15%). This material behaves like a sample of loose particles with "bridges of binder" connecting them. It is pre-destined for applications where thermal or acoustic insulation is important. For intermediate levels of content, the level of performance increases with the quantity of binder. The behavior of the material becomes progressively more similar to that of materials with high binder content. The behavior is guided by the binder matrix. For high binder content, the material is comparable to a continuous binder matrix in which the plant particles are buried. The mechanical performances increase and tend towards those of the pure binder. However, the high deformations that this concrete is capable of dealing with make it advantageous as a filling material. It can undergo differential compression, contract or dilate with no apparent cracking.

The mechanical strengths and the elasticity moduli obtained during the course of this experimental work are poor in comparison to other lightweight concretes such as wood-based concretes of cellular concrete. Thus, at present, it appears crucial to use load-bearing structures to support these concretes. However, compacting and therefore rearranging plant particles with anisotropic properties seems to be one way of using them differently. It seems interesting to apply stress to them in the direction perpendicular to their compaction, i.e. the direction in which their rigidity is greatest. However, this implies increasing the shear stress at the particle/matrix interface, which at present is still a weak point of the material. Thus, it is helpful to optimize the distribution of the binder so as to place as much strain on it as possible, and improve its interaction with the plant material. Indeed, these interactions are fundamental both for the improved cohesion and then, after it has set, for the mechanical behavior of the concrete.

5.5. Bibliography

[ARN 01] ARNAUD L. and CEREZO V., Qualification physique des matériaux de construction à base de chanvre, Final research report CNRS 0711462, Lyon: ENTPE–DGCB–URA, 91 pp., 2001.

[ARN 06] ARNAUD L., HUSTACHE Y. and BOYEUX B., "Construire en chanvre", *Le chanvre industriel, production et utilisations*, Editions France Agricole, pp. 347–372, 2001.

[ARN 07] ARNAUD L. and HUSTACHE Y., Validation des techniques de mise en œuvre du chanvre dans des applications bâtiment, research report DGUHC, Construire en chanvre, 38 pp. 2007.

[ARN 12] ARNAUD L. and GOURLAY E., "Experimental study of parameters influencing mechanical properties of hemp concretes", Construction and Building Materials, vol. 28, pp. 50–56 2012.

[BUT 04] BUTSCHI P.Y., Utilisation du chanvre pour la préfabrication d'éléments de construction, Doctoral Thesis, University of Moncton, Canada, 2004. http://membres.multimania.fr/butch11/These.pdf.

[CER 05] CEREZO V., Propriétés mécanique, thermique et acoustique d'un matériau à base de particules végétales: approche expérimentale et modélisation théorique. Thesis, MEGA-ENTPE. Lyon : INSA de Lyon, 2005.

[DEB 09] DE BRUIJN P.B., JEPPSSON K., SANDIN K. and NILSSON C., "Mechanical properties of limehemp concrete containing shives and fibres", *Biosystems Engineering*, vol. 103, August. 2009, p. 474–479.

[DAR 09] DARTOIS S., NADOT-MARTIN C., HALM D. and DRAGON A., "Discrete damage modelling of highly-filled composites via a direct multiscale "morphological approach"", *J. of Multiscale Modelling*, Vol. 1, nos. 3 & 4, pp. 347–368.

[DIQ 12] DIQUELOU Y., GOURLAY E., ARNAUD L. and KUREK B., "The impact of hemp shiv on cement setting and hardening : the influence of extracts and study of the interface", *IIBCC2012 Int. Inorganic-Bonded Fiber Composites Conference Proceedings*, Canberra, Australia, 12-13 Sept. 2012, Edt T. Cooke & H. Thygesen, 2012, pp. 41–50.

[ELF 08] ELFORDY S., LUCAS F., TANCRET F., SCUDELLER Y. and GOUDET L., "Mechanical and thermal properties of lime and hemp concrete ("hempcrete") manufactured by a projection process", *Construction and Building Materials*, vol. 22, October 2008, pp. 2116–2123.

[GLÉ 11] GLÉ P., GOURDON E. and ARNAUD L., "Acoustical properties of materials made of vegetable particles with several scales of porosity", *Applied Acoustics*, vol. 72, p. 249–259 2011.

[GOU 09] GOURLAY E., De la chènevotte au béton de chanvre, optimisation des propriétés mécaniques et hygrothermiques. Research Masters Dissertation, Vaulx-en-Velin : MEGA ENTPE, 66 p. 2009.

[GOU 11] GOURLAY E., GLÉ P., ARNAUD L. and GOURDON E., "Propriétés multiphysiques des bétons de chanvre", *Matériaux et techniques*, vol. 99, p. 625–631 2011.

[HUS 08] HUSTACHE Y. and ARNAUD L., *Synthèse des connaissances sur les bétons et mortiers de chanvre*, Association construire en Chanvre, 2008.

[JOR 04] JORGE F.C., PEREIRA C. and FERREIRA J.M.F., "Wood-cement composites: a review", *Holz Als Roh-Und Werkstoff*, vol. 62, pp. 370–377 2004.

[NGU 09a] NGUYEN T.T., Contribution à l'étude de la formulation et du procédé de fabrication d'éléments de construction en béton de chanvre, Doctoral Thesis, University of South Brittany, 2009. http://web.univ-ubs.fr/

[NGU 09b] NGUYEN T.T., PICANDET V., AMZIANE S. and BALEY C., "Influence of compactness and hemp hurd characteristics on the mechanical properties of lime and hemp concrete", *European Journal of Environnemental and Civil Engineering*, vol. 13, 2009, pp. 1039–1050.

[NGU 10] NGUYEN T.T., PICANDET V., CARRE P., LECOMPTE T., AMZIANE S. and BALEY C., "Effect of compaction on mechanical and thermal properties of hemp concrete", *European Journal of Environnemental and Civil Engineering*, vol. 14, 2010, pp. 545–560.

[MAG 10] MAGNIONT C., Contribution à la formulation et à la caractérisation d'un écomatériau de construction à base d'agroressources, Doctoral Thesis, University of Toulouse III, 2010.

[MOU 09] MOUNANGA P., POULLAIN P., BASTIAN G., GLOUANNEC P. and KHELIFI H., "Influence de la composition et du mode de mise en œuvre sur le développement des propriétés mécaniques du béton de chanvre", *Actes du 27ème congrès de l'AUGC*, Saint-Malo, 3–5 June 2009. http://www.augc09.univ-rennes1.fr/ (14/09/10)

[RP2C 07] RP2C, Règles professionnelles d'exécution d'ouvrages en béton de chanvre. Commission "Règles Professionnelles Construction Chanvre" RP2C, 60 p. 2007.

[SAM 08] SAMRI D., Analyse physique et caractérisation hygrothermique des matériaux de construction : approche expérimentale et modélisation numérique, PhD Thesis, Ecole doctorale MEGA Lyon 2008.

[VAN 94] VAN BALEN K. and VAN GEMERT D., "Modelling lime mortar carbonation", *Materials and Structures*, vol. 27, pp. 393–398 1994.

Chapter 6

Hygrothermal Behavior of Hempcrete

6.1. Introduction

As stressed in Chapters 2 and 4, hempcrete is a highly hygroscopic material: owing to the presence of an open porous network with staggered characteristic sizes ranging from a nanometer to a centimeter (hemp, binder, arrangement of particles), it exhibits a very great capacity to exchange water vapor with the surrounding air. This vapor can condense into liquid water and come to rest at the surface of the pores when the hygrometry of the air is high. This phenomenon is reversible when the air becomes dry.

The liquid water, which is highly conductive, contained in the hempcrete influences the thermal properties of the material: among other things, its presence modifies the specific heat and the heat conductivity of the concrete. Indeed, the liquid water exhibits a heat conductivity ($\lambda_{water} = 0.6$ W.m^{-1}.K^{-1}) 20 times greater than that of dry air ($\lambda_{air} = 0.026$ W.m^{-1}.K^{-1}). Yet the changes are greater still, because these hygroscopic transfers lead to phase-changing phenomena (vaporization/ condensation) which are accompanied by releases or absorption of energy which influence the variations in temperature and lend the material a greater insulating capacity [SAM 08]. These phenomena have a very noticeable influence on the thermal performances, particularly in comparison with the airtight materials used in building. However, today, these hygrothermal transfers are hardly taken into account – and often not at all – in software for simulating thermal performances, which is a significant handicap for these innovative construction solutions.

Chapter written by Laurent ARNAUD, Driss SAMRI and Étienne GOURLAY.

In this chapter, the thermal characteristics of hempcretes are presented and compared to those of other similar materials in the domain of construction. As an example, we give the key values for these materials (Table 6.1). Then, the projects currently under way to precisely describe transfer phenomena and the influence of the primary materials are discussed, and finally, the results of the numerical modeling currently under way are presented.

Material	ρ_0	c	λ	a	E_{ff}
	[kg.m^{-3}]	[J.kg^{-1}. K^{-1}]	[W.m^{-1}. K^{-1}]	[10^{-7} m^2.s^{-1}]	[J.m^{-2}. K^{-1}. s$^{-1/2}$]
HLC1	**430**	**1000**	**0.105**	**2.45**	**213**
HLC2	**317**	**1000**	**0.082**	**2.62**	**162**
AAC	**421**	**850**	**0.11**	**3.07**	**207**
VPB [LAC 03]	**814**			**3.3**	**405**
Indoor coating	**761**			**1.1**	**417**
Outdoor coating	**1870**			**2.54**	**611**
Fired earth	**1600**	**960**	**0.5**	**3.07**	**831**
Solid wood	500	1880		1.38	350
Extruded polystyrene	25	1450	0.028	9.38	35
Normal heavy concrete	2400	910	1.83	7.1	2020
Immobile air	1	1000	0.026	250	5

Table 6.1. *Values of heat diffusivity **a** and thermal effusivity E_{ff} of dry materials and a number of common construction materials*

6.2. Heat conductivity

Conventionally, the behavior of hempcretes has been studied primarily to obtain their values of conductivity and specific heat – values which today are used almost exclusively to qualify materials in this domain [ARN 01; CER 05]! The work showed that this preliminary approach was insufficient to evaluate the thermal performances of constructions built using hemp-based hemp-based materials.

6.2.1. *Measurement of the conductivity*

The conductivity measurements taken in the lab were carried out on dry samples (previously dried at 50°C for 24 hours) and in a permanent regime with guarded hot-boxes between 5 and 20°C on samples with very different densities.

These samples were then tested after a time in an environmental test chamber with different conditions in terms of relative humidity (50 and 75% RH) (Figure 6.1).

The level of heat conductivity λ of hempcretes varies, depending on the dose of binder, from 64 to 90 mW/(m.K). This porous, lightweight material is therefore a good thermal insulator, although not an "exceptional" one. It seems similar or equivalent to other building materials of similar densities. Such is the case, in particular, for cellular concrete and blocks of fired earth which have also been tested in the same conditions.

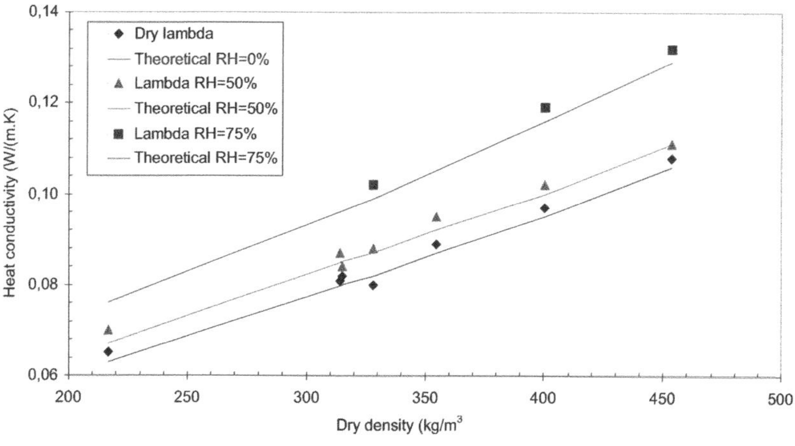

Figure 6.1. *Evolution of the heat conductivity of hempcrete depending on the density and under different conditions of relative humidity*

However, the high sensitivity to the relative humidity of the hemp particles appears unfavorable, because water is an excellent heat conductor. Indeed, a significant change in the heat conductivity is measured: it can go up to 30% when a sample has been left in a humid environment. The heat conductivity of the liquid water which condenses in the pores (capillary condensation) accounts for this noticeable increase (Figure 6.1). All porous materials are sensitive to these hygroscopic exchanges with the surrounding air and, consequently, their thermal performances are inescapably altered.

The evolutions of three different construction materials were compared (Figure 6.2). Samples were subjected to different strain jumps both in terms of temperature and relative humidity [ARN 05; SAM 06]. At the center of walls 30 cm thick, we clearly observe a lower temperature for the wall made of hempcrete. This demonstrates better insulation although the conductivities of these materials are similar!

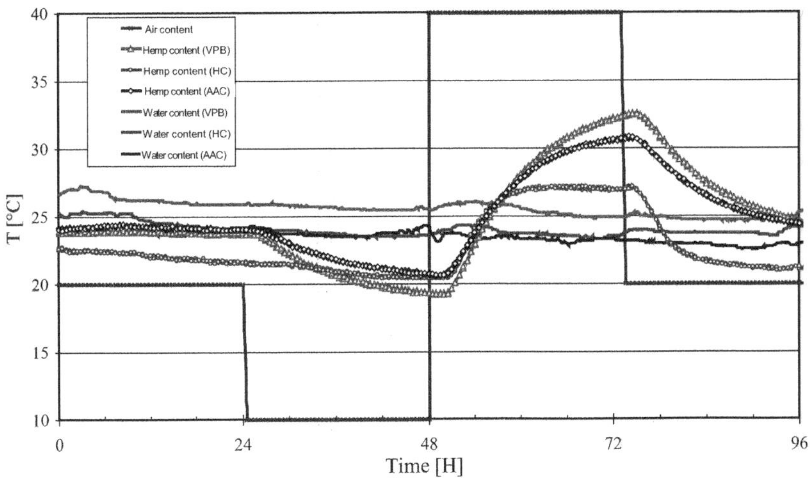

Figure 6.2. *Response at different temperature platforms: comparison of walls made of autoclaved aerated concrete (AAC), vertically perforated bricks (VPB, made of fired earth) and hempcrete (HC). The solid lines represent the change in the temperature of the laboratory, considered to be constant over time and for all the materials*

6.2.2. *Modeling of the heat conductivity in dry and humid conditions*

Given that heat conductivity plays a major role in heat transfers, it seems necessary to measure its value and to be able to model its variations depending on the mixtures used, on the one hand, and the water content of the concretes, on the other.

The research carried out by V. Cérézo for her doctoral thesis in 2005 [CER 05] was able to use self-consistent homogenization to establish reliable models both for dry and damp concretes. The results are recapped below – briefly, because these models are now well known.

Hempcrete is made up of three media: the model with tricomposite additives composed of an air bubble (medium a), surround by plant particles (medium b), which themselves are surrounded by binder (medium c) yields the following very effective model:

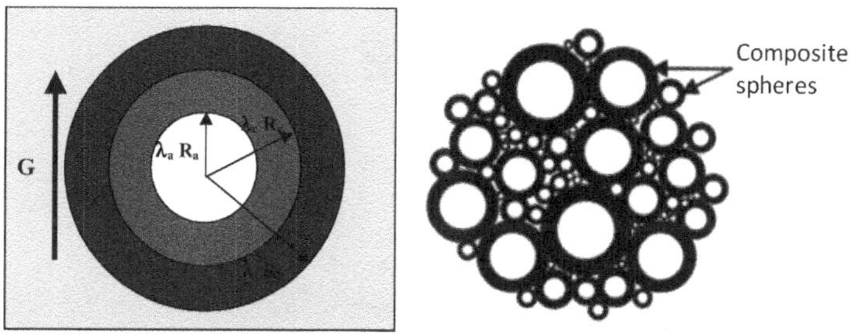

$$\frac{\lambda_{eq}}{\lambda_c} = 1 + \frac{\varepsilon_v}{\left(\frac{1-\varepsilon_v}{3} + \frac{3+\delta(\lambda_a/\lambda_b-1)}{3(\lambda_a/\lambda_b-1)-\delta(\lambda_a/\lambda_b-1).(2\lambda_b/\lambda_c+1)}\right)} \text{ where } \varepsilon_v = \left(\frac{R_b}{R_c}\right)^3 \text{ and}$$

$$\delta = 1 - \left(\frac{R_a}{R_b}\right)^3$$

By positing that $m_{hp} = k \times m_b$ where m_{hp} is the mass of the hemp particles and m_b is the mass of binder, we get:

$$\varepsilon_v = \left(\frac{R_b}{R_c}\right)^3 = 1 - \frac{1}{k+1} \cdot \frac{\rho}{\rho_b} \text{ and}$$

$$\delta = 1 - \left(\frac{R_a}{R_b}\right)^3 = \frac{\rho}{\rho_{hp}} \cdot \frac{k}{k+1} \cdot \frac{1}{1 - \frac{\rho}{\rho_b(k+1)}}$$

The definition of k is based on the hypothesis that that variations in properties between the powdered binder and the hydrated binder cause no major change in the heat conductivity of the hempcrete. This has been verified by [COR 99]. Using the value of the heat conductivity of the hemp particles alone, $\lambda_{hp} = 0.048$ W.m^{-1}.K^{-1}, and values given for the binder PF70: $\lambda_b = 0.188$ W.m^{-1}.K^{-1} and $\rho_b = 860$ kg.m^{-3} [COR 99], the heat conductivity of the hempcrete is obtained on the basis of the formulation and the dry density of the concrete. Thus, for a dry density ranging from 400 and 700 kg.m^{-3}, the conductivity varies between 0.09 and 0.16 W.m^{-1}.K^{-1}. The values for dry products (λ_0) of HLC1 (hemp lime concrete 1) and HLC2 chosen are: $\lambda_{0\ HLC1} = 0.105$ W.m^{-1}.K^{-1} and $\lambda_{0\ HLC2} = 0.082$ W.m^{-1}.K^{-1}.

As regards wet concretes, many configurations have been studied in which the position of the water from capillary condensation and its interactions with the other constituents varied. A mixed model by double homogenization is gained by comparing the amount of water adsorbed by the hempcrete under certain hygrometries and the amount of water adsorbed by each of the constituents taken separately under the same hydric conditions. We use the symbol u_b to denote the mass water content of the binder, and u_{hp} for the mass water content of the hemp particles. Thus, the volume concentrations of water in the binder and the particles are equal to $\varepsilon_{hw} = \frac{1}{1+\frac{1}{u_{hp}}\cdot\frac{\rho_w}{\rho_{hp}}}$ and $\varepsilon_{bw} = \frac{1}{1+\frac{1}{u_b}\frac{\rho_w}{\rho_b}}$.

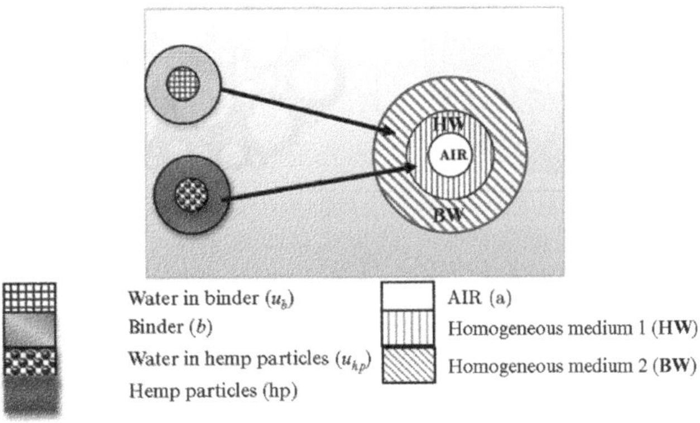

The heat conductivity of wet hempcrete of density ρ_{HLC} is then [CER 05]:

$$\frac{\lambda_{eq}}{\lambda_{bw}} = 1 + \frac{\varepsilon}{\left(\frac{1-\varepsilon}{3} + \frac{3+\delta(\lambda_a/\lambda_{hw}-1)}{3(\lambda_a/\lambda_{bw}-1)-\delta(\lambda_a/\lambda_{hw}-1).(2\,\lambda_{hw}/\lambda_{hw}+1)}\right)}$$

where $\varepsilon_{hw} = 1 + \frac{1}{1+k_3}\frac{\rho_{HLC}}{\rho_{bw}}$ and $\delta = k_3.\frac{\rho_w}{\rho_b}\left(\frac{1}{\varepsilon}-1\right)$ where $k_3 = k\frac{1+u_{hp}}{1+u_b}$.

Using the adsorption/desorption isotherms of the hemp and binder as well as their characteristics, we can deduce the evolution of the heat conductivity of hempcretes HLC1 and HLC2 on the basis of their mass water content *u [%]*:

$$\lambda_{HLC1} = 0.105 + 0.35.u \text{ and } \lambda_{HLC2} = 0.082 + 0.15.u.$$

The values of heat conductivity obtained for hemp mortars are in line with the experimental values obtained by [CER 05]. For HLC1, comparison with the results from [COL 04], who considers a three-phase model with air, water and the solid phase, shows good consistency, because the leading coefficient of the line is 0.35 as compared to 0.29.

Finally, for hemp-lime coating, the approach is identical to that pursued for HLC1 and HLC2. Based on the adsorption-desorption isotherms of the material, the evolution of the heat conductivity of indoor coating based on its mass water content *u [%]* is written $\lambda_{ind\ coating} = 0.204 + 0.25.u$.

6.2.3. *Heat transfers*

Hempcrete's excellent capacity to store relative humidity, far greater than that of other materials, was measured using sorption testing (Figure 6.5). This property accounts for the atypical thermal behavior described above. How, though, are we to explain it, and above all, is it possible to analyze and model it by these transfers of mass and energy by phase changing? The aim of the next sections is to answer these questions.

For this purpose, two hemp-based materials were tested: HLC2, a material used in surfaces of 30 cm in thickness, and HLC1, and formulation for walling, a 15 cm-thick sample of which was tested. These two materials exhibit different formulations, specified in Figure 6.3.

V (m³)		*M (kg)*	*V (m³)*		*M (kg)*
0.715	Air	0	0.79	Air	0
0.174	PF70	266.6	0.12	PF70	190
0.111	Hempshiv	163.4	0.09	Hempshiv	127
1.000	ρ_{0HLC1} **= 430 kg.m^{-3}**	430	1.00	ρ_{0HLC2} **= 317 kg.m^{-3}**	317

Figure 6.3. *Formulations of the hempcretes HLC1 and HLC2 tested – dry materials*

The performances of these hempcretes were compared with those of commercial AAC and VPB. The physical characteristics of these materials are given in Figure 6.4.

V (m³)		*M (kg)*	*V (m³)*		*M (kg)*
0.84	Air	0	0.60	Air (alveoli)	0
0.16	Solid matrix	421	0.10	Air microscopic	0
1.00	ρ_{0AAC} **= 421 kg.m^{-3}**	421 kg	0.3	Shard	837
			1.00	ρ_{VPB} **= 837 kg.m^{-3}**	837 kg

Figure 6.4. *Constitution of AAC and VPB [SAM 08]*

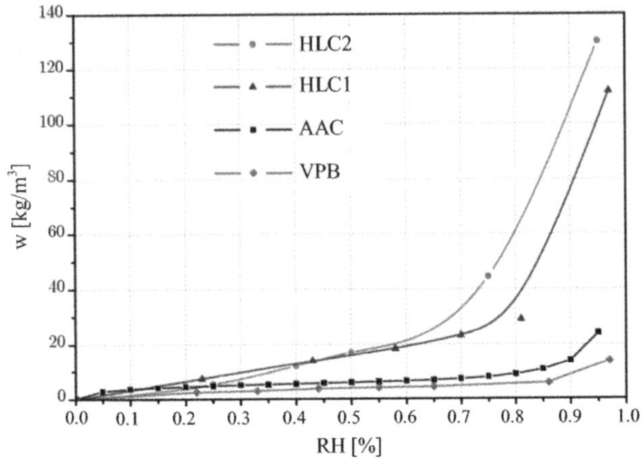

Figure 6.5. *Adsorption isotherms of the materials HLC1, HLC2, AAC and VPB*

6.3. Hygrothermal transfers

6.3.1. *Experimental device*

In order to study the phenomena of mass and energy transfers, an original experimental device was installed in the laboratory [SAM 08]. This device is able to fulfill the following five functions:

– Measure the temperature and relative humidity over time and thereby determine and analyze the distributions of these two variables through the sections of concrete depending on the strains applied in the chamber.

– Measure the water uptake of the materials depending on the relative humidity of the surrounding air in the environmental chamber and the laboratory, using samples placed on balances.

– Characterize the hygrothermal behavior of the materials in the "static" and "cyclic" regimes: demonstrate the role and the impact of each variable of state on the materials' performances, compare their thermal performances, characterize and compare the capacity of each material to compensate for fluctuations in relative humidity (the notion of the "hygric capacity").

– Evaluate the impact of a coating on the hygrothermal performances of the hemp-based materials.

– Demonstrate the phenomenon of phase-changing on the temperature profiles: indeed, the condensation of water vapor or evaporation of liquid water result in an increase or decrease in the temperature of the material.

In order to achieve these goals, a FIRLABO CV 700 FHP2 N20 environmental chamber was used (see Figure 6.6). The door was taken off and replaced by a sample-holder with enhanced insulation. The section available to place the surfaces to be tested measures 101 × 80 cm. The internal volume is approximately one cubic meter.

Thus, it is possible to impose cycles of temperature and relative humidity on the surface in contact with the chamber. The surfaces of materials are then subjected to gradients between the interior of the environmental chamber and the ambient atmosphere in the laboratory. Note that as regards these ambient environmental conditions, the T and the RH are not controlled but are relatively stable because they are regulated in the laboratory. Thus, a new device coupling two independent chambers was developed.

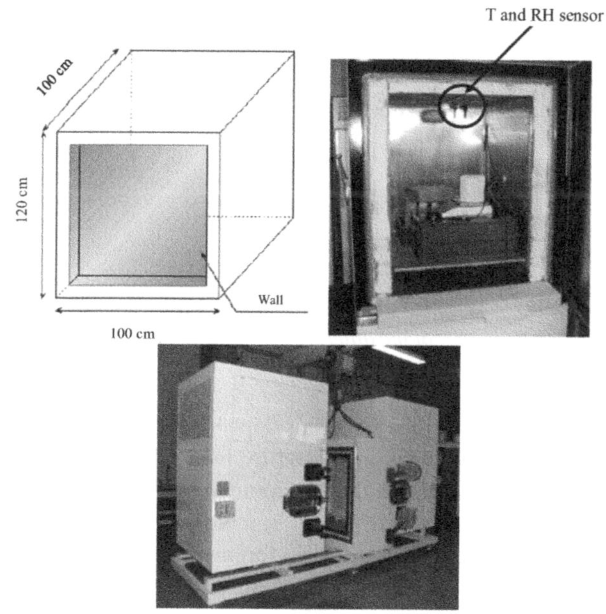

Figure 6.6. *Top: diagram and photo of the experimental chamber used [SAM 08]; bottom: the new device coupling two such chambers [GOU 11b]*

So as to fulfill the set goals, the instrumental setup comprises five temperature and relative humidity sensors, made by HYGROCLIP. The probe is a high-precision

Pt100 sensor, with a measuring capacity ranging from -40°C to +60°C. The sensor measuring the relative humidity of the ambient air is a capacitive sensor: the RH measurement deduced from the voltage at the extremities of a dielectric yields certain and reliable results in a porous medium [PAD 99; PEU 03]. Five sensors are positioned at A, B, C, D and E (see Figure 6.7), where the point C denotes the center of the sample.

By measuring T and RH on each of the two surfaces – one in contact with the environmental chamber (point B) and the other with the ambient atmosphere in the laboratory (point D) – we can get around the inaccuracies of measurement caused by edge effects (particularly due to the phenomenon of convection between the ambient air and the surface of the material).

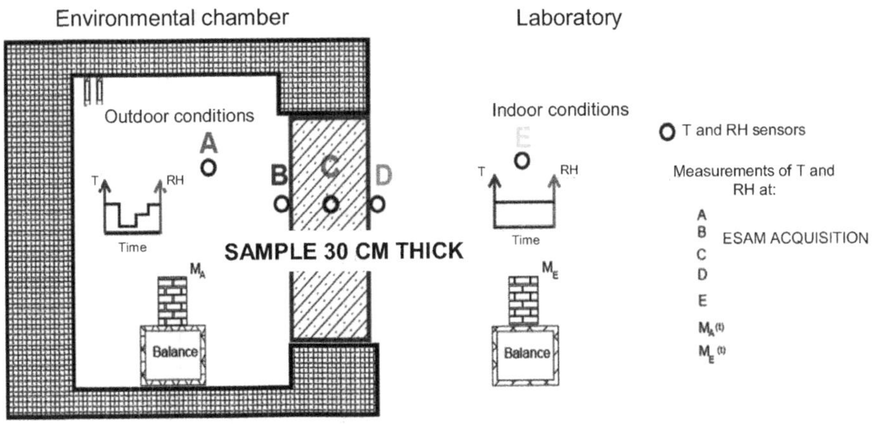

Figure 6.7. *Diagram of the experimental device used, showing the position of the different instruments*

Finally, the water uptake of the materials depending on the conditions in terms of T and RH (sorption/desorption) is measured by two balances (made by PRECISA, accurate to 0.1 g): one is placed in the environmental chamber and the other in the laboratory. The change in mass of a "small" sample of material, either cylindrical or cubic in shape, is monitored over time. The aforementioned dimensions were chosen so that the volume would be greater than the representative elementary volume (REV) of each material. This is around 125 cm^3, if we refer to the determination of REV for hemp-based mortars [COL 04].

The sensors are connected to conditioners directly linked to an electrical box for constant supply at 15 V, and to an acquisition center, by a coaxial cable. All data acquisition is done using the software ESAM-3.

6.3.2. *Stresses*

The strategy of measurement is based on the application of gradients of temperature and relative humidity on both sides of the layer of material. In view of the configuration of the experimental device, the flows of heat and relative humidity are considered to be mono-dimensional across the wall.

The samples were subjected to two types of stress: so-called static stress, corresponding to changes in staggered 24-hour stages, and cyclical, simultaneous stresses of temperature and relative humidity, variable by 24-hour cycles. Such stresses were intended to mimic genuine daily conditions of temperature and relative humidity encountered during the different seasons.

For the static stresses, results are presented with STAT1, comprising four 24-hour stages, for a total duration of 96 hours. The values recorded for T and RH at each stage are indicated in Figure 6.8. The conditions of T and RH at A are identical at the beginning and end of each stress phase in order to examine whether or not the final level of temperature measured at point C over the past 24 hours is identical to that measured there initially.

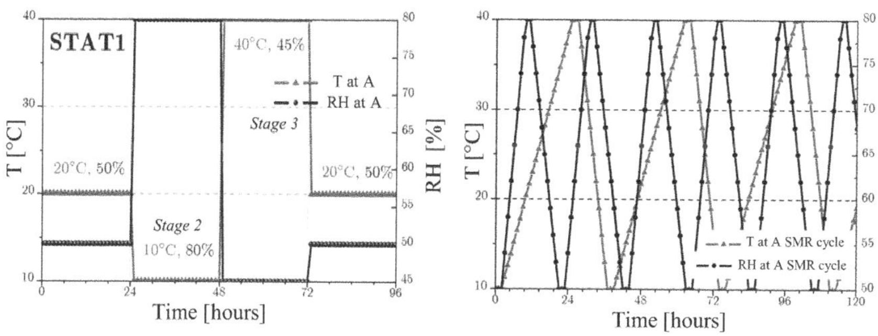

Figure 6.8. *Left: principle of static stress STAT1; right: cyclic variation for the summer cycle – stresses used with the experimental device*

For the cyclic stresses, the cycle labeled SMR (denoting "summer") is characterized by the outdoor temperature at point A, which varies between 10°C and 40°C, and the relative humidity between 50% and 80% (see Figure 6.8). With high RH values, a problem began to manifest itself with the function of the environmental chamber: it became apparent that the RH value oscillated significantly ($\Delta RH = \pm 7\%$). Thus, the frequency of evolution of the relative humidity is 21 hours, whereas for the temperature, it is 37 hours. Over a period of 120 hours, therefore, three cycles of T and five cycles of RH take place.

6.3.2.1. *Repeatability of the measurements*

Figure 6.9 presents the raw results of comparison for STAT1 stress for a sample of AAC, 30 cm in thickness, denoted AAC30. The difference in the average temperature at point E, compared between the two types of stress, is $\Delta T_E = 3°C$, because of the uncontrolled boundary conditions at point E. Over the period between 48 hours and 96 hours, a 2°C adjustment (purely of the ordinate value on the "Y" axis) of the temperature at point C for STAT1 enables perfect consistency of the results (Figure 6.9). This adjusted value of ΔT_C remains smaller than the measured value for ΔT_E and therefore proves excellent repeatability.

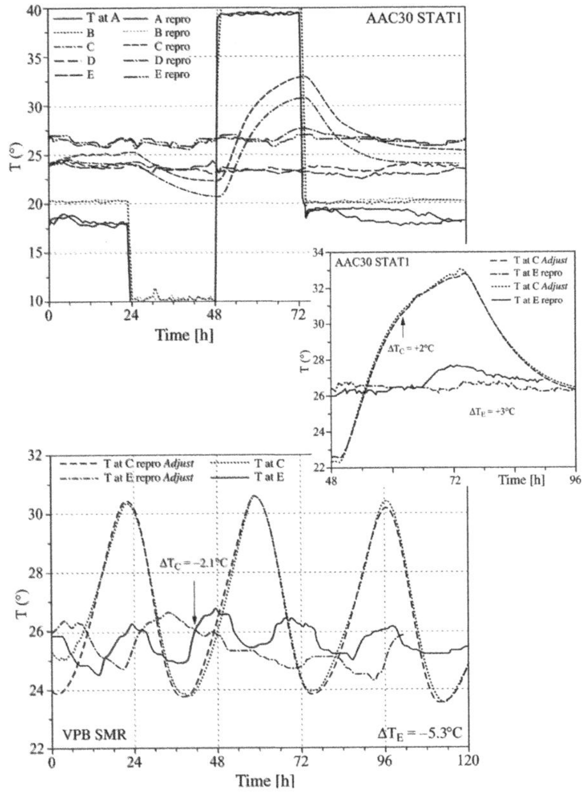

Figure 6.9. *Reproducibility tests for AAC under static stress STAT1 (top) and for VPB under cyclic stress in summer (bottom)*

Another result is obtained for VPB under cyclic stress (Figure 6.9). Figure 6.9 presents the comparison of the temperature changes for two cycles during summer. However, the temperature levels in the laboratory are distinct, with average temperatures at E of 25.4 and 30.7°C, equating to an average difference of

$\Delta T_{E\,SMR} = 5.3°C$. A shift by a value of 2.1°C (smaller than the aforementioned difference) offers a very good superposition of the results for the three cycles of T. The amplitude and the phase shifting of the response at point C are perfectly similar in both cases.

For these two materials – AAC and VPB – in spite of the difference of the boundary conditions at point E, we obtain excellent reproducibility of the measurements.

6.3.3. *Phase changes*

For hempcrete, two examples are presented in order to show that the behavior appears to differ.

In the two tests carried out in similar conditions, the average difference of the temperature at point E is very small ($\Delta T_E = 0.3°C$). However, unlike the materials AAC and VPB, the adjustment needed in order to superpose the temperature changes at point C is far greater, because $\Delta T_C = 2.5°C$ (Figure 6.10). In view of the heat balance equation, one possible explanation is that the original term is not equal to zero. Thus, the hygrothermal performance of the material is different to that of autoclaved aerated concrete and of vertically perforated bricks.

The peculiarity of the material's thermal behavior can be seen on the two STAT1 tests. Indeed, for t between 48 and 96 hours, the temperature at point C increases and decreases rapidly. What is specific to this particular material is this temperature dynamic, marked by the appearance of a very clear period of stabilization, which occurs only 12 hours after the temperature change imposed in the environmental chamber. The evolution of the relative humidity at point C closely follows this trend, where the amplitude of variation is around 20%.

Thanks to these observations, we can envisage the use, in the heat balance equation $\frac{\partial}{\partial x}\left(\lambda\frac{\partial T}{\partial x}\right)+\dot{q}=\rho c\frac{\partial T}{\partial t}$, of a source term $\dot{q}$ which corresponds to the phenomenon of phase-change. This term establishes a direct link between the change in T and RH at point C. Consider the third stage, at t between 48 and 72 hours. The increase in relative humidity at C is due to vaporization of the liquid water that is free or weakly bound in the material. As this phenomenon is endothermic, it absorbs a great deal of energy. That is why the increase in RH at C is accompanied by a drop in the T at C.

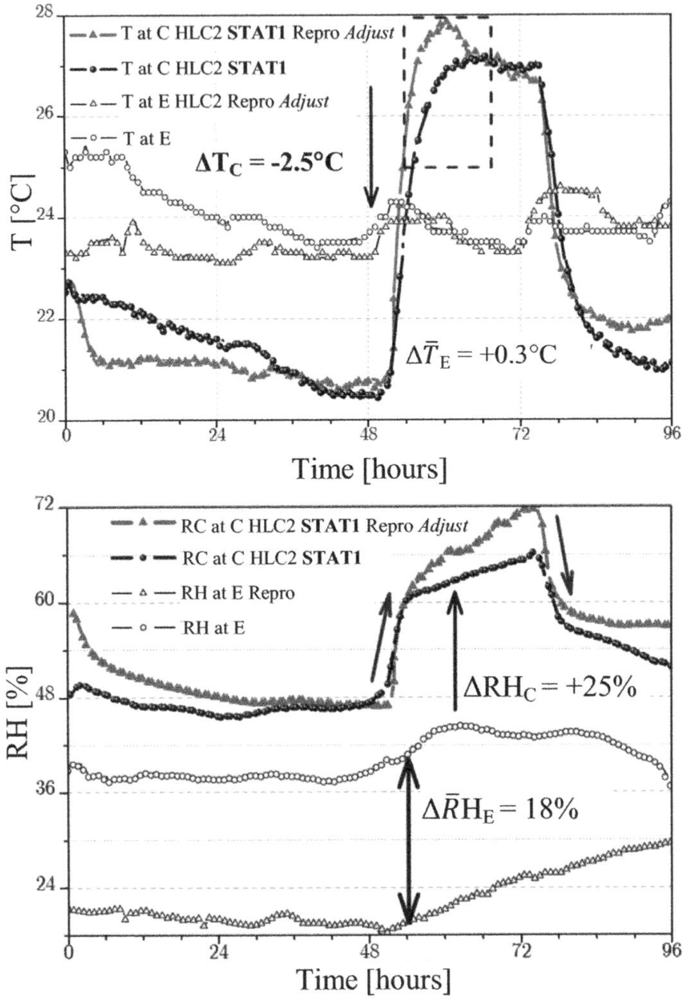

Figure 6.10. *Reproducibility tests for hempcrete under static stress STAT1 (top); temperature and relative humidity graphs after upward adjustment (bottom)*

Superposing the trends of T at C for materials 30 cm in thickness under the influence of STAT1 enables us to appreciate the impact of the phenomenon. Instead of continuing to increase, as happens for AAC30 and VPB, the increase in T at C is significantly slowed and then stabilized with HLC2.

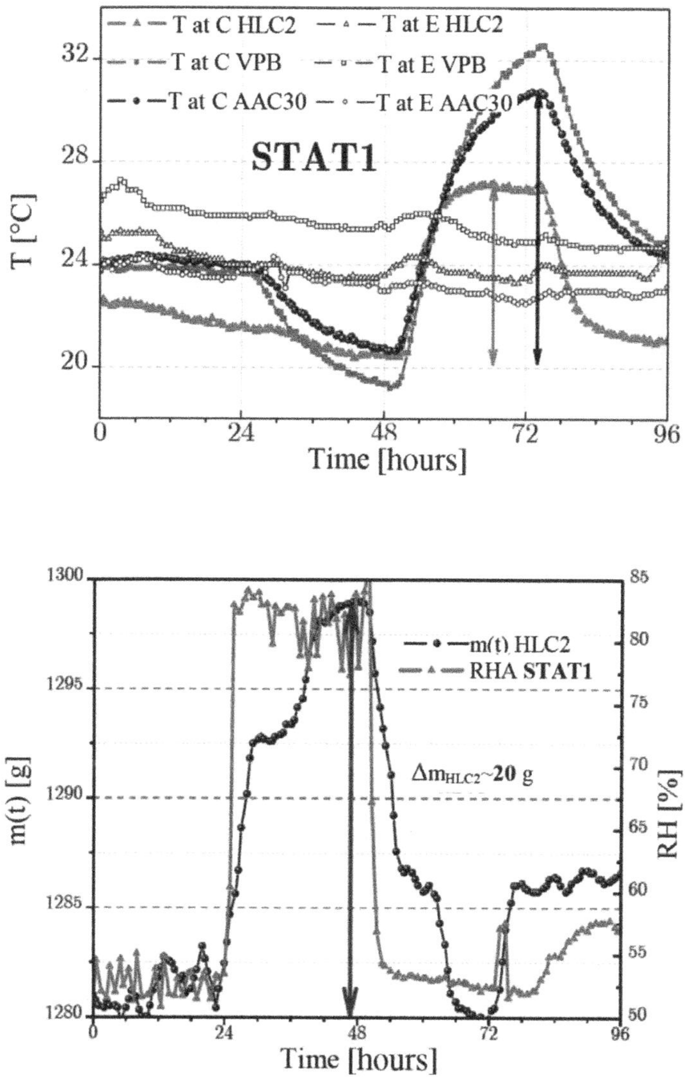

Figure 6.11. *Top: comparison of the temperature changes at C and E for STAT1 (samples 30 cm in thickness: HLC2, AAC30 and VPB). Bottom: variation of the reference mass for hempcrete*

6.3.4. *Hygrothermal transfers*

For irrefutable proof of the role played by these hygrothermal transfers, the same sample of HLC2 hempcrete was subjected to the same gradients of T and RH, under STAT1 stress, in two configurations: the first conventional, as presented above; then for the second, the sample was covered with a very fine, watertight layer of cellophane on each of its surfaces. Thereby, all convection transfer is prevented: only by conduction can heat transfer occur within the material [GOU 09]. The measurements taken are then compared to those taken from the same sample in similar conditions but without the watertight film. The changes in T and RH in the environmental chamber (position A), at the center of the sample (position C) and in the laboratory (position E) are compared in Figures 6.12 and 6.13.

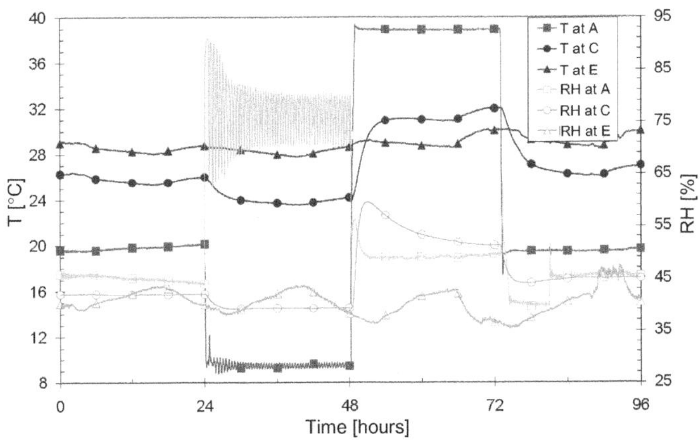

Figure 6.12. *Evolution in temperature and relative humidity at points A, C and E over the course of time for the HLC2 hempcrete wall covered in cellophane*

In Figure 6.12 we note that, during the second stage, the relative humidity at point C decreases: this is due to the condensation of some of the water vapor present in the sample. At the beginning of the third stage, the relative humidity at C increases sharply, before decreasing and then stabilizing. The sudden increase in T by 7°C at point A leads to the vaporization of some of the liquid water in the sample, and therefore the RH at the center of the sample increases abruptly. Within the concrete surface, there is a RH gradient created between the "warm" area of the sample where the water vaporizes and the "cold" area where it does not change stage. The water vapor then migrates toward the "cold" area, which accounts for the decrease in relative humidity observed in the center of the sample. Finally, during the fourth stage, the RH at point C decreases greatly before slightly recovering and then stabilizing. The drop in temperature at point A causes the condensation of the water vapor, which in turn precipitates a decrease in RH at point C. Then a RH gradient arises within the wall and the water vapor migrates to compensate, which accounts for the increase in RH observed at the center of the sample.

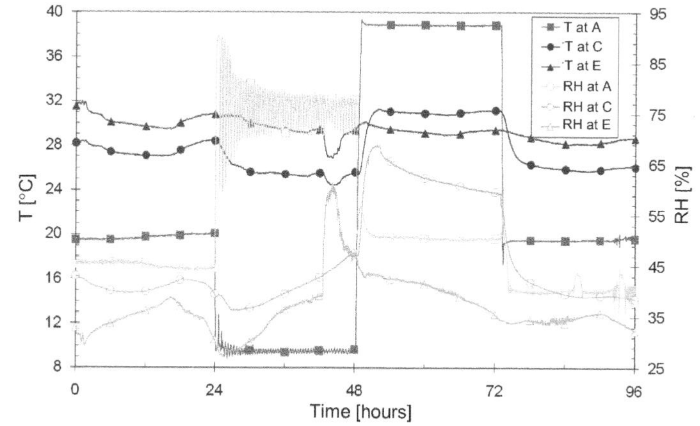

Figure 6.13. *Evolution in temperature and relative humidity at points A, C and E over the course of time for the HLC2 hempcrete wall without cellophane*

By comparing the graphs from Figures 6.12 and 6.13, we note that the amplitude of the change in temperature measured at the center of the wall covered in cellophane between the second and third stages – around 7°C – is greater than that measured at the center of the same wall without cellophane (around 5.5°C). This is attributable to the fact that the cellophane prevents exchanges of water vapor between the wall and the outside world. Indeed, during the second stage, some of the water vapor contained in the sample condenses: the release of energy resulting from this phase-change then halts the drop in temperature in the wall.

In the absence of cellophane, the decrease in the amount of water vapor present in the sample, caused by the phenomenon of condensation, is compensated by an intake of water vapor from the outside, which favors condensation and offsets the drop in temperature. Thus, the drop in temperature at the center of the wall during the second stage is less attenuated when the sample is covered with cellophane.

During the third phase, some of the liquid water contained in the hempcrete vaporizes: the energy absorption caused by this phase-change halts the increase in temperature in the wall. However, during the second stage, because of the contributions of water vapor from the outside and because of the phase-changing that occurs, the wall not covered with cellophane is able to accumulate a larger reserve of liquid water: during the third stage, the phenomenon of vaporization is therefore greater in this wall, which contributes to better counteracting of the rise in temperature at point C. Finally, the increase in temperature at point C during the third stage is less attenuated when the wall is covered with cellophane.

In conclusion, it appears clearly that hempcrete behaves as a natural Phase Change Material (PCM). Work is currently being carried out to study the influence of various types of binder and different kinds of hemp wood on hygrothermal transfers within the material.

6.3.5. *Role of coating products applied to hempcrete*

In view of the hygrothermal behavior of hempcretes, it seems necessary and interesting to examine surfaces treated with coating. Multi-layer envelopes account for the vast majority of real-life situations of use of hemp mortars for new builds, renovation or repair.

We use the notation HLC1E to denote the wall where the material HLC1 constitutes the central part and HLC2E to denote a coated surface which includes the material HLC2. Comparison of the behavior of these walls should enable us to analyze the impact of interior and exterior coating on the hygrothermal behavior of hemp-based materials. In particular, it is a question of determining whether the presence of such coatings alters the physics of the phenomena occurring in HLC2 and HLC1 for which phase-changing is predominant.

Thus, in this configuration, the hemp/lime coating is in direct contact with the interior ambience in the laboratory. The sensor in position C is located at a greater depth in relation to the exterior conditions of T and RH (position A) (7.5 cm + 1.8 cm = 9.3 cm for HLC1E, and 15 cm + 1.8 cm = 16.8 cm for HLC2E – see Figure 6.14).

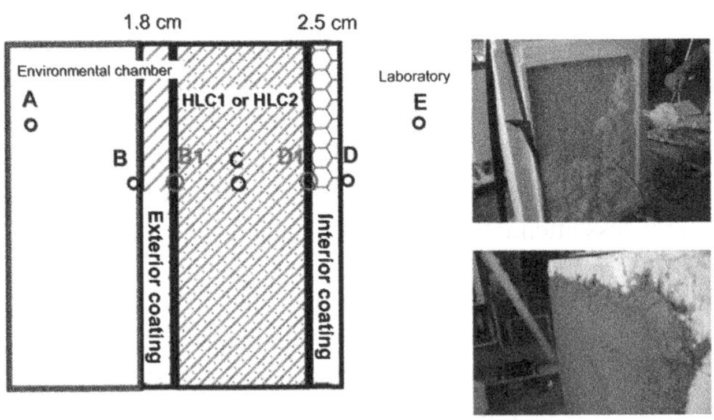

Figure 6.14. *Configuration for the tests on coated walls and photos of the realization of these walls*

Similarly as for the tests above, superposition of the evolutions in T at points C and E between 0 and 96 hours, during the STAT1 tests, was tested on coated walls. The results are presented in Figure 6.15. The average difference of temperature at point E is $\Delta T_E = 0.9°C$. The optimal adjustment value for C is similar: $\Delta T_c = 0.8°C$. We observe excellent repeatability of the thermal behavior at each stage, both in positions C and D1.

The response of the material HLC1E differs from that for HLC1. Let us first turn our attention to the change at point C. Over the stage (48–72 hours), when $T_A = 40°C$, the rise in temperature is less swift than for HLC1 and we see a change similar to that for AAC15. However, the temperature at point C stabilizes before the end of each stage. The presence of the two layers of coating completely alters the response of the material. The sensitivity of the temperature at C to the variations in temperature at E is lesser, because of the layer of hemp/lime coating, which is now in direct contact with the atmosphere in the laboratory. On the one hand, the porosity of the coating surface is different to that of the hempcrete: convection exchanges of heat and humidity at the surface are different, and on the other hand, position C is further from the atmosphere in the laboratory.

The evolution of T at point D1 is identical to the evolution of T at point C for the HLC2 wall. In addition, the similarity is also valid for the RH at D1 (Figure 6.10), increasing by around 30% for *t* between 48 and 72 hours, for both tests. At the interface between the HLC1 material and the coating, therefore, there is a vaporization of liquid water, which accounts for this change at this point. The shape of the graph of temperature change at point C seems to indicate that the phase-change remains localized at D1, without any real impact at point C.

On the basis of the changes in RH at points B1 and C, one is led to believe that the significant increase in RH at D1 is due, in part, to a redistribution of water vapor by diffusion, from A towards E, in the direction of the high pressure gradient of the water vapor that exists at the boundaries. Only by modeling the transfers will we be able to vindicate or debunk this hypothesis. The microstructural difference between HLC1 and the hemp/lime coating results in distinct transfer properties.

We shall see in the next section that hemp/lime coating offers greater resistance to the passage of water vapor than HLC1 does. A large proportion of the water vapor which migrates from the HLC1 toward the coating layer has greater difficulty in evacuating and therefore accumulates at D1.

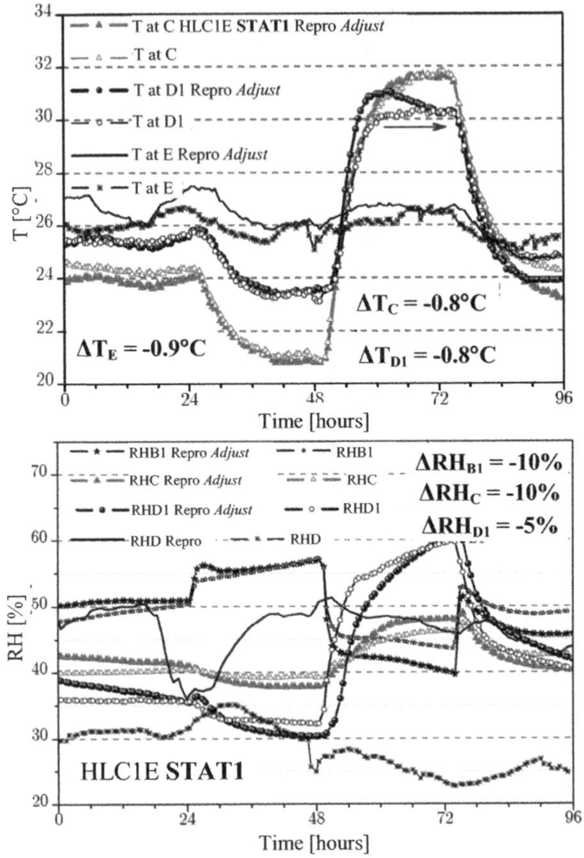

Figure 6.15. *Adjustment of T (top) and RH (bottom) at points C, D1 and E for the reproducibility test under STAT1 (ΔT_C = -0.8°C), t ∈ [0-96 hours] (HLC1E)*

For HLC2E, as can be seen in Figure 6.16, we again see the adjustment of temperature for the three positions C, D1 and E. The average temperature difference in the laboratory is $\Delta T_E = 5.5°C$. With an adjustment $\Delta T_C = 3.5°C$, the changes coincide perfectly throughout the test. This is also verified at the D1 interface, where, because of its position nearer to the ambient atmosphere in the laboratory, the superposition of the trends occurs for $\Delta T_{D1} = 5°C$. At points C and D1, the two adjustment values are lesser than the average difference in temperature at point E. Thus, in view of the heat balance equation, and if we compare it with the HLC2 wall, the source term is no longer predominant.

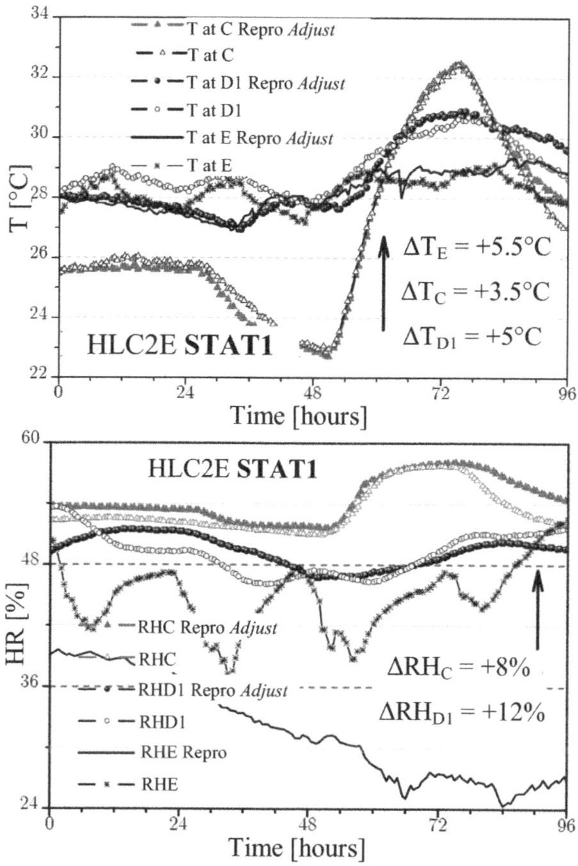

Figure 6.16. *Adjustment of T (top) and RH (bottom) at C, D1 and E for the reproducibility test for STAT1 (ΔT_C = 3.5°C), t ϵ [0-96 hours] (HLC2E)*

Thus, it becomes clear that the coating products tested appreciably alter the thermal behavior of HLC2. This is confirmed by a comparison of the temperature changes at point C for the HLC2 and HLC2E samples (Figure 6.16). The third stage, when the temperature at point A spikes from 10 to 40°C, is very revealing of the change in behavior. For HLC2E, the response at point C is marked by a much-slowed evolution (a less steep slope) and the stage of stabilization is no longer to be witnessed: the temperature at C continues to increase. The amplitude of the temperature change at C (9°C) is thus greater than for HLC2 (7°C). The vaporization phase-change becomes lesser and no longer has the same impact on the heat transfer in hempcrete measured between two layers of coating. Indeed, for *t* between 48 and 72 hours, the change in RH at point C displays a more moderate rise than for HLC2.

However, the variation is of around 10%, and in reference to the adsorption isotherm for HLC2, this implies a not-insignificant change in the material's water content, thus directly influencing the thermal characteristics of the material.

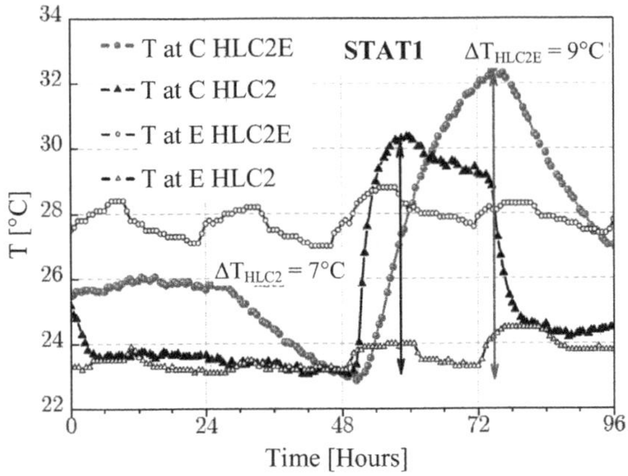

Figure 6.17. *Comparison of the changes in temperature at point C for STAT1, t ϵ [0-96 hours] (HLC2E & HLC2)*

In conclusion, it is helpful to stress that the interfaces between the hempcrete and the interior and exterior layers of coating impose restrictive conditions on the flows of water vapor and/or liquid water going in and out of HLC2 hempcrete (particularly a discontinuity of the capillary pressure). The result of this is renewed management of water in the material. The connection between T and RH is thus far less pronounced in a multi-layer wall. However, work needs to be done to adapt the characteristics of coating products to the transfer capacities of hempcretes in order to optimize the hygrothermal performances and the phenomena of phase-change in these concretes.

6.3.6. *Conclusions*

This work clearly demonstrates that HLC1 and HLC2 walls have hygrothermal behavior which is very different from those of the other materials. Indeed, the comparison of the adjustment values at points C and E implies the existence of an internal source of heat production or the presence of another heat flow (notably convection), introduced explicitly in the heat balance equation in the form of a source term. The superposition of the changes in T and RH at point C appears to indicate that this term refers to latent heat given off by phase-changing. Because the phase-change enthalpy is transported by diffusion of water vapor, mass transfers have a significant impact on the heat transfer. These flows are directly dependent on the material's transfer characteristics – i.e. its permeability to water vapor and air and its hydraulic diffusivity.

For walls made of AAC, VPB, HLC1E and HLC2E, there is a proportional relationship between ΔT_C and ΔT_E, globally conserved for each type of test. The heat balance equation does indeed show an "equilibrium" between the injection of heat and the warming of the material. The predominant mode of transfer in these materials is heat conduction.

These conclusions demonstrate the need to adapt the performances of coating so as not to negate the increase capacity for regulation that may result from the phenomena of hygrometric transfers and phase changes in these hempcretes.

The next part of this chapter offers a comparison of the results obtained against a model of pure heat conduction, in order to give a more precise analysis of the role of conductive and convective transfer, by radiation or by phase-changing in the materials.

6.4. Thermal characterization of various construction materials

Hemp-based materials are distinguished by the appearance of liquid/vapor phase-changes, by condensation or evaporation, which play a significant role in the hygrothermal performance of these walls. In order to precisely evaluate the impacts of these phenomena, it is necessary to quantify them.

Remember that a material collects and reflects heat: this phenomenon depends on the two physical characteristics linked to a combination of its density ρ, its heat conductivity λ and its mass heat capacity c: the heat diffusivity $\mathbf{a} = \lambda/(\rho.c)$ and the thermal effusivity $E_{ff} = (\lambda.\rho.c)^{1/2}$.

In the case of simple walls, these two parameters of the material's thermal behavior were determined by inverse analysis on the basis of the two types of temperature stresses, which consists of applying temperature stresses to the surface of the material such as sudden and staggered spikes in temperature or in the form of cyclic evolutions [SAM 08]. Without going into detail about the process, let us recall that the configuration of the tests is modeled in the following manner: the first layer represents the wall of material (15 or 30 cm in thickness), to which we add a layer of air, 3 m thick (see Figure 6.18). Hence, the boundary conditions are as follows:

– on the surface inside the environmental chamber (point B): Dirichlet condition with the surface temperature measured experimentally (at $x = 0$, $T = T_{B\ \mathrm{measured}}$);

– on the surface exposed to the laboratory: Dirichlet condition, at $x = 3.15$ or 3.30 m (the thickness of the layer of air + that of the material), a temperature equal to the average temperature in the laboratory over the course of the test (point E). Also, a Newtonian condition is applied to the interface between the material and the air (discontinuity of the heat flows). We impose a heat transfer coefficient by convection between the material and the ambient air (the surface temperature will be taken to be equal to the average temperature measured at point D).

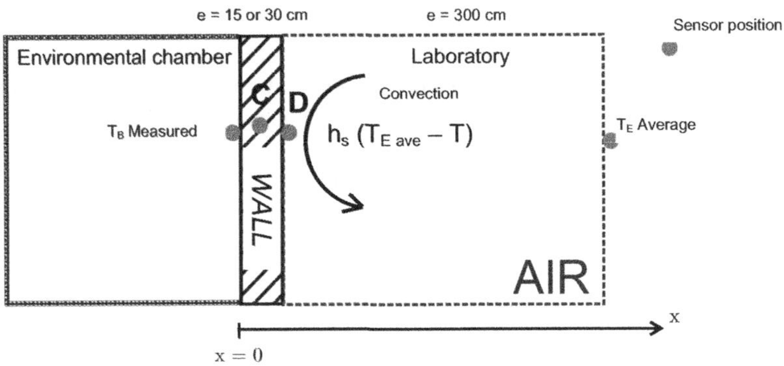

Figure 6.18. *Configuration used for application of the model of heat conduction*

Thus, we are in the context of *unidirectional heat conduction, with no change of state*. In [SAM 08], the equations governing this heat conduction problem were solved numerically. In the present work, these calculations will not be repeated: only a few examples are presented in order to illustrate the possible case of solution and the limitations of the existing models. Finally, an alteration of the existing model sis put forward, in order to take account of the effects of hygrometric transfers and phase-change phenomena.

6.4.1. *Autoclaved aerated concrete*

The temperature changes measured at points C and D with rapid (staggered) stress are presented in Figure 6.19, and compared to the model of heat conduction. For both thicknesses, the same values of heat were injected into the model. These values are specified below in Table 6.2.

	λ [W/(m.K)]	ρ [kg/m^3]	c [J/(kg.K)]	a_{AAC} [m^2/s]
AAC	0.118	429.5	915.3	2.99×10^{-7}

Table 6.2. *Values for the thermophysical performances of autoclaved aerated concrete*

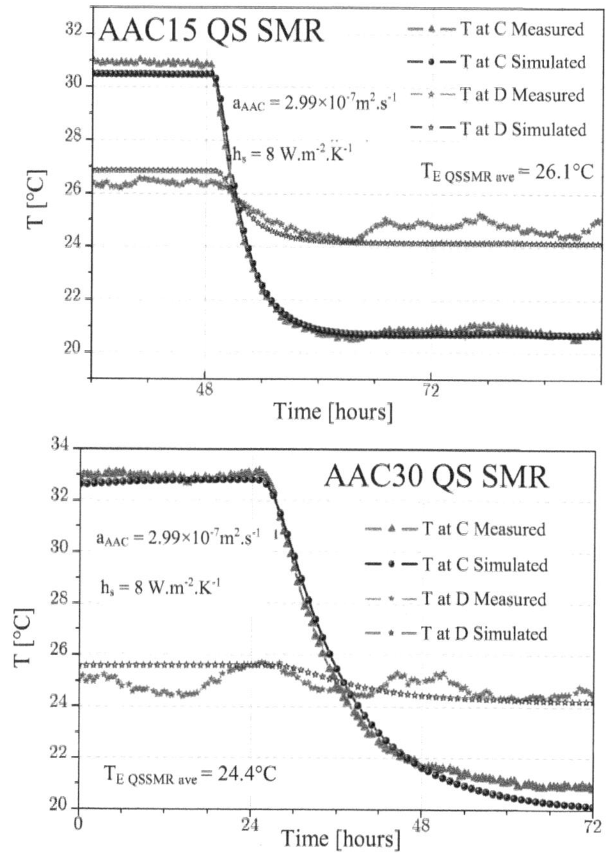

Figure 6.19. *Evolutions in T at points C and D, both theoretical and experimental, for staggered stresses (top: AAC15; bottom: AAC30)*

These values offer very good correlation between the simulated and measured changes of T at points C and D. If we compare the value of the parameter a_{AAC} with the dry value given in Table 6.1, we note that they are very similar, even though in the latter case the material has an average water content of $w = 9$ kg.m^{-3}. For this test, the mass transfer (essentially diffusion of water vapor) therefore does not affect the rate of propagation of heat transfer. Hence, the results are in line with the analysis performed on the material at 15 and 30 cm thickness. Finally, an identical convective heat exchange coefficient was determined for both thicknesses: $h_S = 8$ W.m^{-1}.K^{-2}. This is the value which minimizes the difference with the measurement at D (after several tests of values ranging from 5 to 15 W.m^{-1}.K^{-2}). In addition, it is consistent with the value generally encountered for the interior of a room.

In addition, other static and dynamic stresses have successfully been used to validate the values of λ, ρ, c and h_S. For instance, the simulated and measured changes in T at points C and D for cyclical variations are compared for AAC15 in Figure 6.20. The values of λ, ρ and c, which determine the damping and phase-changing of the signal at points C and D in relation to the thermal wave at point B, lead to variations – both increase and decrease in temperature – identical to the measured values. An appropriate concordance at point D vindicates the adjusted value for h_S. A slight phase-change of the simulated responses at point C is observed in relation to the experiment (approximately two hours). Indeed, the maximum values do not always coincide on the cycles, with differences ranging from 0.5°C to 1°C. The influence of mass transfers on the thermal behavior of AAC, which is not taken into account here, can therefore be taken to be moderate, even in a cyclic regime.

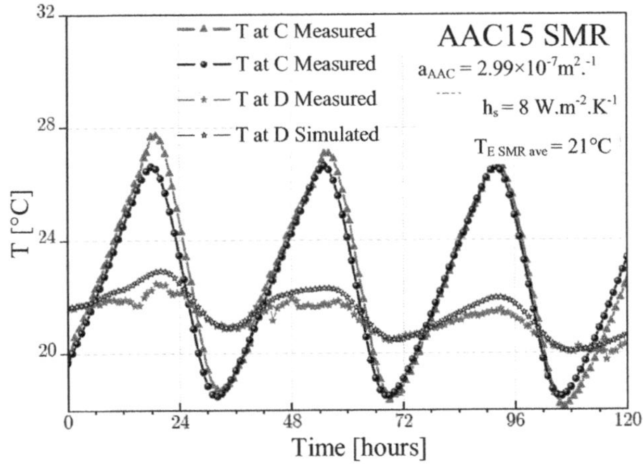

Figure 6.20. *Changes in T and points C and D, both simulated and measured for the cycle during summer (AAC15)*

6.4.2. *Vertically perforated brick*

The same approach is followed for the characterization of vertically perforated bricks of fired earth. Without going into detail about the results [SAM 08], comparison between the measurements and the model gives us the adjusted heat diffusivity $a_{VPB} = 2.98 \times 10^{-7}$ m^2.s^{-1}. The equivalent adjusted value of heat diffusivity a_{VPB} is relatively close to $a = 3.3 \times 10^{-7}$ m^2.s^{-1} as measured by [LAC 03]. The proper consistency of the results is satisfied for the three positions C, C2 and D, even when the temperature at B changes inversely between 10°C and 40°C.

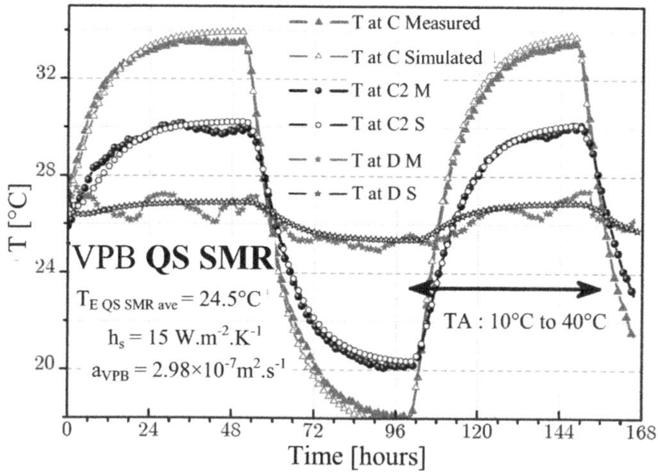

Figure 6.21. *Changes in T at the center of the wall, both simulated and measured, for a staggered test on fired earth VPBs*

In addition, this result shows that in the static regime, the hypothesis of *purely conductive* heat transfer through VPB is confirmed globally. The material VPB is therefore comparable to a homogeneous material.

The value of the convective exchange coefficient imposed at the surface is h_S = 15 W.m^{-2}.K^{-1} – a value which is far greater than the adjusted coefficient for AAC. In reality, the influence of exchanges by high-wavelength radiation (in the far infrared (FIR) spectrum, around 10 μm) is integrated into the high value of the parameter h_S. In order to avoid this bias, a radiative exchange coefficient of 6 W.m^{-2}.K^{-1} is taken into account for an average temperature at point E of 26°C and by subtraction, the convective exchange coefficient is adjusted to h_S = 9 W.m^{-2}.K^{-1}. These values are also validated for cyclic stresses.

6.4.3. *Hempcrete*

For a wall made of HLC2, 30 cm in thickness, static stresses offer good correspondence between the simulated and measured results with an adjusted value of heat diffusivity, $a_{HLC2} = 7.5 \times 10^{-7}$ m^2.s^{-1} (see Figure 6.22). In comparison with the dry value (Table 6.1), this value is three times greater. The water content of the material, which alters the values of λ, ρ and c, is not enough: the heat diffusivity of the material does not go above 2.6×10^{-7} m^2.s^{-1}. The heat conduction model is therefore not adapted to the specificity of the thermal behavior of HLC2. Taking account of the material's water content is not enough to account for the high value of the adjusted apparent heat diffusivity. It confirms the need for a source term in the heat balance equation, justifying a rapid heat transfer (high values of dT/dt). The notable fall in RH at point C, from 65% to 35% (see Figure 6.22) appears to result from significant condensation. Given that this phase-change is exothermic, in relation to the conduction model, it leads to a significant difference in temperature at point C, of around 4°C – the result of a large amount of heat released into the material. It is therefore essential to take account of the mass transfers within HLC2 surfaces. A model integrating these mass transfers must be used in order to accurately express the thermal behavior of the hemp-based material. These observations are also valid for the study of HLC1 hempcrete. Thus, it is not possible to use this method to characterize this material.

The same is not true for coated walls, and the low permeability of the coating products tested, as highlighted above. Therefore, we have modified the approach (see the diagram in Figure 6.23) in order to analyze the entirety of a multi-layer wall.

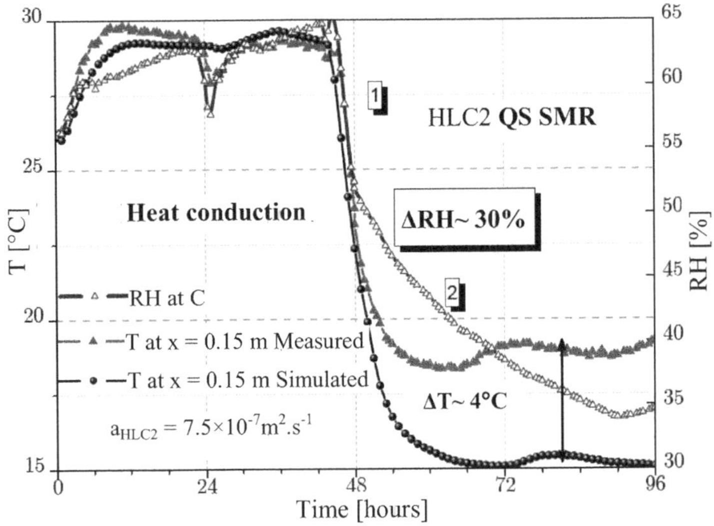

Figure 6.22. *Simulated and measured changes in T at point C, superposed on the changes in RH at point C for the staggered test on hempcrete (HLC2)*

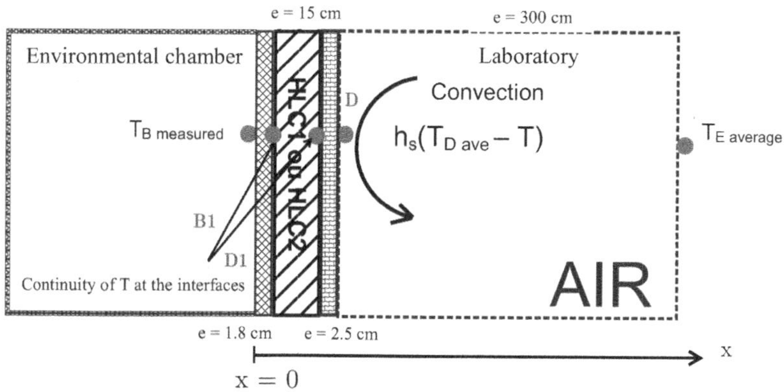

Figure 6.23. *Configuration used for the application of the heat conduction model to coated walls*

The boundary conditions of the system (exterior coating + Hemp Material + interior coating) are identical to the model above: we see an imposed temperature at point B equal to the measured value of T at B, a temperature at point E equal to the average temperature in the laboratory, and at point D, a Newtonian condition where the exchange coefficient is imposed. The presence of the two layers of coating on either side of the HLC wall imposes two new conditions at the interfaces.

In the model, the values of λ, ρ and c of the interior and exterior coating products used for the experiment are indicated in Table 6.3. For the exterior coating, the thermal values were adjusted by the same approach: they were first contrasted against the measured changes of T at point B1, and then compared with other static stresses in order to confirm the reliability of the values. As regards the interior coating, the values used are taken from [COL 04], even though the value of the conductivity, in particular, seems high.

	λ [W/(m.K)]	ρ [kg/m^3]	c [J/(kg.K)]	a [m^2/s]
Interior coating (hemp/lime)	0.21	761	1090	2.5×10^{-7}
Exterior coating (sand/lime)	0.2	1870	1000	1.1×10^{-7}

Table 6.3. *Thermophysical values of coating products [COL 04]*

For the materials HLC1 and HLC2 only, we use the values of λ, ρ and c given in Table 6.1, also integrating the influence of the material's water content. For this purpose, we evaluate the average value of the RH at point C, and using the adsorption isotherm, deduce from it a value of the average water content w. By applying the models obtained by self-consistent homogenization, we determined the new values for the thermal variables (see section 6.2.2).

To begin with, for the HLC1E wall, the result is shown in Figure 6.24. The average value of RH at point C is 65% (with variations of ±10% RH). The adsorption isotherm of HLC1 gives us an average value for the water content of the material: $w = 19$ kg.m^{-3}. From this, we deduce the values of the thermophysical properties of HLC1, indicated in Table 6.4.

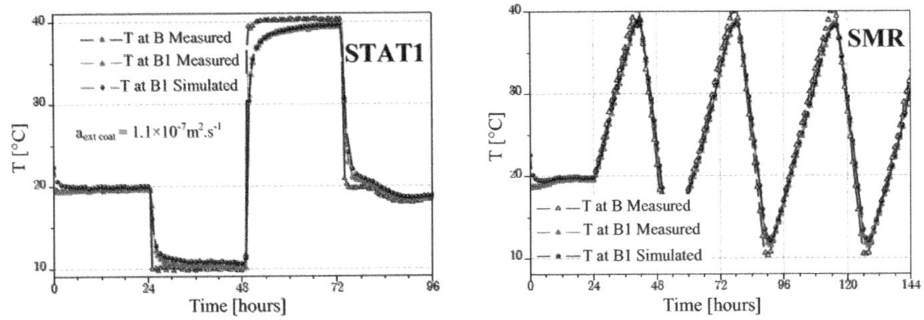

Figure 6.24. *Simulated and measured changes of T at point B1 for the STAT1 and SMR tests*

At this stage, the comparison between the simulated and measured changes of T at points C and D no longer requires any adjustment of a value for the convective heat exchange coefficient at point D. When $h_S = 5$ W.m^{-2}.K^{-1}, the changes at point C match perfectly. Such is also the case for STAT1 stresses, in particular (see Figure 6.25).

For the HLC2E wall, the same procedure yields the thermophysical values given in Table 6.4 for HLC2 hempcrete.

	λ [W/(m.K)]	ρ [kg/m^3]	c [J/(kg.K)]	a [m^2/s]
HLC1 (RH = 65%)	0.12	452	1150	2.3×10^{-7}
HLC2 (RH = 55%)	0.09	332	1150	2.36×10^{-7}

Table 6.4. *Values of the thermophysical properties of HLC1 for $w = 19$ kg.m^{-3} and HLC2 for $w_{ave} = 16$ kg.m^{-3}*

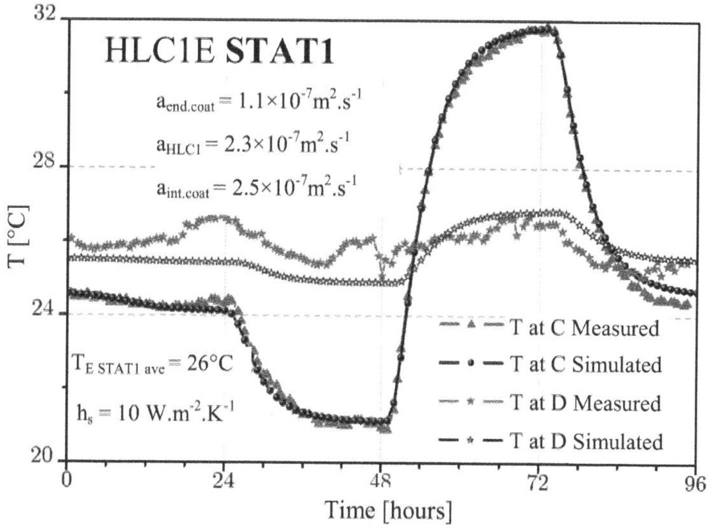

Figure 6.25. *Simulated and measured changes in T at points C and D for the STAT1 test (HLC1E)*

In order to quantify the robustness of the model for coated walls, the temperature changes at points C and D were compared for the summer cycle. The values of λ, ρ and c for the interior and exterior coating products (Table 6.3) are the same as in the static regime. For the materials HLC1 and HLC2, we always take account of the average water content of the porous medium. For the HLC1E wall, the simulated and measured temperature changes at points C and D during the summer cycle are illustrated in Figure 6.26. For the summer cycle, the average water content of the material is $w = 17$ kg.m^{-3} (RH = 50%). The new values of $\lambda = 0.117$ W.m^{-1}.K^{-1}, $\rho = 447$ kg.m^{-3} and $c = 1120$ J.kg^{-1}.K^{-1} for HLC1 yield a relatively similar value for the heat diffusivity $a_{HLC1} = 2.4 \times 10^{-7}$ m^2.s^{-1}. Similarly, at point D, we impose an identical convective heat exchange coefficient $h_S = 5$ W.m^{-2}.K^{-1}.

For the HLC2E wall, the average water content of the material, during the summer cycle, is $w = 16$ kg.m^{-3} (RH$_{C\ ave}$ = 55%, and over the course of the whole test, the variations of RH at C settle at ±3% of this value), just like for the QS SMR test. The thermal parameters used are therefore the same as in the static regime.

Similarly to in the static regime, the heat conduction model offers accurate simulations of the changes in temperature at point C. These results vindicate the conclusions drawn in the static regime – i.e. altered thermal behavior of the hemp-based material HLC1, where the presence of the coating "limits" the influence of the mass transfers on the heat transfer.

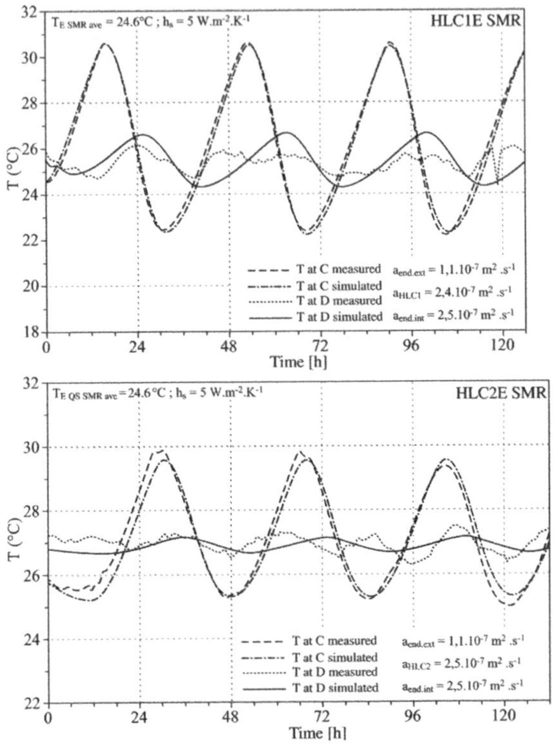

Figure 6.26. *Simulated and measured temperature changes at points C and D for cyclic stresses in summer (top: HLC1E and bottom: HLC2E)*

6.4.4. *Conclusions*

For all types of stress, the *heat conduction* model delivers reliable results regarding the thermal characterization of walls made of various materials. The values of λ, ρ, c, h_S and a for the materials tested were therefore validated by means of a common approach: inverse analysis followed by tests of the reliability and robustness of the model.

	λ	ρ	c	a	h_S
AAC	0.118	429.5	915.3	2.99×10^{-7}	8
VPB				2.98×10^{-7}	15
HLC1	0.12	452	1150	2.3×10^{-7}	
HLC2	0.09	332	1150	2.36×10^{-7}	
Interior coating	0.21	761	1090	2.5×10^{-7}	5
Exterior coating	0.2	1870	1000	1.1×10^{-7}	

The case of walls coated with hemp mortars HLC1 and HLC2 is worthy of a number of comments. The limitations of the pure heat conduction model have been demonstrated. The hygrothermal performance of hemp-based materials is very different from that of other materials. With this model, it proves impossible to retranscribe the particular dynamic of the temperature at point C, with inaccuracies reaching up to 4°C. In addition, the measured changes of RH at C account for the existence of phase-changes by evaporation (increase in RH) or condensation (drop in RH) in the material. Given that the former phenomenon is endothermic and the latter exothermic, it is crucial to integrate them as a source term in the heat balance equation. For this reason, a modeling procedure with coupled phenomena is presented in the next section.

Finally, for HLC1E and HLC2E hempcrete coated walls, it has been possible to adjust values of λ, ρ, c and h_S for all three materials. The simulated results conform quite closely to the experimental, measured results, and confirm that the coating products tested act as thermal and hydric filters. Thus, it seems important in the future to take account simultaneously of the definition of the filler materials and the coating applied to them.

6.5. Modeling of coupled heat- and mass transfers

6.5.1. *Introduction*

The pure heat conduction model used above proves insufficient to describe the hygrothermal performance of hemp-based materials HLC1 and HLC2. A strong correlation has been shown between mass transfer and heat transfer. It becomes essential to take account of the relative humidity within the envelope of materials, because it has a very noticeable impact on the energy consumption.

The theoretical description of mass transfers in porous media reveals the complex phenomena which are involved: diffusion in vaporous and liquid phases, adsorption and capillary condensation [PHI 57]. These phenomena are particularly difficult to model in the case of a building where the materials are very different and the properties are often poorly defined.

This final section gives a very succinct presentation of the transfer laws for vapor, liquid and heat on the macroscopic scale, and then a model of the coupled transfers of heat and mass is presented in detail. Thereafter, the nonlinear equations are employed in the numerical simulation software COMSOL 3.2. The results for walls built of hempcrete alone or coated hempcrete are then compared to the experimental results.

6.5.2. *Transfer laws*

The essential difficulty in determining the transfer laws on the macroscopic scale lies in the choice of an appropriate Representative Elementary Volume (REV), the separation of the phenomena involving the different phases (vapor, liquid), the phase-changes which can occur, and also the definition of the macroscopic transfer coefficients evaluated experimentally or on the basis of empirical models, or models bases on the morphology of the porous structure which is difficult to determine. Here we adopt a macroscopic approach which entails imagining the porous medium as an equivalent (fictitious) continuous medium for which the variables and physical parameters are taken to be the average values for a certain volume of the porous medium.

6.5.2.1. *Vapor transport*

On the macroscopic scale, Fick's law tells us that in isothermal conditions, the density of flow of vapor g_v [kg.m^{-2}.s^{-1}], dependent upon the gradient of concentration of the vapor, is written [KRU 96] $g_v = -\frac{D_a}{\mu}\nabla\rho_v$ where the coefficient μ denotes the factor of resistance to water vapor ($\mu = \frac{D_a}{\delta_v}$, with δ_v [m^2.s^{-1}] denoting the permeability to water vapor under the gradient of ρ_v). Thus, the diffusion of vapor in a porous material is interpreted as the diffusion of vapor in air, but reduced because of the geometry of the porous medium. Thus, it is dependent upon the porosity and the tortuosity. If we introduce the gradient of partial vapor pressure and the perfect gas law, g_v is then written $g_v = -\frac{D_a}{\mu\, R_v\, T}\nabla p_v = -\delta_p\,\nabla p_v$. The permeability to water vapor δ_p [kg.Pa^{-1}.m^{-1}.s^{-1}] (under the gradient of p_v) is therefore given by $\delta_p = \frac{\delta_v}{R_v\, T}$.

In non-isothermal conditions, when a temperature gradient is applied, there are two modes of transport within the porous medium which cause migration of vapor: effusion (Knudsen transport) and thermo-diffusion, related to the difference in density between dry and damp air. Thus, the density of the non-isothermal flow of vapor $g_{T,v}$ [kg.m^{-2}.s^{-1}] based on the temperature gradient is $g_{T,v} = -D_{T,v}.\frac{\partial T}{\partial x}$ where $D_{T,v}$ [kg.K^{-1}.m^{-1}.s^{-1}] is the diffusivity coefficient of vapor under the influence of a temperature gradient. Finally, the equation governing the transport of water vapor when a temperature gradient is imposed is taken from [PHI 57]: it combines the diffusion of vapor under the influence of a mass water content gradient w [kg.m^{-3}] and thermo-diffusion: $g_v = -D_{w,v}.\frac{\partial w}{\partial x} - D_{T,v}\frac{\partial T}{\partial x}$, where $D_{w,v}$ is the diffusivity of vapor under the effect of a mass water content gradient. In general, for applications relating to the physical behavior of the building, thermo-diffusion is negligible in comparison to the other forms of transport [KRU 96]. Consequently, with an increase in the temperature gradient through a damp material, that material's permeability to water vapor will increase.

6.5.2.2. *Liquid transport*

The transport of water vapor decreases progressively as the water content of the medium increases, becoming negligible when saturation point is reached.

In isothermal conditions, the transfer of liquid water is described, in macroscopic terms, by Darcy's law, even in the case of a non-saturated porous medium [PHI 57]: the density of the liquid flow is given on the basis of the capillary pressure gradient p_c: $g_l = -K\frac{\partial P_c}{\partial x}$, where g_l is the density of liquid flow [kg.m^{-2}.s^{-1}], p_c is the capillary pressure [Pa], the difference between total pressure P and liquid pressure P_l under the meniscus, and K represents the hydraulic conductivity [kg.Pa^{-1}.m^{-1}.s^{-1}], which is dependent upon the capillary pressure p_c. This expression is highly nonlinear, because of the nonlinearity of the function $K(p_c)$.

It is practically impossible to determine the hydraulic conductivity experimentally. According to Kelvin's law, the capillary pressure p_c is a function of the relative humidity, which itself is a function of the water content (the inverse of the isotherm). Thus, the isothermal liquid transfer can be written as a function of the mass water content w [kg.m^{-3}], $g_l = -D_{w,l}\frac{\partial w}{\partial x}$, where $D_{w,l}$ [m^2.s^{-1}] is the coefficient of isothermal water diffusivity.

When $w < w_{cr}$ (the critical water content), the mass transport is comparable, in general, to the diffusion of water vapor. Furthermore, the transport of liquid water exists as capillary suction. The minimum value of water content at which this process takes place is difficult to determine, because it depends greatly on the porous structure of the material.

In [KRU 96], the mass water content w is used as the gradient causing the density of flow by surface diffusion $g_{s,l}$ [kg.m^{-2}.s^{-1}]: $g_{s,l} = -D_{s,l}\frac{\partial w}{\partial x}$, where $D_{s,l}$ is the coefficient of surface diffusion [m^2.s^{-1}]. If a continuous potential, such as relative humidity RH, is used, the equation of surface diffusion is expressed as a function of the relative humidity, as a driving potential: $g_{s,l} = -D_\varphi\frac{\partial RH}{\partial x}$, where D_φ [kg.m^{-1}.s^{-1}] denotes the hydraulic conductivity in liquid form. According to [KÜN 94], it is written $D_\varphi = D_{s,l} \times \xi_w$ where ξ_w is the tangent to the adsorption isotherm of the material, $\xi_v = \frac{\partial w}{\partial HR}$ [kg.kg^{-1}]. The mass water content w is converted to relative humidity RH by means of the adsorption isotherm, which is highly nonlinear. This implies therefore that at all times, the material is in hygroscopic equilibrium with the surrounding air.

When we introduce a temperature gradient, on the one hand, the viscosity of the liquid water is altered, and on the other, the liquid flow is due to a temperature gradient (Soret effect). According to [DEV 58], the density of liquid flow in non-isothermal conditions is written as $g_l = -D_{w,l} \cdot \frac{\partial w}{\partial x} - D_{T,l} \frac{\partial T}{\partial x}$ [kg.m^{-2}.s^{-1}], where $D_{T,l}$ [kg.K^{-1}.m^{-1}.s^{-1}] is the coefficient of water diffusivity under the influence of a temperature gradient.

6.5.2.3. *Total humidity flow*

The mass transfer combines the mechanisms of transport of water vapor and liquid water. The total flow g may not necessarily be caused by a single potential, but rather may be a linear combination of a number of gradients. The range of variables to choose from in order to model the coupled transport of heat and mass in a porous material is extensive: temperature T, relative humidity RH, partial vapor pressure p_v, capillary pressure p_c or indeed the concentration of water vapor ρ_v.

In the hygroscopic domain (RH < 95%), the total flow g [kg.m^{-2}.s^{-1}] through a porous material is expressed as the sum of a component dependent on the vapor pressure gradient g_p and another linked to the temperature gradient g_T [PEU 03]: $g = g_p + g_T = -\delta_p \frac{\partial p}{\partial x} - D_T \frac{\partial T}{\partial x}$ where $D_T = D_{T,v} + D_{T,l}$ [kg.K^{-1}.m^{-1}.s^{-1}], the total diffusivity coefficient under the influence of a temperature gradient. The flow proportional to δ_p is supposed to include the liquid transfer for high RH values. The mass transfer equation was used by [GAL 98].

In non-isothermal conditions, the water content gradient may be at the root of a mass transfer, in the same direction or the opposite direction to the partial vapor pressure gradient. Thus, the total flow of humidity can be written as the sum of a component g_p, dependent on the vapor pressure gradient, and a component g_w, which is dependent on the water content gradient: $g = g_p + g_w = -\delta_p \frac{\partial p}{\partial x} - D_{w,l} \frac{\partial w}{\partial x}$. This equation was employed notably by [KRU 96] on a number of different mineral-based materials such as AAC, natural sandstone, gypsum, etc. If a continuous potential across the material-material or material-air interfaces is preferable [KÜN 94], the relative humidity gradient is used instead of the water content gradient, so that $g = g_p + g_\varphi = -\delta_p \frac{\partial p}{\partial x} - D_\varphi \frac{\partial HR}{\partial x}$. Only the pure vapor transport is taken into account in δ_p, whereas D_φ corresponds to the flow in the liquid phase – i.e. transport by surface diffusion and capillary suction.

6.5.2.4. *Isothermal water diffusivity*

Dynamic mass transport can be described by Fick's second law which, if we discount the source term and consider a uni-directional transfer, is written $\frac{\partial w}{\partial t} = \frac{\partial}{\partial x}\left(\delta_p \frac{\partial p}{\partial x}\right)$. In isothermal conditions, the driving force is given as a function of the mass water content w. The connection between w and the vapor pressure gradient is given by the derivative to the adsorption isotherm – in other words, the term ξ [kg.m^{-3}], expressing a given material's capacity to store humidity, $\xi = \frac{\partial w}{\partial RH}$. Thus, we get $\frac{\partial w}{\partial t} = \frac{\partial}{\partial x}\left(\delta_p \frac{p_{sat}}{\xi}\frac{\partial w}{\partial x}\right) = \frac{\partial}{\partial x}\left(D_w \frac{\partial w}{\partial x}\right)$, where D_w [m^2.s^{-1}] is the isothermal water diffusivity. This equation clearly reveals the similarity that exists between Fourier's law for heat diffusion and Fick's law for mass diffusion. Thus, the water diffusivity D_w is the counterpart to the heat diffusivity a. It corresponds, therefore, to a measurement of the rate of change of the water content in the material when its surface is subjected to cyclic variations in RH. Finally, the isothermal water diffusivity is evaluated on the basis of the permeability to water vapor δ_p, by the formula $D_w = \frac{\delta_p . p_{sat}}{\xi}$.

6.5.2.5. *Heat transfer*

Combining all of these phenomena, De Vries [DEV 87] suggests that the macroscopic heat flow be written in the form $g_q = -\lambda_0 \nabla T + L_v g_v + c_l.(T - T_0).g_{tot}$, where g_q is the macroscopic heat flow density [W.m^{-2}]; λ_0 is the pure conductivity of the porous medium [W.m^{-1}.K^{-1}]; L_v is the latent heat of vaporization [J.kg^{-1}]; c_l is the mass heat of the liquid [J.kg^{-1}.K^{-1}]; T_0 is the reference temperature [K]; g_v is the mass density of vapor flow [kg.m^{-2}.s^{-1}]; and g_{tot} is the total mass density of the flows (liquid + vapor) [kg.m^{-2}.s^{-1}].

It is estimated that convective movement through a porous medium begins to manifest itself when the Rayleigh number becomes greater than or equal to $4\pi^2$, with the Rayleigh number Ra being the product of the Grashof number Gr and the Prandtl number Pr. We shall take a value of Pr = 0.71 for air. In terms of determining a local Grashof number Gr, the calculation is performed after having chosen a macroscopic temperature gradient ΔT, a characteristic length and thickness e of material. The characteristic length considered is the median diameter of the pores, D, gained by mercury porosimetry testing. For the particular materials under study here, D is given in Table 6.5, on the basis of [COL 04; BEL 92; GOY 07]. In the expression of Gr, $\frac{G\beta\rho^2}{\mu^2} = 1.329 * 10^8$ and T_p-T_∞=D×$\Delta T/e$, where e = 5 cm and ΔT = 10 K.

Material	AAC	HLC	VPB	Hemp/lime coating
Median diameter of the pores (μm)	10.37	0.78	0.324	1.10
Ra	3.1×10^{-10}	7×10^{-15}	2×10^{-16}	3×10^{-14}

Table 6.5. *Values of the Rayleigh numbers for the materials under examination*

The Rayleigh numbers thus obtained are below the boundary value $4\pi^2$. It seems that, for these materials and in these conditions, natural convection can be discounted, and the expression of the heat flow is written thus, on the basis of the volume water content θ (instead of w): $g_q = -\lambda_0 \nabla \mathrm{T} + \rho_l L_v . (D_{Tv} \nabla T + D_{\theta v} \nabla \theta)$.

6.5.3. *Transfer model: the Künzel model*

As previously indicated, mass transport takes place in both the liquid and vaporous phases. In the vaporous phase, two gradients are involved simultaneously: the concentration gradient and the temperature gradient. Transfers in the liquid phase occur under the influence of capillary forces, and the gradient is the difference in capillary pressure inside the pores. The choice of forces is crucial in the formulation of the equations. Galbraith [GAL 98] uses thermodynamics of irreversible phenomena to show that the temperature/relative humidity or temperature/vapor pressure "couples" represent true potentials for humidity transfers. However, many authors use potentials derived from the relative humidity and the vapor pressure. These potentials, obtained by way of the relations between the thermodynamic properties of the substances, often cause discontinuities at the boundaries of the material. Thus, the criterion chosen is the accessibility of the potential when measuring. Therefore, our choice moved towards a system of coupled equations, already used in a simulation computer tool developed by the *Fraunhofer Institut für Bauphysik* (IBP): WUFI [KÜN 94]. The equations governing the coupled transports of heat and mass are as follows:

$$\begin{cases} \rho_0 c \dfrac{\partial T}{\partial t} = \dfrac{\partial}{\partial x}\left(\lambda \dfrac{\partial T}{\partial x}\right) + L_v \dfrac{\partial}{\partial x}\left(\delta_p \dfrac{\partial}{\partial x}(RH.p_{sat})\right) \\ \dfrac{\partial w}{\partial RH} . \dfrac{\partial RH}{\partial t} = \dfrac{\partial}{\partial x}\left(D_\varphi \dfrac{\partial RH}{\partial x} + \delta_p \dfrac{\partial}{\partial x}(RH.p_{sat})\right) \end{cases}$$

where RH is relative humidity [-]; t is time [s]; ρ_0 is the dry apparent density [$kg.m^{-3}$]; c is the mass heat of the material [$J.kg^{-1}.K^{-1}$]; w is the water content of the material [$kg.m^{-3}$]; p_{sat} is the saturation vapor pressure [Pa]; λ is the thermal conductivity of the material [$W.m^{-1}.K^{-1}$]; D_φ is the liquid conductivity [$kg.m^{-1}.s^{-1}$] ($D_\varphi = \xi_w \times D_w$, D_w is the water diffusivity of the material [$m^2.s^{-1}$]); δ_p is the material's permeability to water vapor [$kg.m^{-1}.s^{-1}.Pa^{-1}$]; and L_v is the latent heat from phase-changing [$J.kg^{-1}$].

These two equations contain storage terms and transport terms. The heat flows comprise heat transfer by conduction, a function of the water content (included in the thermal conductivity) and a source term resulting from the processes of evaporation/condensation of the water. The latter term expresses the effects of phase-changing (latent heat), the driving factor behind which is the density of vapor flow g_v.

Transport in the liquid phase (capillarity, surface diffusion) is governed by the gradient in terms of relative humidity (RH). The diffusion of vapor is governed by the partial vapor pressure p_v, which depends on the relative humidity RH and the saturating vapor pressure p_{sat}. The interpolation of the saturation pressure can be obtained using the Clausius–Clapeyron relation, which describes the behavior of vapor in equilibrium with the liquid phase.

The liquid transport coefficient D_φ greatly increases with the water content, whereas the influence of the temperature remains limited to its effect on the viscosity of the water. The coefficient brings together the effects or surface diffusion and capillary transport.

The permeability to water vapor δ_p is slightly dependent upon the temperature and the water content, while the saturating vapor pressure increases exponentially with increasing temperature.

The storage of humidity is described by the derivative of the adsorption isotherm, ξ. This function is not heavily dependent on the temperature. The heat storage term corresponds to the mass heat $\rho_0 \times c$. The influence of the water content is directly integrated into this term. Note that the model does not take account of the flow of air by difference of total pressure. In addition, the total pressure, on both sides of the wall, has not been measured. The mutual effects of the temperature and the water content on the different coefficients lead to strong pairing between the two nonlinear equations. This is why the software COMSOL Multiphysics [COM 06] was used to solve these equations.

6.5.4. *Determination of the transfer coefficients*

6.5.4.1. *Permeability to water vapor*

The permeability to water vapor δ_p is obtained by imposing a partial vapor pressure gradient Δp_v, constant and uni-directional, through a sample of thickness e ("steady-state cup" measurements). This method is described in the international standards [EN-ISO 01]. By measuring the density of vapor flow g_v in the permanent regime, we can get $\delta_p = \frac{e}{\Delta p_v}$. We can then determine the factor of resistance to water vapor $\mu_p = \frac{\delta_a}{\delta_p}$, δ_a denotes the permeability to vapor in the air which is a function of the temperature and is written $\delta_a = 2.10^{-7}\, T^{0.81}/_{P_L}$, where P_L is the ambient atmospheric pressure [Pa] and T is the ambient temperature [K]. When T = 23°C, we get $\delta_a = 2.10^{-10}$ kg.m^{-1}.s^{-1}.Pa^{-1}.

When attempting to measure permeability to water vapor, we come up against numerous difficulties. Indeed, this assumes a perfect mixture of damp air on both sides of the sample. In reality, this type of boundary condition is problematic. It is therefore helpful to take account of the water resistances of the inner and outer layers of the material by introducing a factor Z_p [Pa.m^2.s.kg^{-1}], equal to $1/\beta_p$, where β_p is the convective mass exchange coefficient, and of the layer of air between the saline solution and the sample. $Z_{p,in}$ and $Z_{p,out}$ are determined on the basis of the convective heat exchange coefficients. On the inner surface of the sample, the velocity of the air is zero. On the outer surface, the convective heat exchange coefficient is directly dependent upon the velocity of the surrounding air. In general, we consider $h_{s,out} = 5.82 + 3.96 \times v$, where v is the velocity of the air. The equation is written in the following form: $\delta_p = \frac{e}{\frac{\Delta P_v}{g_v} - (Z_{p,in} + Z_{p,out} + Z_{p,air})}$. This approach is particularly important for highly permeable materials, such as HLC1 and HLC2 hempcretes. [DUF 06] also shows that the influence of the surface humidity exchange coefficient β_p is powerful for the most permeable materials such as thermal insulation. Today, this effect is overlooked. However, it leads to a systematic under-estimation of the measured vapor permeability – all the more so if the material is permeable and the surface exchange coefficient is small.

Finally, it should also be noted that the vapor permeability depends on the temperature, the degree of saturation and the relative humidity of the material.

6.5.4.2. *Vapor permeability values of materials*

For hempcrete, very few values are available in the existing body of literature, and the values for dry and humid permeability obtained by these writers (particularly [COL 04] and [EVR 06]) are contrasted. It is very difficult to know the exact conditions in which the measurements are taken, and it is not clear whether all the adjustments and precautions set out above have been taken into account. In spite of this, for our purposes, we shall use the values obtained that are shown in the following tables and compared to those of other enveloping building materials.

It becomes obvious that hempcretes are set apart from other materials by their high permeability to vapor: hence, they are sometimes referred to as “breathable” materials.

	Measure 0-50%	**Measure 0-85%**
$10^{11}\times\ \delta_{p\mathrm{HLC1}}$ [kg.m^{-1}.s^{-1}.Pa^{-1}]	1.7	2.3
μ_{HLC1}	11.5	8.7

Table 6.6. *Permeability to water vapor of HLC1 hempcrete – measurements at 0-50% and 0-85%, according to [COL 04]*

Material	HLC1 [COL 04]	AAC [GAL 99]	Fired earth [DWI 94]	Hemp/Lime coating [COL 04]	HLC2 [EVR 06]
μ_{dry}	11.5	10	16	13	3.6
μ_{humid}	8.7	8.3	10	10.6	
Material	Mineral wool	Loose cellulose fibers	Wood-shaving concrete	Clay/Straw mortar [COL 04]	Lime coating mortars
μ_{dry}	1	2	15	13.3	10
μ_{humid}	1	2	10	11.8	6

Table 6.7. *Water vapor resistance factor of the materials under examination and a number of construction materials*

6.5.4.3. *Isothermal water diffusivity D_w*

The isothermal water diffusivity coefficient D_w is determined by the relation $\frac{\partial w}{\partial t} = \frac{\partial}{\partial x}\left(D_w \frac{\partial w}{\partial x}\right)$. While the value of the vapor permeability is known, we still need to determine $\xi = \frac{\partial w}{\partial HR}$, obtained from the derivation of the adsorption isotherm. Smoothing the adsorption isotherms for the materials AAC, HLC1 and HLC2 (GAB smoothing, as per [KÜN 94]) give us access to the evolution of the parameter as a function of the RH (Figure 6.27).

We get the evolution of D_w for small and medium values of mass water content by exploiting the results from the adsorption isotherms of the materials, which cover between 0 and 95% relative humidity.

For autoclaved aerated concrete, the permeability value used to calculate D_w takes account of the change in δ_p on the basis of the RH. The result coincides with that obtained by [PEU 03].

As regards HLC1 material, the water vapor permeability value obtained at 0-50% relative humidity is used for the initial calculation at the dry point, and that obtained at 0–85% for the other points. As we did not have reliable values for HLC2, the calculation was not carried out in this case. Therefore, the D_w coefficient will be obtained by calibration during the modeling phase.

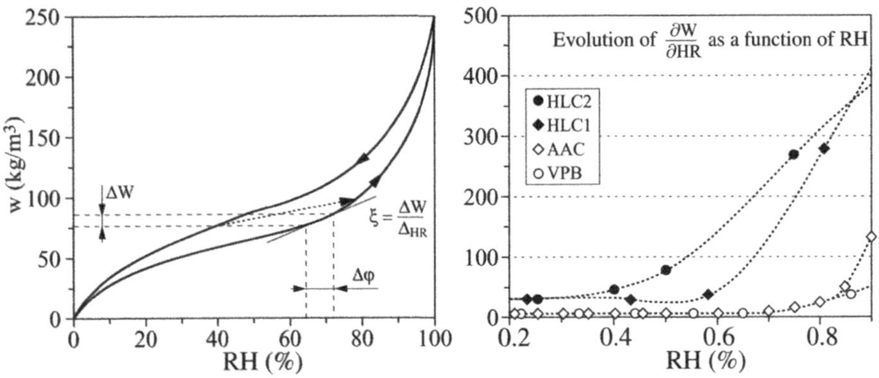

Figure 6.27. *Diagrammatic representation and change of ξ as a function of the RH, for the materials under investigation*

The isothermal water diffusivity coefficient, for both HLC1 and AAC, is between 10^{-10} and 10^{-8} $m^2.s^{-1}$. The result given by [PEU 03] on AAC, for RH between 40 and 60%, indicates a value of D_w of 8.10^{-9} $m^2.s^{-1}$. This value expresses an influence of the phenomena of evaporation-condensation (liquid/vapor series). With the model advanced by [KÜN 94], a constant value of D_w is chosen in the hygroscopic domain: initially this choice would seem to be vindicated, by similar values throughout the range of RH values, around an average of 4.10^{-9} $m^2.s^{-1}$.

Finally, one last comment about this coefficient relates to the crucial distinction that must be drawn between the processes of water absorption and drying. Indeed, these configurations yield different values for D_w. For instance, for AAC, we note that the D_w associated with the drying of the material is always smaller than the D_w associated with the imbibing phase. This difference is all the more pronounced as we begin to approach the capillary water content. For our purposes, in view of the water state of AAC walls ($w/w_{cap} < 0.2$), with liquid transport only occurring in the form of surface diffusion, the difference between adsorption and desorption is very slight. For hemp-based materials, the quantities of water present in the material become significant when the RH surpasses 60%: the difference between the D_w coefficients associated with the drying and imbibing phases could be more pronounced when the w/w_{cap} ratio is greater than or equal to 0.2. For us, during the tests, the water content of the materials, estimated on the basis of the change in RH, remains low in comparison to the capillary water content.

In conclusion, therefore, we shall use exactly the same water diffusivity coefficient D_w for both adsorption and desorption.

6.5.4.4. *Vapor diffusivity associated with a temperature gradient* D_{Tv}

The coefficient D_{Tv} corresponds to the transport of vapor under the influence of a thermal gradient. It is evaluated on the basis of $g_v = -D_{w,v}.\frac{\partial w}{\partial x} - D_{T,v}.\frac{\partial T}{\partial x}$, taking the water content w to be constant. This results in constant relative humidity, so long as the impact of the temperature on the adsorption isotherm remains small. For [PHI 57], taking account of the mass transfer caused by the temperature gradient leads to mass transport ratios which are between 1.3 and 3 times greater than when the pressure gradient of the vapor alone is considered. [DEV 58] and [BEL 92] propose a method whereby the transfer coefficients can be deduced by heat conductivity tests using the impulse method with a single-rod probe. The change in the coefficient D_T (= D_{Tv} + D_{Tl}) with changing mass water content for AAC is traced in Figure 6.28 [BEL 92], over the whole range of values of mass water content and for four different temperatures: 26°C, 30°C, 34°C and 42°C. The coefficient increases slowly and continuously with increasing water content. However, the influence of the temperature is not apparent, though it should appear clearly in view of the fact that the kinetics of drying is faster at higher average temperatures.

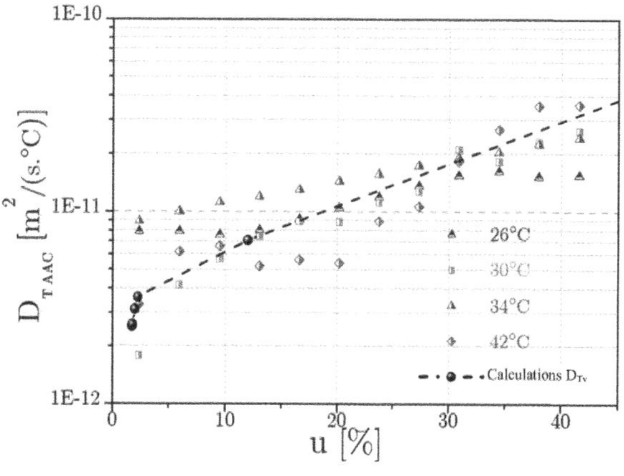

Figure 6.28. *Diffusivity coefficient for a temperature gradient D_T [BEL 92], compared to the calculations*

As we have seen, this coefficient is very difficult to determine. As a general rule, the impact of the temperature is taken into account only by way of the saturation vapor pressure $p_{sat}(T)$, with an exponential increase.

For AAC and HLC, the dependency of the transfer coefficients on the water content of the material and the temperature may be significant. As we do not have precise data relating specifically to hemp-based materials, the expression $g_v = -D_{w,v}.\frac{\partial w}{\partial x} - D_{T,v}.\frac{\partial T}{\partial x}$ is used, with constant values for the coefficients δ_p and D_w, calibrated for an average water content of the material measured during the test.

6.5.5. *Numerical modeling*

A computer program was written specially to numerically solve the partial differential equations, while being able to integrate several coupled equations simultaneously. The program is based on the Finite Element Method (FEM), which consists of discretizing the partial differential equations on a spatial matrix in order to turn them into a system of nonlinear algebraic and differential equations [COM 06].

Ω denotes the 1-dimensional domain formed by the wall represented by the sequential association of a segment 15 or 30 cm in thickness corresponding to the material, and a segment 3 m in length representing the layer of air in the laboratory. The terms e_a and d_a are matrices.

$$\begin{cases} \rho_0 c \dfrac{\partial T}{\partial t} = \dfrac{\partial}{\partial x}\left(\lambda \dfrac{\partial T}{\partial x}\right) + L_v \dfrac{\partial}{\partial x}\left(\delta_p \dfrac{\partial}{\partial x}(HR.p_{sat})\right) \\ \dfrac{\partial w}{\partial HR}.\dfrac{\partial HR}{\partial t} = \dfrac{\partial}{\partial x}\left(D_\varphi \dfrac{\partial HR}{\partial x} + \delta_p \dfrac{\partial}{\partial x}(HR.p_{sat})\right) \end{cases}$$

$$\rightarrow \begin{cases} e_a \dfrac{\partial^2 u}{\partial^2 t} + d_a \dfrac{\partial u}{\partial t} + \nabla.\Gamma = F \text{ on } \Omega \\ -n.\Gamma = G + \left(\dfrac{\partial R}{\partial u}\right)^T \mu \quad \text{on } \partial\Omega \\ 0 = R \text{ on } \partial\Omega \end{cases}$$

The identification of the Künzel model and of the equations drawn from [COM 06] is based on:

– u denoting the vector $\left|\begin{matrix} T \\ RH \end{matrix}\right.$;

– $e_a = \begin{bmatrix} 0 & 0 \\ 0 & 0 \end{bmatrix}$ because the model does not contain a second time differential;

– $d_a = \begin{bmatrix} \rho c & 0 \\ 0 & \xi \end{bmatrix}$, a mass matrix comprising the terms of humidity- and heat storage;

– $\Gamma = \begin{bmatrix} L_v \delta_p \frac{\partial}{\partial x}(RH.p_{sat}) & 0 \\ 0 & D_\varphi \frac{\partial RH}{\partial x} + \delta_p \frac{\partial}{\partial x}(RH.p_{sat}) \end{bmatrix}$ denoting the source term: this expression therefore brings together the phase-change enthalpy for the heat flow and the liquid transport / vapor diffusion for the mass transfer;

– the last two equations are the boundary conditions: at points B and E, Dirichlet conditions in terms of T and RH, of the form $R = T - T_i = RH - RH_i = 0$ ($i = B$ or E); and for D, there is discontinuity of the heat flow and the mass flow. Therefore, we introduce convective heat exchange (h_s) and mass exchange (β_p) coefficients. The configuration is illustrated in Figure 6.29.

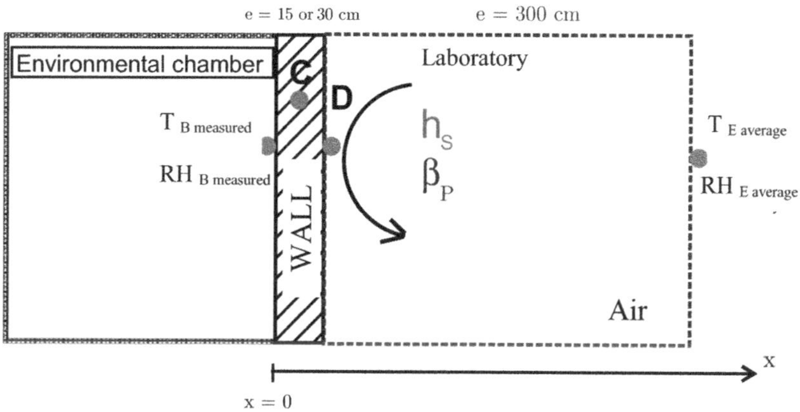

Figure 6.29. *Configuration used for the application of the combined model for heat and mass*

6.5.5.1. *Application to aerated autoclaved concrete walls*

With the aim of validating our model, to begin with, we tested it for AAC15 and AAC30 walls, to vindicate its reliability and robustness, before turning our attention to hempcrete.

The values of the transfer coefficients for aerated autoclaved concrete (AAC) are taken from the existing body of work. To begin with, the transfer coefficients are considered constant, determined on the basis of the average value of the RH at point C measured during the test. It is also possible to evaluate the influence of the water content on the permeability to water vapor δ_p, the isothermal water diffusivity D_w or indeed the heat conductivity λ of AAC.

The heat conductivity of the material is evaluated depending on its average water content. This is determined in relation to the average value of RH at point C measured during the course of the test. As the link is established by the adsorption isotherm of the material, we have direct access to the mass water content w_{ave}. The values of specific heat and heat conductivity are then calculated. The variations of the thermal values of AAC are given by:

$$\lambda_{AAC} = 0.11.(1 + 3.45.u) = 0.11 + 0.414.u = 0.11 + 9.10^{-4}w,$$

$\rho_{AAC}.c_{AAC} = \rho_0.c_0 + w.c_w$, where $\rho_0 = 421$ kg.m^{-3}, $c_0 = 850$ kJ.kg^{-1}.K^{-1} and $c_w = 4.18$ kJ.kg^{-1}.K^{-1}.

Similarly, the latent heat of vaporization is equal to $L_v = 2.5 \times 10^6$ J.kg^{-1}, and – let us remember – the dependency of the temperature on the saturation vapor pressure is given by the empirical relation valid for T between 0 and 80°C, with a precision of ±0.15% [PEU 03] $p_{sat} = e^{23.5771 - \frac{4042.9}{T-37.58}}$, where T is expressed in Kelvin [K].

The value of the permeability to vapor chosen is set on the basis of the average value of RH at point C measured during the test. For the term ξ, the hysteresis between adsorption and desorption is discounted. We choose a uniform storage capability $\xi = 20$ kg.m^{-3}, obtained for 60% RH. For the isothermal water diffusivity D_w, over the range of RH values between 0 and 80%, we choose a constant value in the hygroscopic domain: $D_w = 1.1 \times 10^{-10}$ m^2.s^{-1}.

Finally, the values of the convective heat- and mass exchange coefficients are adjusted in relation to an inverse analysis test. The 15 cm-thick wall under static stress with a mass water content $w = 8.5$ kg.m^{-3} yields the values of permeability and diffusivity of the material: $\delta_p = 5.10^{-11}$ kg.m^{-1}.s^{-1}.Pa^{-1} and $D_w = 1 \times 1.10^{-10}$ m^2.s^{-1}. The convective heat exchange coefficient is the same as that adjusted for the conductive model: $h_S = 8$ W.m^{-2}.K^{-1}, and the Lewis relation then gives $\beta_p = 4.10^{-8}$ kg.Pa^{-1}.m^{-2}.s^{-1}.

Once these values are fixed, we need to validate the robustness of the combined heat and mass model. The results obtained with a number of different stresses are compared against the model's results. An initial example can be taken from the AAC15 wall in the cyclic regime in summer, and another from the AAC30 wall. For the 30 cm-thick wall, the hygrothermal parameters adopted – with the exception of the water diffusivity D_w – are the same as in the static regime. A value for the diffusivity D_w is chosen as equal to 7.10^{-9} m^2.s^{-1} – a value which is ten times greater than that in the static regime. This value coincides with the peak observed, for a value of RH between 40% and 60%, corresponding exactly to the zone of variation of the RH at point C during the summer cycle. The comparisons in temperature and relative humidity at point C, in Figure 6.31, are consistent.

Data	λ [W/(m.K)]	ρ [kg/m^3]	C [J/(kg.K)]	a [m^2/s]
AAC15	0.118	429.5	915.3	2.99×10^{-7}
	δ_p [kg.m^{-1}.s^{-1}.Pa^{-1}]	D_w [m^2.s^{-1}]	h_S [W.m^{-2}.K^{-1}]	β_P [kg.Pa^{-1}.m^{-2}.s^{-1}]
	5×10^{-11}	1.1×10^{-10}	8	4×10^{-8}

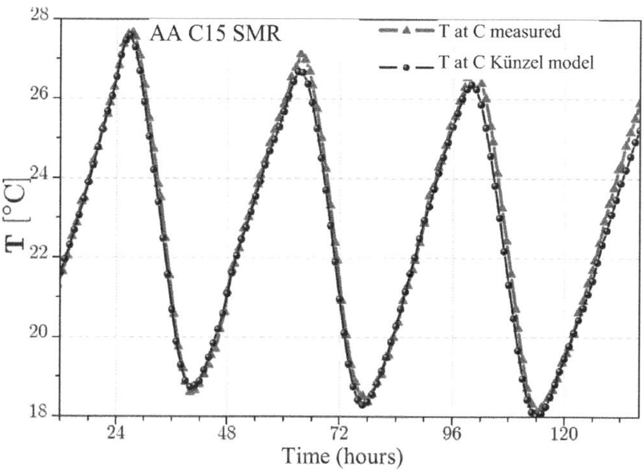

Figure 6.30. *Evolutions of T at point C – the simulated (Künzel model) and measured values, $h_S = 8$ W.m^{-2}.K^{-1} and $\beta_p = 4 \times 10^{-8}$ SI (summer cycle AAC15)*

The Künzel model can be used to obtain satisfactory results on the temperature cycles and stabilize the differences at the maximum values attained, present for the conductive model. With the 30 cm sample, the simulated results obtained in terms of the change in relative humidity at C are very similar to the measurement, thereby validating the model for samples of this thickness.

Data	λ [W/(m.K)]	ρ [kg/m^3]	c [J/(kg.K)]	a [m^2/s]
AAC30	0.118	429.5	915.3	2.99×10^{-7}
	δ_p [kg.m^{-1}.s^{-1}.Pa^{-1}]	D_w [m^2.s^{-1}]	h_S [W.m^{-2}.K^{-1}]	β_P [kg.Pa^{-1}.m^{-2}.s^{-1}]
	5×10^{-11}	7×10^{-9}	8	4×10^{-8}

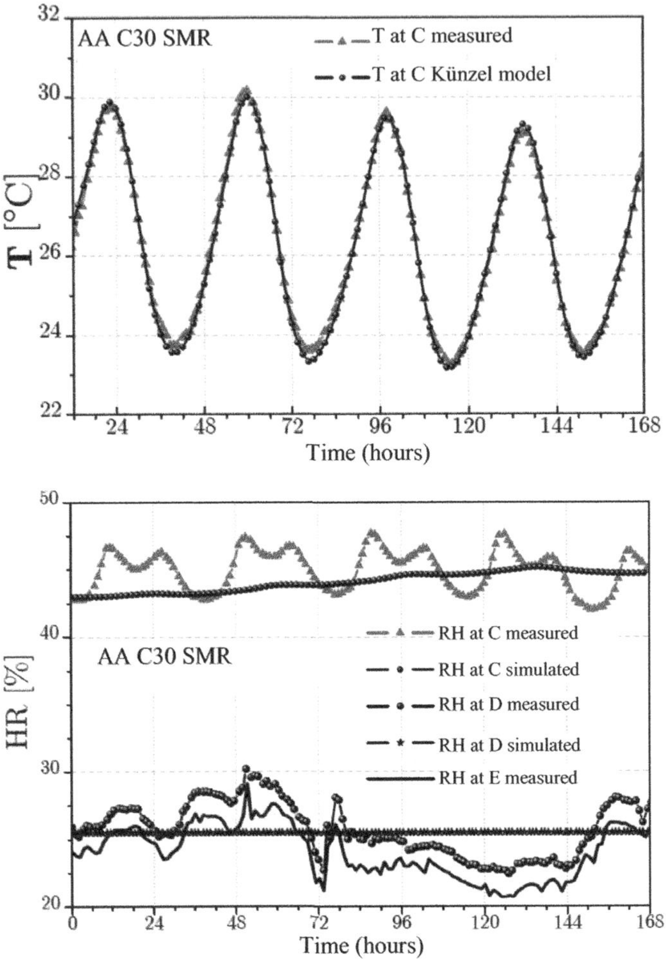

Figure 6.31. *Change in T at point C (top) and RH at points C and D (bottom), comparison of the measured and simulated values, $h_S = 8\ W.m^{-2}.K^{-1}$ and $\beta_p = 4\times10^{-8}$ SI (summer cycle AAC30)*

6.5.5.2. *Application to hempcrete walls*

The application of the model has been validated for aerated autoclaved concrete. We now turn our attention to a hempcrete wall (HLC2) 30 cm in thickness. The heat conduction model was not able to express the specificity of this material's thermal behavior, and in particular the influence of hygroscopic transfers.

When taking account of the influence of the mass water content w on the materials thermal parameters λ, ρ and c, we based our calculations on the models seen in section 6.2.2:

$$\lambda_{HLC2} = 0.082 + 0.15.u = 0.082 + 4.7 \times 10^{-4}.w$$

$$\rho_{HLC2}.c_{HLC2} = \rho_0.c_0 + w.c_w,$$

where $\rho_0 = 317$ kg.m^{-3}, $c_0 = 1000$ kJ.kg^{-1}.K^{-1} and $c_w = 4.18$ kJ.kg^{-1}.K^{-1}.

As before, to begin with, the transfer coefficients are deemed to be constant. The values of heat conductivity and heat capacity are determined on the basis of the average value of the RH at point C which, along with the adsorption isotherm, can be used to calculate the average water content of the material around point C. The change in the parameter ξ depending on the RH is introduced directly into the model, but this result is obtained only on the basis of the adsorption isotherm: the hysteresis is discounted. We do not have a value for the vapor permeability nor for the isothermal water diffusivity of the material. Hence, these values are obtained by adjusting for a static stress (here STAT1). These same values are then used for all the other stresses; the HLC2 wall is a case apart, because the average water content at points C is very different from one form of stress to another (there is a great change in the RH at C).

Static stresses: the repeatability of the tests for the STAT1 stress demonstrated a very clear plateau of stabilization of the temperature at C, for t between 48 and 72 hours, when the T at point A changes from 10°C to 40°C. There is a direct link between the change in T at point C and the change in RH at C, which increases by almost 20%, owing to significant vaporization (endothermic phenomenon). The average relative humidity at point C is 50% so, according to the adsorption isotherm, $w_{ave} = 20$ kg.m^{-3}. Yet the third stage exhibits a very particular dynamic. However, for t between 48 and 72 hours, the average value of RH is 65%: thus, we shall take $w_{ave} = 34.8$ kg.m^{-3}. Figure 6.32 presents a comparison between the changes in T at points C and D – both simulated and measured values – for the STAT1 test. The values of the adjusted coefficients δ_p and D_w are respectively equal to 8.10^{-10} kg.Pa^{-1}.m^{-1}.s^{-1} and 4.10^{-8} m^2.s^{-1}. Similarly, the convective heat exchange coefficient h_S and convective mass exchange coefficient β_p are equal to 15 W.m^{-2}.K^{-1} and 5.10^{-8} kg.m^{-2}.s^{-1}.Pa^{-1} respectively. Comparison with the heat conduction model shows that the Künzel model provides a response which corresponds more closely with the measured results at the two points of comparison C and D. In the third stage (t between 48 and 72 hours), we again see the rapid dynamic followed by the plateau of temperature stabilization. This plateau coincides with an increase in the relative humidity at C, which is reflected in the model. Noticeable differences appear, however, in the dynamic of RH at C. We can draw a direct link between this difference and the difference in temperature over the same period, between 60 and 72 hours. Two hypotheses can be put forward to explain these differences. To begin with, comparing the simulated and measured trends in RH at C, one might be led to think that a larger amount of water is vaporized, so more heat is absorbed and the temperature at point C is therefore lower, because the plateau begins earlier on. Secondly, the difference in temperature at C may suggest that we under-estimate the amount of heat required to vaporize the water in the hempcrete. Just like with wood, it may be helpful to add extra energy to the latent heat of vaporization of the liquid contained in the matrix (sorption heat). The correction to be made to the latent heat would be around 244 kJ/kg for a pore radius of 1 nm and 0.002 kJ/kg for a pore radius of 100 μm.

Thus, these results confirm that HLC2 hempcrete is very different from AAC, with transfer coefficients that are ten times greater. Bear in mind that these values of permeability to water vapor and water diffusivity must be considered as apparent values, because they directly include the influence of the relative humidity and temperature.

The next step is to verify the reliability of the model and the values for δ_p and D_w with other static stresses. Hence, Figure 6.33 presents the simulated and measured trends in temperature at point C, for QS SMR stress. The average relative humidity at point C is 45%, which corresponds to an average water content $w_{ave} = 19$ kg.m^{-3}. Earlier, we used the heat conduction model to highlight the impact of the mass transfers particularly on the value of the material's thermal diffusivity. With the Künzel model, coherent results are obtained for $\delta_p = 8.10^{-10}$ kg.Pa^{-1}.m^{-1}.s^{-1} and $D_w = 8.10^{-6}$ m^2.s^{-1}. The value of the material's vapor permeability is identical to that employed for STAT1.

Data	λ [W/(m.K)]	ρ [kg/m^3]	c [J/(kg.K)]	a [m^2/s]
HLC2 STAT1	0.099	352	1315	2.13×10^{-7}
	δ_p [kg.m^{-1}.s^{-1}.Pa^{-1}]	D_w [m^2.s^{-1}]	h_S [W.m^{-2}.K^{-1}]	β_P [kg.Pa^{-1}.m^{-2}.s^{-1}]
	8×10^{-10}	4×10^{-8}	15	5×10^{-8}

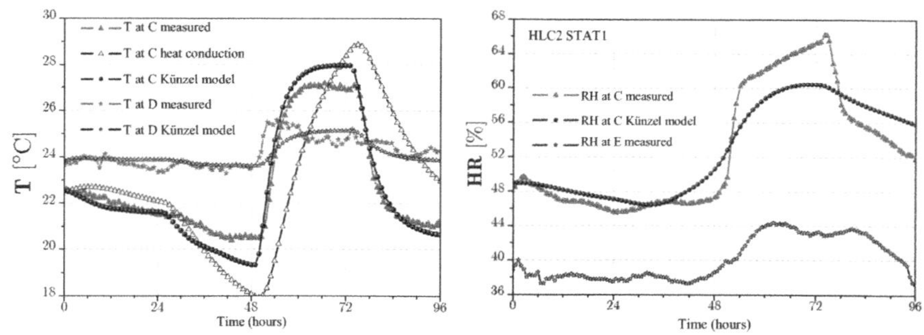

Figure 6.32. *Changes in T at points C and D, and in RH at point C – simulated (Künzel model) and measured values. $h_S = 15$ W.m^{-2}.K^{-1} and $\beta_p = 5 \times 10^{-8}$ SI (STAT1 HLC2)*

Data	λ [W/(m.K)]	ρ [kg/m^3]	c [J/(kg.K)]	$\boldsymbol{a}$ [m^2/s]
HLC2 QS SMR	0.091	336	1180	2.30×10^{-7}
	δ_p [kg.m^{-1}.s^{-1}.Pa^{-1}]	D_w [m^2.s^{-1}]	h_S [W.m^{-2}.K^{-1}]	β_P [kg.Pa^{-1}.m^{-2}.s^{-1}]
	8×10^{-10}	8×10^{-6}	15	5×10^{-8}

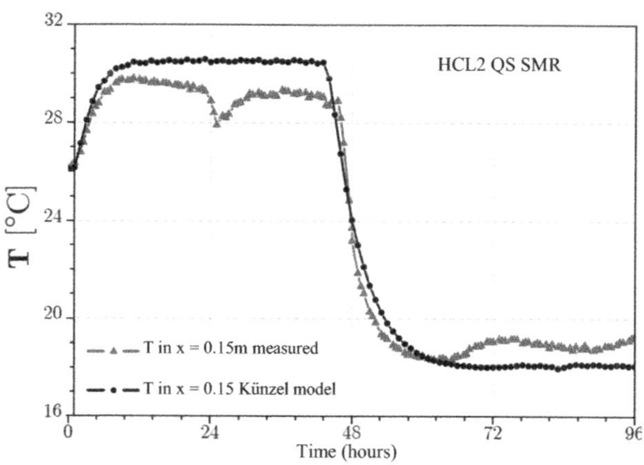

Figure 6.33. *Evolution of T at point C – experimental value versus simulated value (Künzel model) $h_S = 15$ W.m^{-2}.K^{-1} and $\beta_p = 5 \times 10^{-8}$ SI (QS SMR HLC2)*

Cyclic stresses: we now consider a typical summer stress. The average water content in the material is $w_{ave} = 19$ kg.m^{-3} during the summer cycle. We use *the same values for the parameters λ, ρ, c, δ_p and D_w*. The only difference lies in the adjusted value of the convective mass coefficient β_p. The value adopted is ten times lesser than that used in the static regime. The numerical values of the parameters are indicated in the table above Figure 6.34. Thus, we superpose the evolutions of T at point C and those of T at point D – both measured and simulated – over the three periods of the summer cycle.

The result is satisfactory over the first two periods, from 0 to 72 hours, but beyond this point significant differences begin to appear at points C and D. The hypothesis of constant mass transfer coefficients is no longer enough to account for the hygrothermal performance of the hempcrete wall under cyclic stress conditions. The dynamic regime does indeed present a more complex problem, because the RH at point C is time-dependent. However, the phenomenon of hysteresis is not taken into account here. Thus, the water content of the material changes successively with the adsorption or desorption isotherm. As the differences between the two curves are significant beyond a value of 50% RH, failing to take account of hysteresis could lead to considerable errors.

In addition, the peculiarity of hempcrete is that it has two components – the hemp shiv and the binder – which have different hydric transfer properties.

Data	λ [W/(m.K)]	ρ [kg/m^3]	c [J/(kg.K)]	a [m^2/s]
HLC2 SMR	0.091	336	1180	2.3×10^{-7}
	δ_p [kg.m^{-1}.s^{-1}.Pa^{-1}]	D_w [m^2.s^{-1}]	h_S [W.m^{-2}.K^{-1}]	β_P [kg.Pa^{-1}.m^{-2}.s^{-1}]
	8×10^{-11}	4×10^{-7}	5	2×10^{-8}

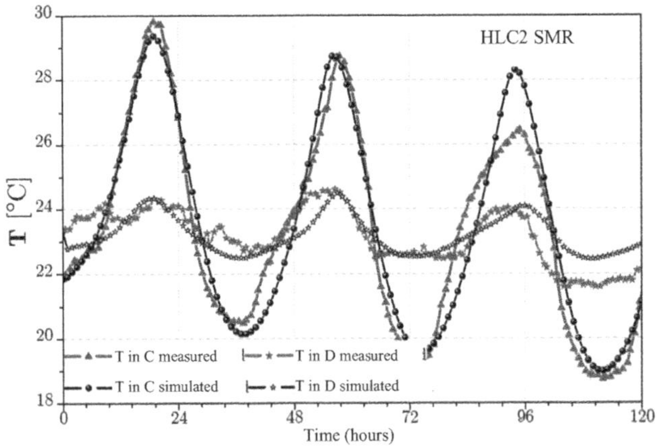

Figure 6.34. *Measured versus simulated (Künzel model) changes in T at points C and D, h_S = 15 W.m^{-2}.K^{-1} and β_p = 5 × 10^{-9} SI (SMR HLC2)*

These comparisons prove that the hygrothermal performance of the hempcrete wall is a complex phenomenon. However, this initial approach has the merit of clearly demonstrating the peculiarity of the material's behavior, characterized by high values of permeability and water diffusivity. Similarly, the value of the convective heat exchange coefficient is high, expressing the fact that the ambient atmosphere in the laboratory has a not-insignificant influence. The porosity of the surface of the hemp material plays an important part. Hence, it is crucial to examine the role of surface coating products.

6.5.5.3. *Coated hempcrete wall*

We now apply the Künzel model to a coated hempcrete (HLC2) wall. Remember that the heat conduction model yielded accurate results, showing that the influence of mass transfers on heat transfer is limited.

The boundary conditions are the same as for the HLC2 wall. To these are added the conditions at the interfaces between HLC2 and the two layers of coating, for which continuity of temperature and relative humidity is assumed at points B1 and D1. Hence, the configuration from Figure 6.29 is modified as shown in Figure 6.35.

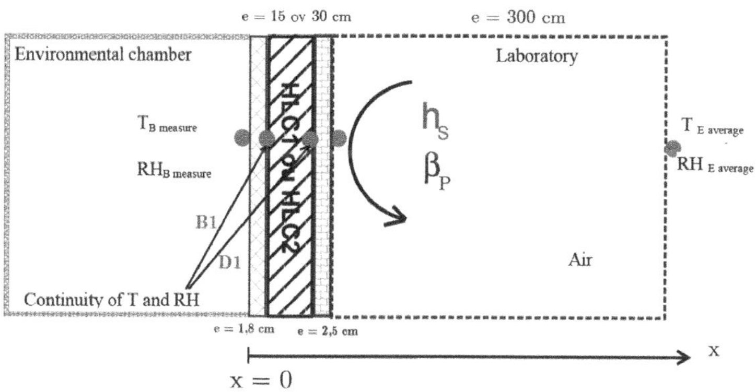

Figure 6.35. *Setup used for the application of the combined heat and mass model to the coated HLC2E wall*

For the HLC2 material, we assume the mass transfer coefficients δ_p and D_w to be constant for an average value of relative humidity at point C. The values of the thermal parameters λ, ρ and c for the interior and exterior coating products are those given in section 6.4.3. The values of δ_p and D_w for the hemp/lime coating are taken from [COL 04]. As regards the interior coating, the values of the coefficients will be validated by comparison between the simulated and measured changes in RH at point B1. The hygrothermal data for the coating products are given in Table 6.8.

Mass transfer coefficients	ξ	δ_p	D_w	a
Exterior coating (sand/lime)	10	4×10^{-11}	5×10^{-10}	1.1×10^{-7}
Interior coating (hemp/lime)	30	1×10^{-11}	8×10^{-9}	2.5×10^{-7}

Table 6.8. *Values of the transfer coefficients for the interior and exterior coating products*

Static stress: the model is applied for STAT1 stress. The average relative humidity at point C is equal to 53%, meaning the average water content in the HCL2 material of $w_{ave} = 20$ kg.m^{-3}. This average value is very close to that calculated for the HLC2 wall under QS SMR stress (Figure 6.33). Hence, the same values are used for λ, ρ and c. As regards the mass transfer coefficients δ_p and D_w, they are adjusted from this first stress so as best to correspond to the measured evolution of T at point C.

The results obtained are presented in Figure 6.36. We have chosen to keep a value of the isothermal water diffusivity in the HLC2 material identical to that of the uncoated wall – i.e. $D_w = 4 \times 10^{-7}$ m^2.s^{-1}. For the permeability to water vapor, the simulated temperature changes at point C are given for two values: the value adjusted for the HLC2 wall – $\delta_p = 8 \times 10^{-10}$ kg.m^{-1}.s^{-1}.Pa^{-1} – and the same value divided by 10. The overall consistency of the results is obtained for $\delta_{p\,HLC2} = 8.10^{-11}$ kg.Pa^{-1}.m^{-1}.s^{-1}. There is still a slight difference of around 1 hour between the simulated and measured values, which can be clearly seen during the last temperature cycle (t between 72 and 96 hours).

The impact of the presence of the coating products is thus primarily to be seen in the coefficient of permeability to water vapor of the HLC2 material. An identical value to that used for the HLC2 wall ($\delta_{p\ HLC2} = 8 \times 10^{-10}$ kg.Pa^{-1}.m^{-1}.s^{-1}) leads to significant differences – particularly between 48 and 72 hours – with a swifter response in terms of temperature. It is this parameter which has the greatest impact on heat transfer in the material. A decrease in the value of δ_p demonstrates frustrated migration of water vapor through the HLC2 material, and therefore the phase-change in the material is no longer of consequence. Thus, the layers of coating play the role of a buffer for the mass transfer.

This is confirmed by the simulated changes in RH at points B1 and C deduced from the (see Figure 6.36). They are similar to the measured changes, with the exception of the slight jump of around 5% recorded at C, between 48 and 72 hours. We may ask ourselves whether this difference is behind the difference in T at point C noted for the same period.

Note, finally, that the adjusted convective heat and mass exchange coefficients h_S and β_p are respectively 5 W.m^{-2}.K^{-1} and 2×10^{-8} kg.m^{-2}.s^{-1}.Pa^{-1}. In comparison with the uncoated HLC2 wall, the presence of the hemp/lime coating in direct contact with the ambient atmosphere in the laboratory alters the value of h_S, decreasing it threefold. This again highlights the fact that the heat exchange coefficient h_S depends greatly on the porosity of the surface of the material in contact with the ambient atmosphere in the laboratory.

Data	λ [W/(m.K)]	ρ [kg/m^3]	c [J/(kg.K)]	a [m^2/s]
HLC2 STAT1 in HLC2E	0.091	336	1180	2.3×10^{-7}
	δ_p [kg.m^{-1}.s^{-1}.Pa^{-1}]	D_w [m^2.s^{-1}]	h_s [W.m^{-2}.K^{-1}]	β_P [kg.Pa^{-1}.m^{-2}.s^{-1}]
	8×10^{-11}	4×10^{-7}	5	2×10^{-8}

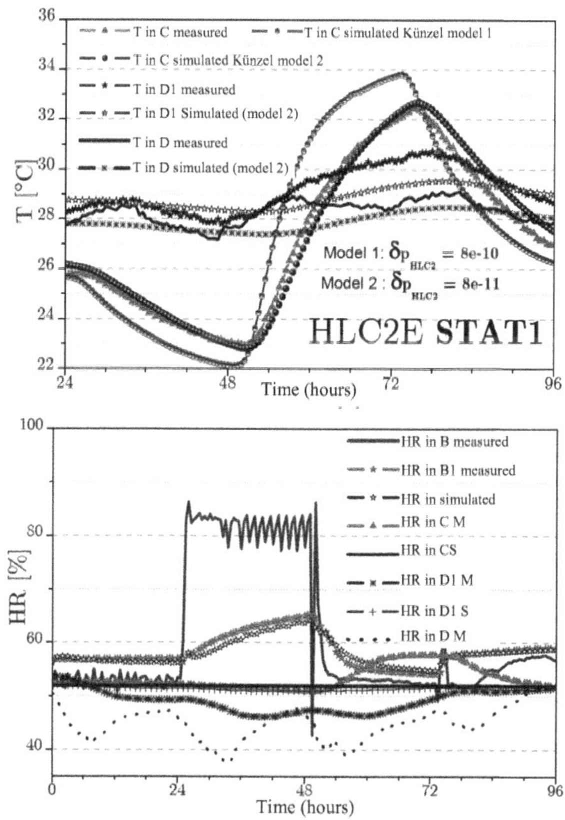

Figure 6.36. *Simulated (Künzel model) versus measured changes in T at points C, D1 and D (top) and in RH at points B1, C, and D1 (bottom). $h_S = 5$ W.m^{-2}.K^{-1} and $\beta_p = 2 \times 10^{-8}$ SI (STAT1 HLC2E)*

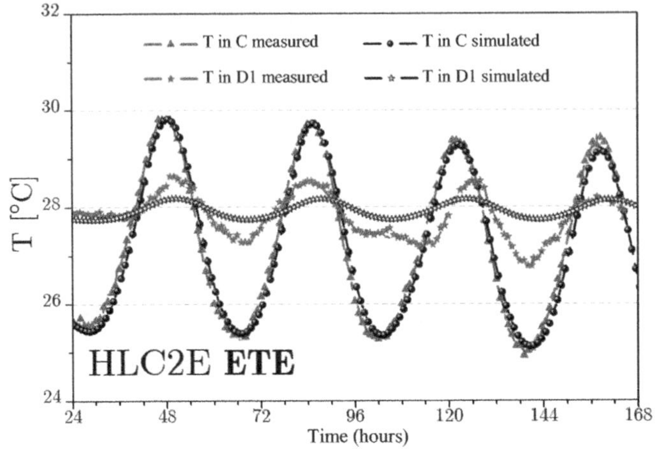

Figure 6.37. *Simulated (Künzel model) versus measured changes in T at points C and D1 $h_S = 5$ W.m^{-2}.K^{-1} and $\beta_p = 2.10^{-8}$ SI (SMR HLC2E)*

Cyclic stresses: another comparison is performed, with a cyclic stress in summer. The average value of RH at C, during the summer cycle, is 53%, just like for static stresses STAT1 and STAT2. The application of the Künzel model is therefore done with the same values for the hygrothermal parameters. Figure 6.37 demonstrates that the results correspond fairly closely to the measured performance. The adjustment of the simulations with a coefficient δ_p, ten times reduced in relation to the HLC2 alone, gives a good account of the attenuation of the transfers by evaporation and condensation. The capacity of the coated wall to deal with the fluctuations of RH, therefore, is greatly improved.

Whether in the static or cyclic regime, the layers of coating act as hydric filters in regard to mass transfers, which therefore alters the response of the hemp-based material. The key parameter is the permeability to vapor, which exerts the greatest influence on the heat transfer.

A slight phase-shift is observed between the simulated and measured responses at point C (Figure 6.36). This could cast doubt on the hypothesis of steady RH at the interfaces B1 and D1. Indeed, the values of the hydric transfer coefficients in the interior hemp/lime coating and the HLC2 hempcrete are too far removed from one another to be able to consider a perfect hydraulic contact (continuity of the capillary suction). We can therefore put forward an improvement to the simulations: modifying the setup by introducing a thin layer of air at the D1 interface, which is tantamount to introducing a hydric and thermal resistance.

6.6. Conclusions

The building of dwellings or renovation work including hemp-based materials often draw initial comments from users that are very positive in regard to the thermal and hygrothermal conditions. The aim of this chapter was to objectively analyze and characterize the hygrothermal performance of construction materials including hempcretes, subjected to conditions of temperature and relative humidity that are similar to the real conditions of use.

Three materials were tested for the purposes of this investigation: aerated autoclaved concrete (AAC), hemp/lime concrete (HLC) and vertically perforated bricks (VPB). The particular reasons behind this choice are the similar physical characteristics of these materials – their high porosity (between 70% and 80%) – meaning they all have excellent insulating properties but of different kinds. In the current context, where improving the energy efficiency of buildings is a matter of priority, the study carried out on these enveloping materials is a fundamental step. In particular, for hempcrete, the studies reported here are crucial in order to adapt the legislation and professional regulations to this new construction material.

This in-depth study presents the physical, thermal and hydric properties of these materials – in particular specifying the apparent densities, porosities and adsorption/desorption isotherms. Significant gains in terms of mass are observed for hemp-based materials in particular with an RH of over 50%. Finally, the thermal values of these dry materials are specified. Self-consistent homogenization is employed in order to establish the change in the heat conductivity as a function of the mass water content u.

The materials were then tested using an innovative device, which is able to measure – in order to characterize and evaluate, and therefore better understand and model the thermal and hydric performances of the construction materials subjected to temperature and relative humidity gradients. Different regimes of stress in terms of T and RH were applied at the surface of the walls of the materials in contact with an environmental chamber, in static and cyclic regimes. This approach corresponds to the "ideal" performance of a wall in a building, where priority is given to the "material" performance.

The main results obtained are as follows:

– the analysis of the reproducibility of the results within the wall and at its surface revealed notable differences in behavior between the materials;

– for the AAC, VPB, HLC1E and HLC2E walls, good consistency of the results is obtained, with the "balance" between the injection of heat and the warming of the material being maintained. Thus, conductive heat transfer proves to be significant in these materials;

– conversely, for walls made of HLC1 and HLC2 hempcrete, the results prove the existence of an internal source (or well) of heat and/or the presence of another flow of heat (notably by convection), directly interfering in the heat balance equation;

– the changes of T and RH within the wall are consistent if we consider it to be a question of latent heat of phase-change. As the enthalpy of phase-change is transported by diffusion of water vapor, mass transfers have a significant impact on the heat transfer in these materials;

– this analysis is supported by experiments, which led us to install a watertight membrane between the hemp-based material and the surround air so as to limit hygroscopic transfers.

Next, a phase of modeling was presented. Digital modeling of the transfers was performed by the software COMSOL Multiphysics in two distinct phases.

Firstly, the comparison of the measured results with the pure heat conduction model confirms the previous analyses. This comparison enables us to provide precise responses regarding the thermal characterization of walls made of AAC, VPB, HLC1E and HLC2E. The values of the thermal parameters λ, ρ, c, h_S and $\boldsymbol{a}$ of the materials are obtained and validated by way of an innovative approach by inverse analysis performed on a simple stress. Tests on the reliability and robustness of the model with these thermophysical values then demonstrated the applicability of the approach to other stresses in both static and cyclic regimes. This first result is interesting because such measurements of thermophysical values are scant in the existing body of literature, and they present the advantage of having been obtained in a laboratory environment on representative samples and for operational ranges close to real functioning conditions.

A second important result relates to the hemp-based materials HLC1 and HLC2. The purely conductive model proves inapt to characterize the thermal behavior of these materials. Mass transfers – particularly phase-changes by evaporation/condensation – have a significant impact on heat transfer in these materials.

Finally, for coated walls made of HLC1E and HLC2E hempcrete, it was possible to adjust values of λ, ρ, c and h_S for all three materials. The simulated results, which fit the experimental results quite closely, confirm that the thermal behavior of HLC1E and HLC2E is different to that of the walls alone. The coating products tested act as "thermal and hydric filters", which may limit mass transfers, which has a direct impact on the amount of latent heat involved. This result demonstrates that it would be helpful to design the enveloping and coating materials together.

The observation made about the hemp-based materials HLC1 and HLC2 therefore led us to model the combined transfers of heat and mass in order to integrate – among other factors – the influence of phase-changing (vaporization or condensation). The Künzel model serves as a reference to describe the coupled transfer laws on the macroscopic scale. This model is adapted in the software suite COMSOL Multiphysics. In particular, the model includes the hydric transfer coefficients – i.e. the permeability to water vapor and water diffusivity of the material. While the thermal values are calculated for an average water content of the material (determined on the basis of the average value of RH at the center of the walls), we take the transfer coefficients to be constant.

The comparison between the simulated and measured values with static stresses yields crucial information. Indeed, the appropriate consistency of the results is obtained by calibrating the values of water diffusivity D_w and apparent permeability to water vapor δ_p. These values, which are difficult to measure, are once again obtained by inverse analysis. The consideration of the mass transfers and phase change, made possible by this model, is crucial in order to model the thermal behavior of hemp-based material. In addition, the high values of D_w and δ_p appear to highlight the existence of a driving gradient other than that of the partial vapor pressure, which greatly influences the vapor diffusion.

In the cyclic regime, the results confirm the appropriateness of the model and the approach. However, tests demonstrate that there are limitations inherent to the use of constant transfer coefficients. The water content of the material changes successively in accordance with the adsorption or desorption isotherm. In view of the differences between the two isotherms with values of RH greater than 50%, it seems unwise to continue to discount hysteresis.

Finally, for a wall coated with HLC2E, good consistency between the simulated and measured results is obtained for a smaller value of the coefficient of permeability to water vapor for HLC2, in relation to the wall alone. The filtering effect of the layers of coating is therefore directly influential on the value of δ_p, which appears to be the parameter which has the greatest influence on heat transfer.

In conclusion, it appears that hemp-based materials naturally behave as phase-change materials (PCM): the thermal behavior reduces the amplitude of the variations in the ambient air temperature, whilst improving the thermal comfort by bringing down the surface heat of the material. Thus, the use of such materials is an excellent means of passively regulating the indoor temperature, and thereby decreasing the building's energy requirements. In addition, the experimental results show a regulation of the relative humidity in the envelope because of constant exchanges of water vapor between the indoor and outdoor environments, modulating sudden changes in temperature. Hence, these materials are able to improve summer and winter comfort, and stabilize the indoor temperature between day and night, whilst preventing the phenomena of condensation and dampness on the walls.

Work is currently being carried out with a view to integrating this model of hygrothermal behavior into the computation software used to predict the energy performances of buildings. It should be stressed that such projections are required by the thermal regulations currently in force (in France, RT 2005 and now RT 2012). It is therefore urgent to react because to date, none of these software tools takes account of the genuine dynamic hygrothermal behavior of the building materials and, as demonstrated in this chapter, this leads to a significant under-estimation of the actual performances of hemp-based materials. Therefore, these under-appreciated, high-performance solutions are often disregarded – and wrongly so!

6.7. Bibliography

[ARN 93] ARNAUD L., Pâte crue de béton cellulaire, approches théorique et expérimentale : l'homogénéisation - le vibroscope, thesis, Mécanique ECL, Paris : Ecole Centrale Paris, 280 p., 1993.

[ARN 00a] ARNAUD L., "Mechanical and thermal properties of hemp mortars and wools: experimental and theoretical approaches", *Third International Bioresource Hemp 2000 Symposium*, Wolfsburg, Germany, September 2000.

[ARN 00b] ARNAUD L., MONNET H., CORDIER C., SALLET F., "Modélisation par homogénéisation autocohérente de la conductivité thermique de bétons et laines de chanvre", *Congrès SFT 2000*, Lyon, France, May 2000.

[ARN 01] ARNAUD L., CÉRÉZO V., Qualification physique des matériaux de construction à base de chanvre. Final research report CNRS 0711462, Lyon: ENTPE–DGCB–URA, 91 p. (+ appendices), 2001.

[ARN 05] ARNAUD L., SAMRI D., "Hygro-thermal behaviour of building porous materials", *3rd Biot Conference on Poromechanics*, Norman, Oklahoma, United States, p. 761–766, May 2005.

[ARN 06a] ARNAUD L., SAMRI D., "Réponse scientifique et technique des matériaux chanvre", *3ème Assises de la construction du chanvre*, Paris, France, September 2006.

[ARN 06b] ARNAUD L., SAMRI D., ELAQRA H., Amélioration des performances mécaniques du béton de chanvre en vue de la préfabrication de briques industrielles, Contrat de recherche N° CNRS 04 W 453/50.0516, Programme EURËKA, Hemp Lime Blocks, April 2006.

[AZI 88] AZIZI S., MOYNE C., DEGIOVANNI A., "Approche expérimentale et théorique de la conductivité thermique des milieux humides I-Expérimentation", *International Journal of Heat and Mass Transfer,* no. 31, p. 2305–2317, 1988.

[BEL 92] BELLINI DA CUNHA NETO J.A., Transport d'humidité en matériau poreux en présence d'un gradient de température - Caractérisation expérimentale d'un béton cellulaire, Doctoral thesis, University of Grenoble 1, 1992.

[BOU 96] BOUTIN C., "Conductivité thermique du béton cellulaire autoclavé : modélisation par méthode auto-cohérente", *Matériaux et Constructions*, no. 29, p. 609–615, 1996.

[BRU 38] BRUNAUER S., EMMETT P.H., TELLER E., "Adsorption of gases in multimolecular layers", *Journal of the American Chemical Society*, no. 60, p. 309–319, 1938.

[CER 05] CÉRÉZO V., Propriétés mécanique, thermique et acoustique d'un matériau à base de particules végétales : approche expérimentale et modélisation théorique, Thesis, MEGA-ENTPE, Lyon: INSA de Lyon, 16 June 2005.

[COL 04] COLLET F., Caractérisation hydrique et thermique de matériaux de génie civil à faibles impacts environnementaux, Doctoral thesis, Rennes: INSA Rennes, 14 December 2004.

[COM 06] Comsol Multiphysics Modeling Guide Version 3.2, Reference Manual, 348 p.

[COR 99] CORDIER C., Caractérisation thermique et mécanique des bétons de chanvre, Rapport de stage de fin d'études, Vaulx-en-Velin : ENTPE, 56 p., 1999.

[CRA 83] CRAUSSE P., Etude fondamentale des transferts couplés de chaleur et d'humidité en milieu poreux non saturé, Doctoral thesis, Toulouse : Institut National Polytechnique de Toulouse, 209 p., 1983.

[CRA 96] CRAUSSE P., LAURENT J.-P., "Influence des phénomènes d'hystérésis sur les propriétés hydriques des matériaux poreux", *Revue générale de Thermique,* no. 35, p. 95–106, 1996.

[DAI 86] DAÏAN J.-F., Processus de condensation et de transfert d'eau dans un matériau méso et macroporeux - Etude expérimentale du mortier de ciment, Doctoral thesis, Université scientifique, Technologique et Médicale de Grenoble, 319 p., 1986.

[DAI 92] DAÏAN J.-F., BELLINI DA CUNHA J.A., "Experimental determination of AAC moisture coefficients under temperature gradients", *Advances in Autoclaved Aerated Concrete*, p. 105–111, 1992.

[DEF 94] DE FREITAS V.P., ABRANTES V., CRAUSSE P., "Moisture migration in building walls - analysis of the interface phenomena", *Building and Environment*, no. 31, p. 99–108, 1994.

[DEV 58] DE VRIES D.A., "Simultaneous transfer of heat and moisture in porous media", *Transaction of the American Geophysical Union*, no. 39, p. 909–916, 1958.

[DEV 87] DE VRIES D.A., "The theory of heat and moisture transfer in porous media revisited", *International Journal of Heat and Mass Transfer*, no. 30, p. 1343–1350, 1987.

[DWI 94] DWI ARGO B., Détermination expérimentale de l'influence de l'hystérésis sur les propriétés hydriques des matériaux du Génie Civil, Doctoral thesis, INSA de Toulouse, 1994.

[EN-ISO 01] European standard EN ISO 12572, Hygrothermal performance of building materials and products – determination of water vapour transmission properties, European Committee for Standardization, 2001.

[EVR 06] EVRARD A., "Sorption behaviour of Lime-Hemp Concrete and its relation to indoor comfort and energy demand", *23rd Conference on Passive and Low Energy Architecture*, Geneva, Switzerland, September 2006.

[EVR 08] EVRARD A., Transient hygrothermal behavior of Lime-Hemp Materials, Doctoral thesis, Université Catholique de Louvain, May 2008.

[FIC 55] FICK A., "Über diffusion", *Annalen der Physik und Chemie*, no. 94, p. 59–86, 1855.

[GAL 98] GALBRAITH G.H., MCLEAN R.C., GILLESPIE I., GUO J.S. AND KELLY D., "Nonisothermal moisture diffusion in porous building materials", *Building Research and Information*, no. 26, p. 330–339, 1998.

[GAL 99] GALBRAITH G.H., GUO J.S., MCLEAN R.C., LEE C.K. AND KELLY D.J., "The temperature dependence of moisture permeability", *CIB W40 Meeting*, Prague, Czech Republic, August/September 1999.

[GAR 00] GARNIER P., Le séchage des matériaux poreux - Approche expérimentale et approche théorique par homogénéisation des structures périodiques, Rapport de DEA de Génie Civil, Vaulx-en-Velin : ENTPE, 100 p., 2000.

[GOU 09] GOURLAY E., De la chènevotte au béton de chanvre, optimisation des propriétés mécaniques et hygrothermiques, Rapport de Master Recherche, Vaulx-en-Velin : ENTPE, 66 p., 2009.

[GOU 10] GOURLAY E., ARNAUD L., "Comportement hygrothermique des murs de béton de chanvre", *Congrès SFT 2010*, Le Touquet, France, May 2010.

[GOU 11a] GOURLAY E., GLÉ P., ARNAUD L., GOURDON E., "Propriétés multiphysiques des bétons de chanvre", *Matériaux & Techniques*, no. 99, p. 625–631, 2011.

[GOU 11b] GOURLAY E., ARNAUD L., "Impact of raw materials on the mechanical and thermal properties of hemp concretes", *International Conference on Advances in Construction Materials through Science and Engineering*, Hong Kong, China, September 2011.

[GOU 12] GOURLAY E., ARNAUD L., "Optimisation des propriétés mécaniques et thermiques des bétons de chanvre - Impact des matières premières", *Annales du Bâtiment et des Travaux Publics*, no. 1–2, p. 58–64, March 2012.

[GOY 07] GOYER S., Caractérisation des propriétés microstructurales et hydriques de bétons légers - Application à la modélisation des transferts hydriques, Research Masters Dissertation, Vaulx-en-Velin : ENTPE, 110 p., 2007.

[HAS 68] HASHIN Z., "Assessment of the Self Consistent Scheme Approximation: Conductivity of Particulate Composites", *J. Composite Materials*, no. 2, p. 284–300, 1968.

[HOL 02] HOLM A.H., KÜNZEL H.M., "Practical application of an uncertainty approach for hygrothermal building simulations – drying of an AAC flat roof", *Building and Environment*, no. 37, p. 883–889, 2002.

[KRU 96] KRUS M., Moisture transport and storage coefficients of porous mineral building materials – Theoretical principles and new test methods, Stuttgart: Fraunhofer IRB Verlag, 1996.

[KÜN 94] KÜNZEL H.M., Verfahren zur ein- und zweidimensionalen Berechnung des gekoppelten Wärme- und Feuchtetransports in Bauteilen mit einfachen Kennwerten (English title: One and two-dimensional calculation of the simultaneous heat and moisture transport in building components, using simple parameters), Doctoral thesis, Fraunhofer Institute for Building Physics, University of Stuttgart, Germany, 1994.

[LAC 03] LACARRIÈRE B., Etude numérique et expérimentale des échanges thermiques en cavités partitionnée : application aux transferts instationnaires dans les briques à perforations verticales, Doctoral thesis, Paul Sabatier University, Toulouse: 140 p., 2003.

[PAD 99] PADFIELD T., The role of absorbent building materials in moderating changes of relative humidity, Doctoral thesis, Department of Structural Engineering and Materials, Technical University of Denmark, Denmark, 1999.

[PEU 03] PEUHKURI R., Moisture dynamics in building envelopes, Doctoral thesis, Department of Civil Engineering, Technical University of Denmark, Denmark, 2003.

[PHI 57] PHILIP J.R., DE VRIES D.A., "Moisture movement in porous material under temperature gradients", *Transaction of The American Geophysical Union*, no. 38, p. 222–232, 1957.

[RES 00] RESTAGNO F., Interactions entre contacts solides et cinétique de la condensation capillaire - Aspects macroscopiques et aspects microscopiques, Doctoral thesis, École Normale Supérieure, Lyon: 151 p., 2000.

[SAM 04a] SAMRI D., Caractérisation hygrothermophysique des matériaux de construction, Rapport de stage de fin d'études, Vaulx-en-Velin: ENTPE, 57 p., 2004.

[SAM 04b] SAMRI D., Contribution au comportement hygrothermique des matériaux de construction, Rapport de DEA, Vaulx-en-Velin: ENTPE, 85 p., 2004.

[SAM 06] SAMRI D., ARNAUD L., "Assessment of heat and mass transfers in building porous materials", *4th European Conference on Energy Performance & Indoor Climate in Buildings*, Lyon, France, November 2006.

[SAM 08] SAMRI D., Analyse physique et caractérisation hygrothermique des matériaux de construction : approche expérimentale et modélisation numérique, Doctoral thesis, Ecole doctorale MEGA, Lyon: 284 p., 2008

Chapter 7

Acoustical Properties of Hemp Concretes

7.1. Introduction

Over the past few years, hemp concrete has become a referential material in the range of environmentally friendly materials used in construction. Made from plant particles, it is characterized by very advantageous properties in terms of its hygrothermic and mechanical performances. Now, these properties are relatively well known, and their function has been modeled [ARN 01; CER 05; SAM 08; ARN 12].

The aim of this chapter is to discuss the properties of this material from an acoustical point of view. The acoustical properties of hemp concretes were first studied by Cérezo [CER 05], who highlighted the extremely interesting performances of this material in terms of absorption on the one hand, and on the other hand function which is complicated to model, given the multiple scales of porosity in the material. Then, new studies were performed, looking into the effect of the various parameters of manufacture of the material on its acoustical properties [GLE 11], and finally, the acoustical behavior of this material was able to be explained by way of a multi-scale approach [GLE 12].

The use of hemp concrete for acoustical purposes is particularly interesting, in that multiple parameters are involved: firstly at the level of the choice of components from a range of hemp particles with different particle size distributions (PSD) and microstructures, and from different types of binders; and secondly at the level of the manufacturing method, with choices to be made between a variety of

---

Chapter written by Philippe GLÉ, Emmanuel GOURDON and Laurent ARNAUD.

formulations and casting techniques. Because they alter the final porosity and microstructure of this composite material, these parameters can be adjusted in order to control the physical performances of the concretes. In particular, the porosity is of crucial importance in the physical processes of sound insulation and absorption. This prior control of the performances in terms of absorption and insulation may facilitate an approach of "made-to-measure material" with regard to the noise issues encountered in situ.

Finally, given that nowadays, the model of hemp concrete and of hemp particles ("shiv") is familiar [GLE 12], it offers the possibility, by inversion on the basis of the measured acoustical properties, to characterize the primary material (the shiv). This possibility is particularly interesting in view of the fact that currently, there are few techniques available to determine the characteristics of hemp aggregates, such as their dimensions, their apparent density or their porosity. The advantages to an acoustic-based characterization method are many: firstly, the acoustical measurement is taken on the basis of a representative volume, which may not necessarily be so with a micro-tomography, and also, the measurement is non-destructive, unlike the measurement of porosity, obtained by a mercury intrusion.

Thus, this chapter is organized into three parts. The first aims to give as broad as possible overview of the mechanisms of control of the acoustical properties of hemp concretes. Secondly, the technique for modeling this material is presented and illustrated by means of a number of examples. In the third and final part, it is explained how, on the basis of acoustical measurements taken from hemp shiv, it is possible to characterize this primary material very precisely

7.2. Acoustical properties of the material on the basis of the main mechanisms

In this section, the acoustical properties of hemp concretes are discussed on the basis of various parameters such as the characteristics of the components and the mode of casting of the material. These results were first presented in [GLE 11].

7.2.1. *Influence of the components*

7.2.1.1. *Particle size distribution*

The PSD is one of the key characteristics of hemp particles. Its effect has been characterized in our published research, both on the acoustical properties of shiv alone, and on the acoustical properties of hemp concretes.

7.2.1.1.1. Effect on the acoustic properties of shiv

Measurements regarding loose hemp particles were made using a B&K type 4106 impedance tube (10 cm in diameter), in a vertical position. The particles were poured into the tube, shaken and slightly compacted if need be, in order to obtain the different desired densities over a thickness of 10 cm. This was repeated three times in order to ensure the reproducibility of the experiment, and to test the influence of the preparation of the samples (influence of the degree of precision on the height and state of surface of the sample, possible differences in positions and directionality of the particles). The three PSDs shown in Figure 7.1 are compared; their average dimensions are given in Table 7.1.

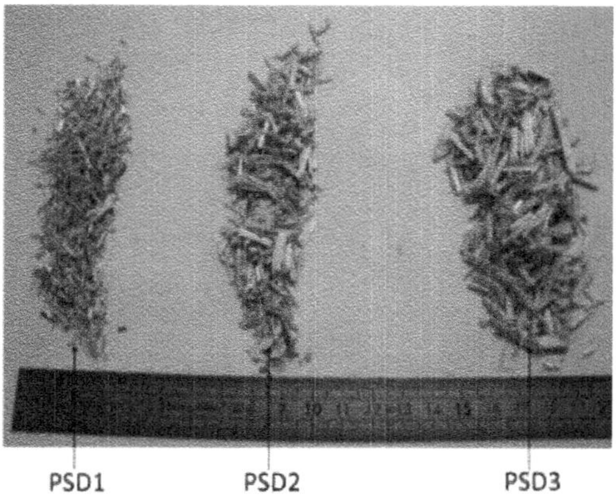

Figure 7.1. *Photo of the particle size distributions*

Shiv	PSD1	PSD2	PSD3
Average length (mm)	9	8	4
Average width (mm)	2.5	2.5	1
Average thickness (mm)	0.5	0.5	0.5

Table 7.1. *Particle size distributions of the types of hemp shiv tested*

The results are presented in Figure 7.2 for a shiv whose apparent density is 102 kg.m^{-3}. To begin with, these results show that the reproducibility is particularly good, and that the preparation of the samples has little bearing at all on the final acoustical properties. In addition, the influence of the PSD is shown to be great. The acoustical properties of hemp shivs PSD1 and PSD2 are very similar, which is unsurprising as they exhibit similar PSDs. However, when it comes to PSD3 shiv, comprising smaller particles, the sound absorption is better, particularly for low frequencies at the level of the first absorption peak. This noticeable difference can be explained by the fact that with small particles, the inter-particle pores are smaller, and therefore, the air-flow resistivity is greater.

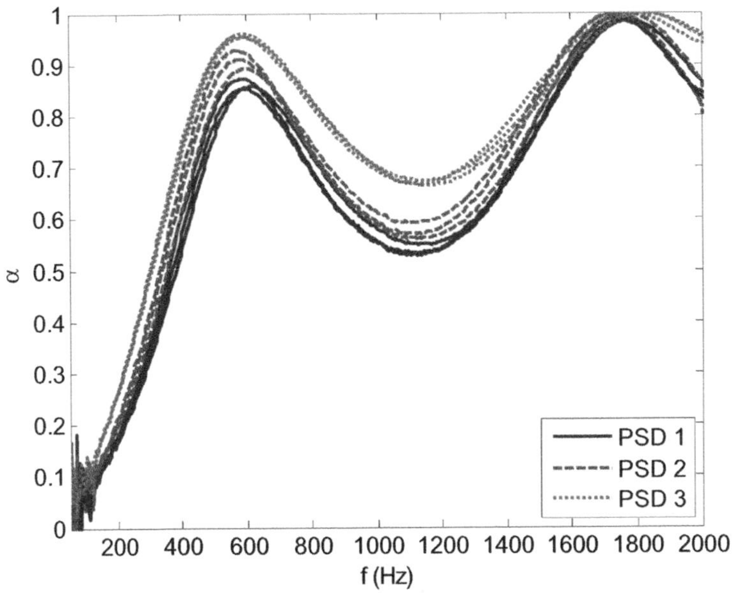

Figure 7.2. *Sound absorption of shiv depending on the PSD – low density*

7.2.1.1.2. Effect on the acoustical properties of the hemp concretes

The effect of the PSD is less marked when the samples of hemp concretes are tested. For instance, Figure 7.3 illustrates the sound absorptions of hemp concretes made with hydraulic lime with a walling formula (a binder-to-shiv mass ratio of two). These measurements were taken with a large Kundt's tube, 5.5 m long with 60×60 cm^2 cross-section; the samples are 10 cm in thickness. The absorptions measured are so similar that it is impossible to distinguish any effect at all of the differing PSDs.

The effect of the PSD is minor, and – as we shall see shortly – can be discounted in comparison to the other parameters of fabrication, which prove to be far more influential.

7.2.1.2. *Nature of the binder*

Secondly, the effect of the binder was studied on samples of pure binder and samples of hemp concrete.

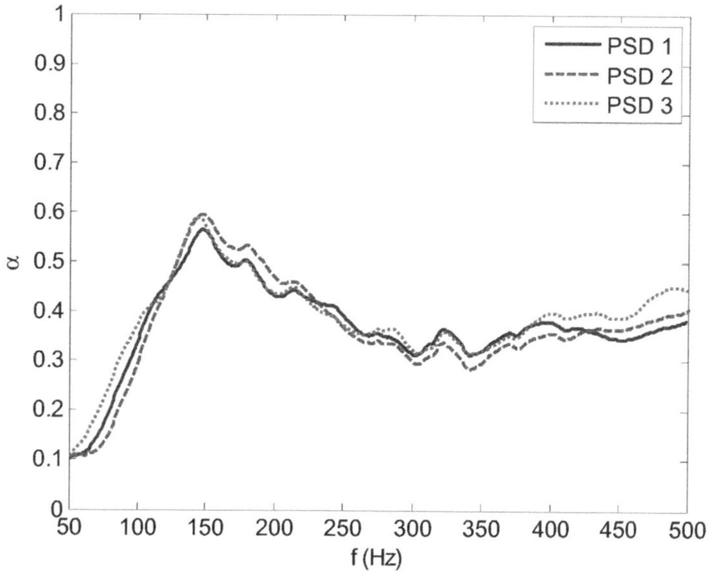

Figure 7.3. *Sound absorption of hemp concretes depending on the PSD – walling formula with hydraulic lime*

7.2.1.2.1. Effect on the acoustic properties of the binders

To begin with, samples of pure binders (aerated lime, hydraulic lime and natural prompt cement) were tested. Their absorptions are presented in Figure 7.4. The values obtained are very poor – less than 10% in the range of frequencies in question, between 150 and 2000 Hz. This is due to the high resistivity of these materials. However, the sound absorption of the aerated lime-based binder is better than the two others, which could be explained by the greater porosity of this binder [GLE 11].

7.2.1.2.2. Effect on the acoustical properties of the hemp concretes

Then, the effect of the binder was observed on samples of hemp concrete. Figure 7.5 compares the sound absorptions of concretes made with the three types of binder and with PSD3 hemp shiv. These measurements, performed with a large impedance tube, demonstrate that aerated or hydraulic lime-based mixtures yield approximately identical acoustical properties, and that these two substances are more advantageous than a hemp concrete made with natural prompt cement. In particular, this is due to the fact that the porosity is more open in lime-based concretes and facilitates greater dissipation of the sound [GLE 11].

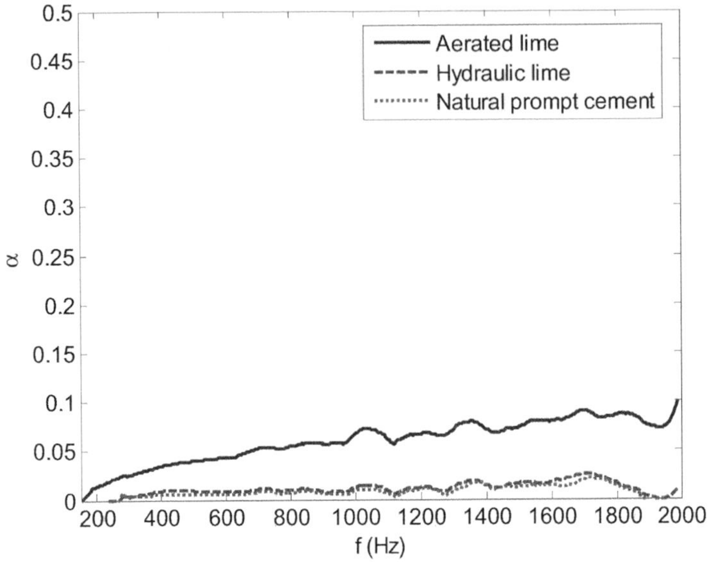

Figure 7.4. *Sound absorption of pure binders*

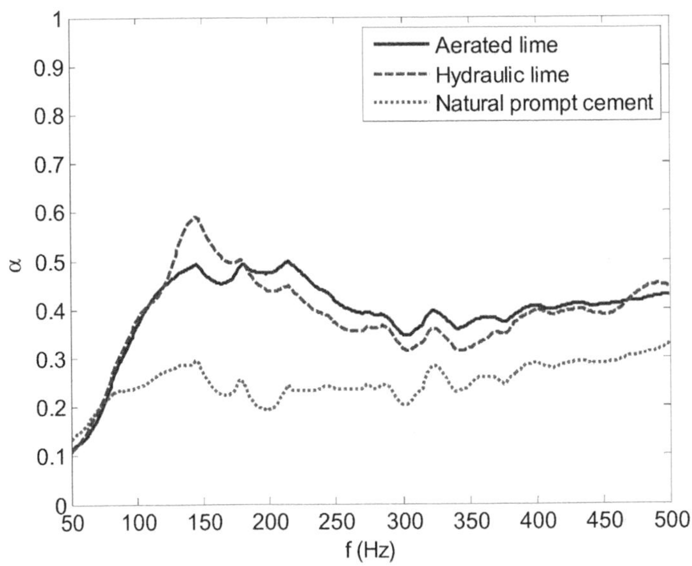

Figure 7.5. *Sound absorption of hemp concretes depending on the type of binder – walling formula with PSD3 shiv*

Hence, the choice of binder may contribute to an optimal formulation of a material, and it is best to choose a lime-based binder in order to maximize sound absorption.

7.2.2. *Influence of the casting method*

7.2.2.1. *Apparent density of the material*

The acoustical properties depend greatly on the mode of casting, beginning with the apparent density. This effect can be seen in shiv and in hemp concretes and is demonstrated in Figure 7.6 for PSD2 shiv. Three densities are compared:

– a low density, 102 $kg.m^{-3}$, for which the particles are simply poured;

– an intermediate density, 127 $kg.m^{-3}$, for which the material is poured and then shaken;

– a high density, 153 $kg.m^{-3}$, for which the material is poured, shaken and slightly compacted.

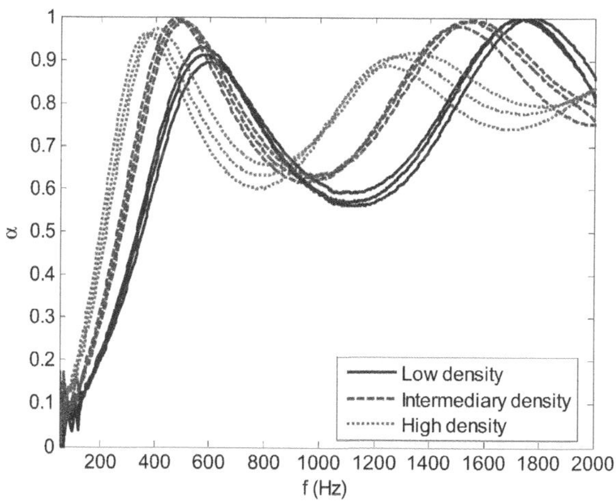

Figure 7.6. *Sound absorption of PSD2 shiv depending on its density*

We note that the density of the shiv has a very pronounced effect on its sound absorption. On the one hand, the sound absorption curve shifts toward low frequencies for the case of high densities, and on the other, the amplitude of the sound absorption is modulated by that density. It is optimal for the intermediate density. We can also see on this figure, in comparison with Figure 7.2, that the sound absorption of the shiv is more sensitive to its apparent density than its PSD.

The mode of emplacement of the particles, and the resulting apparent density of the shiv, therefore appears to be a way to control both the frequential position of the absorption peaks and their level.

7.2.2.2. *Binder content*

The properties of hemp concretes are also very sensitive to the amount of binder used, and particularly to the binder-to-shiv mass ratio (B/S). The acoustical properties of hemp concretes of 5 cm in thickness have been tested for different formulations, varying the B/S mass ratio between 0.5 and 6. The results are presented in Figure 7.7. Here, we can clearly see that with a higher binder content in the hemp concrete, its sound absorption is automatically reduced, with the material becoming less and less porous and more and more impermeable to sound waves.

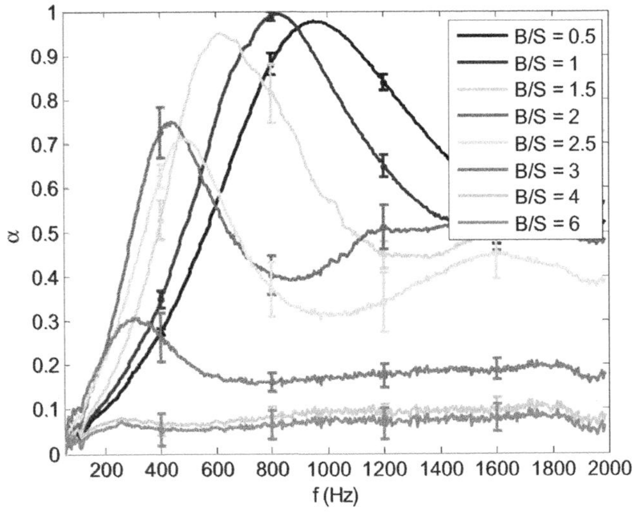

Figure 7.7. *Sound absorption of hemp concretes depending on the binder content*

Consequently, the sound absorption can exhibit very variable levels, spanning all values between 0.1 and 1 depending on the B/S ratio chosen, so this parameter proves to be the most influential in terms of the acoustical properties of the material.

7.2.2.3. *Water content*

Finally, the water content is also an important parameter in the formulation of hemp concretes. In particular, when too much water is used in the mixture, it becomes foamy, which makes it less dense. Furthermore, if too much water is used during the tempering process, we see a creeping of the binder toward the lower face of the samples: a fine layer of the binder may then form, in contact with the mold of the sample. As a result, the sound absorption of the samples is – of course – greatly affected.

Figures 7.8 and 7.9 compare the surface states and sound absorption of the upper and lower surfaces of hemp concretes, with and without excess water, respectively. We can clearly see that with excess water, the porous network is partially blocked by the binder, so the sound absorption of the surface in question decreases.

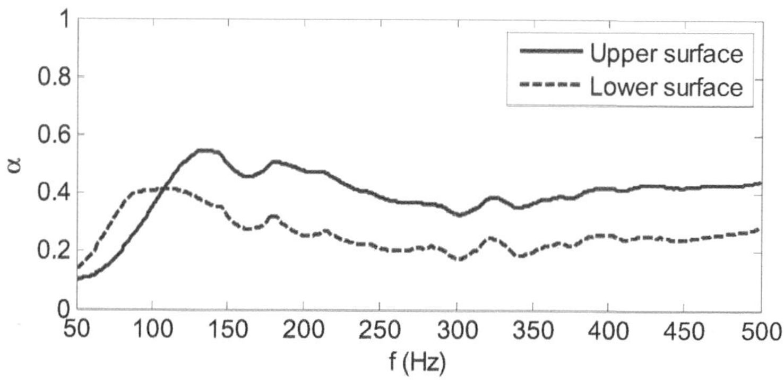

(a)

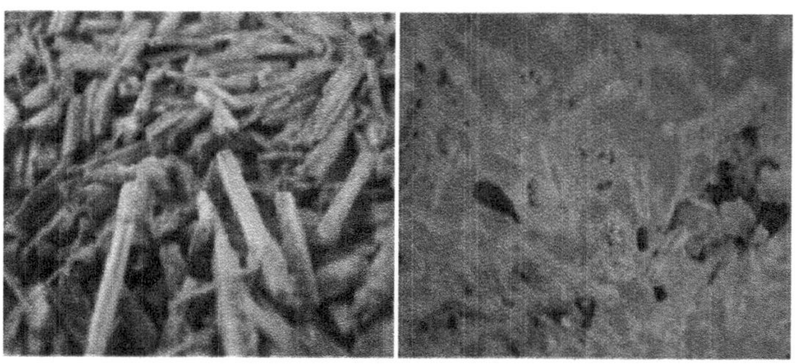

(b) (c)

Figure 7.8. *Effect of excess water on the sound absorption (a) of WALLING formulation hemp concretes – with PSD1-type shiv and an aerated lime-based binder, with photographs of the upper (b) and lower (c) surfaces*

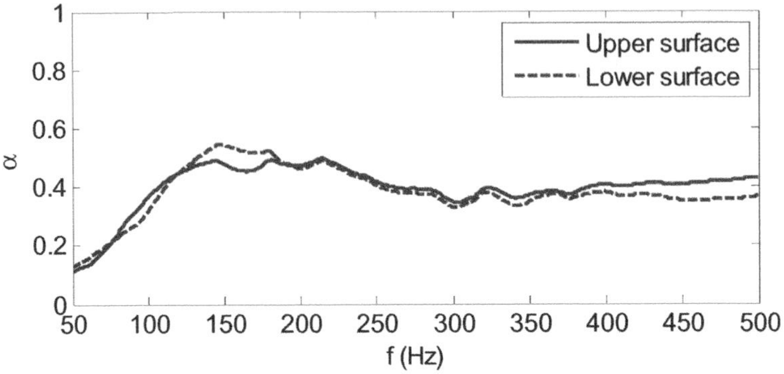

(a)

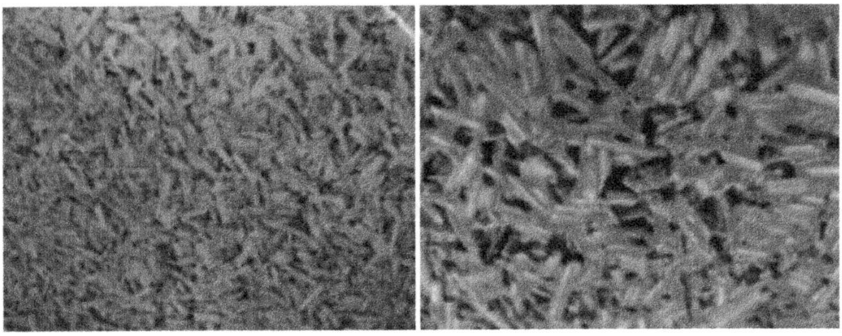

(b) (c)

Figure 7.9. *Effect of excess water on the sound absorption (a) of WALLING formulation hemp concretes with PSD3-type shiv and an aerated lime-based binder, with photographs of the upper (b) and lower (c) surfaces*

7.3. Modeling the acoustical properties

The earliest acoustical studies focusing on hemp concretes and shiv [GLE 11] were based on a model referred to as "single porosity", whereby the whole of the sample of material is described by one average dimension of the size of the pores. In this section, a new model is presented, which is able to yield more accurate predictions because it takes account of the different scales of porosity existing in the material. Later in the chapter, we consider ten different types of shiv. Five of these are examples of raw shiv from different origins, and are labeled as S_A to S_E. The other five are artificial types of shiv, taken from PSD portions of S_B shiv, and studied in order to observe the effect of the particle size. These are labeled S_{B1} to S_{B5}.

7.3.1. *Physical analysis of the acoustical properties being measured*

Among the set of experimental measurements taken on different types of shiv and hemp concretes, one particular feature is always in evidence. This is illustrated in Figure 7.10, for the example of a type of shiv with apparent density ρ = 130 kg.m^{-3}. We can see that the real part of the dynamic bulk modulus is greater than $1/\phi$ and γ/ϕ (where ϕ denotes the open porosity of the material and γ the ratio of the specific heats of the air) above 300 Hz. This dynamic bulk modulus, denoted K_{eq}, is an intrinsic acoustical property of the material, which can be used to describe the heat dissipation that occurs. Similarly, there is the dynamic density ρ_{eq}, which for its part describes the visco-inertial dissipation. However, with porous materials, the real part of K_{eq} is bounded by the asymptotic limits characteristic of the heat exchanges [ALL 09], i.e. behavior which is isothermal at low frequencies ($Re(K_{eq}/P0) = 1/\phi$ for $\omega \ll \omega_t$) and adiabatic at high frequencies ($Re(K_{eq}/P0) = \gamma/\phi$ for $\omega \gg \omega_t$). These behaviors are delimited by the characteristic thermal frequency of the material ω_t [LAF 97]. This peculiarity can also be seen in Figures 19, 20 and 21 from [GLE 11], where the value of this measurement is underestimated by the model.

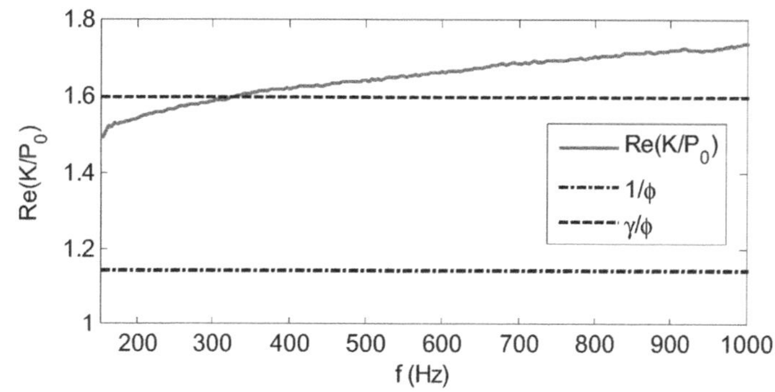

Figure 7.10. *Measurement of the real part of the normalized dynamic bulk modulus compared to its simulated asymptotic limits $1/\phi$ and γ/ϕ for shiv S_A whose apparent density is $\rho = 130\ kg.m^{-3}$*

This difference is equivalent to an overestimation of the open porosity by more than 10%. However, the measurements of open porosity ϕ were taken with a degree of uncertainty of approximately 1%. We therefore came to the conclusion that conventional models are inadequate in terms of predicting the acoustic behavior of such materials. However, if when analyzing the material's behavior, we take account of the different scales of its porosity, and base our experiment on the model developed by Olny and Boutin [OLN 03], everything becomes logical once more.

In the case of porous materials with so-called "double porosity", characterized by the presence of pores on two scales (a microscopic dimension lm and a mesoscopic dimension lp), the intrinsic macroscopic acoustic properties ρ_{eq} and K_{eq} can be expressed as a function of the intrinsic properties of the microscopic and mesoscopic scales. If the contrast of permeability existing between the mesopores and the micropores is sufficiently great, when $lm/lp \approx 10^{-3}$, the model is described by equations [7.1] and [7.2]. The subscripts p and m refer, respectively, to the mesoscopic and microscopic scales.

$$\rho_{eq}(\omega) = \left(\frac{1}{\rho_p(\omega)} + \frac{1-\phi_p}{\rho_m(\omega)}\right)^{-1} \quad [7.1]$$

$$K_{eq}(\omega) = \left(\frac{1}{K_p(\omega)} + \frac{(1-\phi_p)F_d(\omega,\omega_d)}{K_m(\omega)}\right)^{-1} \text{ where } \omega_d \approx \frac{P_0}{\phi_m \sigma_m l_p^2} \quad [7.2]$$

σ is the resistance to the passage of air ($N.m^{-4}.s$), and P_0 is the air pressure at equilibrium (Pa). F_d is dependent upon the frequency and distinguishes three kinds of acoustic behaviors:

– at low frequencies where $\omega \ll \omega_d$, $F_d \approx 1$, the pressure is uniform in the pores and micropores, and both scales are involved from the acoustic point of view;

– at intermediary frequencies such that $\omega \approx \omega_d$, there is coupling between the pores and micropores, and consequently increased sound dissipation;

– at high frequencies with $\omega \gg \omega_d$, $F_d \approx 0$, such that the micropores no longer play any part in sound dissipation.

In the third case, $K_{eq}(\omega) \approx K_p(\omega)$, and consequently, the real part of the dynamic bulk modulus is between $1/\phi_p$ and γ/ϕ_p. This is equivalent to an increase of this value in relation to a singly-porous material, and accounts for the behavior observed in Figure 7.10.

The shiv and hemp concretes are characterized respectively by two and three scales of porosity. For shiv, it is a question of intra-particle pores, the size of which ranges from 10 to 60 µm [GAR 98], and inter-particle pores, ranging from 1 to 10 mm in size, depending on the apparent density and the PSD of the shiv [GLE 11]. In the case of hemp concrete, in addition to the previous two types, we see intra-binder pores, around a micrometer in size. Consequently, the inter-scale ratio between the inter- and intra-particle ports is $\frac{l_m}{l_p} \in [10^{-2}; 10^{-3}]$, which suggests there is a significant contrast in terms of permeability between these two networks.

In the particular case of shiv, we apply the double porosity model with high contrast, looking at the inter-particle pores for the mesoscopic scale and the intra-particle pores for the microscopic scale.

7.3.2. *The adapted double porosity model and its parameters*

The porous model developed by Olny and Boutin [OLN 03] is a model of the equivalent fluid based on the rigid frame hypothesis. Zwikker and Kosten [ZWI 49] showed that above a certain characteristic frequency, the vibrations of the frame and of the fluid in the material are de-coupled, which makes such a hypothesis possible. This de-coupling frequency ω_{dec} was evaluated on the basis of equation [7.3] for shiv S_A, depending on the apparent density. The results are given in Table 7.2.

$$\omega_{dec} = \frac{\sigma\phi^2}{\rho} \quad [7.3]$$

Density ρ $(kg.m^{-3})$	100	110	120	130	140	150
ω_{dec} $(rad.s^{-1})$	11	13	14	19	22	27
f_{dec} (Hz)	1.8	2.0	2.3	3.1	3.5	4.2

Table 7.2. *Decoupling frequencies for shiv S_A for various apparent density values*

The frequencies calculated in the case of this particular type of shiv are similar for all the types of shiv and for most configurations of hemp concretes, and are well below the range of frequencies tested experimentally [150; 2000 Hz]; therefore, we can make the rigid frame hypothesis. Thus, the intrinsic properties ρ_{eq} and K_{eq} are calculated directly using equations [7.1] and [7.2], which are simplified in our case of high contrast of permeability to become equations [7.4] and [7.5].

$$\rho_{eq}(\omega) \approx \rho_{inter}(\omega) \quad [7.4]$$

$$K_{eq}(\omega) \approx K_{inter}(\omega) \quad [7.5]$$

In equation [7.4], the previous equation has been simplified. Indeed, in the context of our hypothesis, we have $\omega \gg \omega_d$, which implies that $\sigma_{intra} \gg \frac{P_0}{\phi_{intra}\omega l_{inter}^2} > 80000\, rad.s^{-1}$.

Consequently, the characteristic viscous frequency of the intra-particle medium ω_v^{intra} defined in [JOH 87] by equation [7.6] is much higher than the experimental range of frequency. This proves that the flow in the material is mainly due to flow through the inter-particle pores [OLN 03], so the material's dynamic density can be approximated by finding that of the inter-particle medium.

$$\omega_v^{intra} = \frac{\sigma_{intra}\phi_{intra}}{\rho_0 \alpha_\infty^{intra}} = O(\sigma_{intra}) \gg 80000 rad.s^{-1} \qquad [7.6]$$

In order to evaluate the intrinsic properties of the inter-particle medium, the model advanced by Johnson *et al.* [JOH 87] can be used to model visco-inertial effects with $\rho_{inter}(\omega)$, and Zwikker and Kosten's model [ZWI 49] can be used to model thermal effects with $K_{inter}(\omega)$. These models have already been used [GLE 11] to model shiv and hemp concretes, looking at the amount of open porosity. The expressions of these models are given in equations [7.7] and [7.8].

$$\rho_{inter}(\omega) = \frac{\rho_0 \alpha_\infty^{inter}}{\phi_{inter}} \left[1 - \frac{j\sigma_{intra}\phi_{inter}}{\omega \rho_0 \alpha_\infty^{inter}} \sqrt{1 + j\frac{4\alpha_\infty^{inter\,2} \eta \rho_0 \omega}{\sigma_{intra}^2 \Lambda_{inter}^2 \phi_{inter}^2}} \right] \qquad [7.7]$$

$$K_{inter}(\omega) = \frac{\gamma P_0}{\phi_{inter}} \left(1 + 2(\gamma - 1) \frac{T(\sqrt{N_{Pr}}\lambda\sqrt{-j})}{\sqrt{N_{Pr}}\lambda\sqrt{-j}} \right)^{-1} \qquad [7.8]$$

T is the ratio of the Bessel functions of order 1 and 0, and λ is defined by $\lambda = \sqrt{\frac{8\alpha_\infty^{inter}\rho_0\omega}{\sigma_{intra}\phi_{inter}}}$.

ρ_0 is the density of air at equilibrium ($kg.m^{-3}$), η its viscosity ($Pa.s$) and N_{Pr} its Prandtl number. Five acoustic parameters characterize the inter-particle medium: the porosity ϕ_{inter}, the resistivity σ_{intra} , the high-frequency limit of the tortuosity α_∞^{inter}, and the viscous characteristic length Λ_{inter} (m). For the rest of this chapter, the subscript $inter$ is omitted from the variables, with the exception of ϕ_{inter}.

The sound absorption α is calculated by $\alpha = 1 - \left|\frac{Z-Z_0}{Z+Z_0}\right|^2$, where Z is the surface impedance of the sample, and Z_0 that of the air. If the sample is touching a rigid termination, $Z = -jZ_c \cot(ke)$, with Z_c being the characteristic impedance of the material ($Z_c = \sqrt{\rho_{eq} K_{eq}}$), k its wave number ($k = \omega\sqrt{\frac{\rho_{eq}}{K_{eq}}}$) and e the thickness of the material (m).

In order to be able to model the dynamic density and the dynamic bulk modulus according to equations [7.7] and [7.8], we then need to know four acoustic parameters for each of the materials. To begin with, the visco-inertial parameters σ, α_∞ and Λ are characterized indirectly on the basis of the real and imaginary parts $Re(\rho_{eq})$ and $Im(\rho_{eq})$ of the dynamic density, using the analytical method described in [PAN 06]. The resistivity is estimated on the basis of the low-frequency limit of the imaginary part of the dynamic density, while the tortuosity and the viscous length are obtained analytically, using Johnson *et al.*'s model [JOH 87]. Finally, the porosity ϕ_{inter} is evaluated so that the Zwikker and Kosten model corresponds to the measurement of the real part of the dynamic bulk modulus. The expressions used in this phase of indirect characterization are given in equations [7.9] to [7.11]. It should be noted that, to begin with, the tortuosity was sought in the form $\frac{\alpha_\infty}{\phi_{inter}}$, because the porosity ϕ_{inter} only becomes known at the end of the process.

$$\sigma = \lim_{\omega \to 0} -\omega Im(\rho_{eq}) \quad [7.9]$$

$$\frac{\alpha_\infty}{\phi_{inter}} = \frac{1}{\rho_0}\left[Re(\rho_{eq}) - \sqrt{Im(\rho_{eq}) - \frac{\sigma^2}{\omega^2}}\right] \quad [7.10]$$

$$\Lambda = \frac{\alpha_\infty}{\phi_{inter}} \sqrt{\frac{2\rho_0\eta}{\omega Im(\rho_{eq})\left[\frac{\rho_0\alpha_\infty}{\phi_{inter}} - Re(\rho_{eq})\right]}} \quad [7.11]$$

7.3.3. *Experimental validation of the model*

The model has been successfully applied to all the types of shiv looked at here and the vast majority of the hemp concretes. The results are presented in Figure 7.11 for three types of shiv with very different densities and PSDs, whose acoustic parameters are given in Table 7.3. From the figure, we can see that the simulations coincide perfectly with the measurements taken. Therefore, the choice of the model is validated.

Hemp shiv	ϕ_{inter}	$\sigma\ (N.m^{-4}.s)$	α_∞	$\Lambda\ (\mu m)$
$S_C\ \rho = 100kg.m^{-3}$	0.79	1530	1.72	389
$S_A\ \rho = 130kg.m^{-3}$	0.75	3090	2.07	215
$S_{B5}\ \rho = 150kg.m^{-3}$	0.77	9290	2.15	101

Table 7.3. *Acoustic parameters used for the modeling*

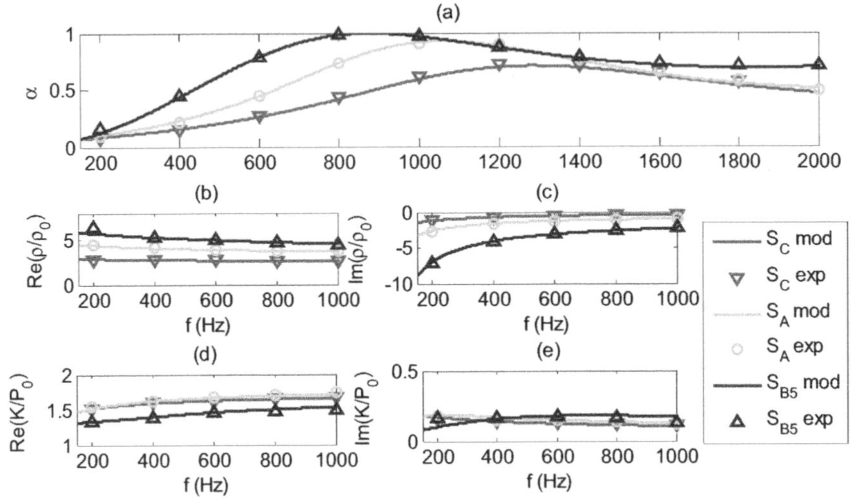

Figure 7.11. *Comparison of the experimental data (exp) and modeled values (mod) for sound absorption (a), for the real (b) and imaginary (c) parts of the dynamic density, and for the real (d) and imaginary (e) parts of the dynamic bulk modulus*

7.4. Application of the model to the acoustical characterization of shiv

A number of studies have been performed with the aim of unmasking the laws governing the evolution of the acoustical parameters depending on the material's basic parameters, which can be recorded as soon as it is manufactured. For instance, Castagnede *et al.* [CAS 00] looked at the change in the acoustical parameters of fibrous materials depending on the degree of compression of the material and its characteristics in a reference state. Here, in the case of hemp shiv, the goal is to be able to draw the connection between the acoustical parameters and the fundamental characteristics of the particles (PSD, intra-particle porosity, apparent density) and the arrangement of the granular mixture (orientation of the particles, inter-particle porosity).

7.4.1. *Porosity of shiv*

7.4.1.1. *Open porosity and density of frame*

For granular materials, the porosity is a linear function of the apparent density of the material, which depends on the density of the frame ρ_{frame} [GLE 11]. This frame density has been characterized for all the types of shiv on three samples, with a mass M_{shiv} of 4 g, in order to ensure the representativeness of the measurement.

The measurement consists of evaluating the volume of the frame V_{frame} of the shiv, using an air porosimeter [LEC 03]. The average densities are given in Table 7.4. On the basis of these values, the porosity is then evaluated using equation [7.12]. For the types of shiv tested, the frame density ranges between 870 and 1350 kg.m^{-3}. These values are similar to those obtained in [GLE 11]. In addition, we note that for shiv S_B, the frame density is the same no matter what the PSD. Hence, this parameter appears to depend solely on the origin of the hemp. In particular, it has been observed that the frame densities are highly sensitive to the retting of the plant, which involves leaving the hemp stems exposed to precipitation and the Sun for a few weeks so that they partially rot and defibration becomes easier. This process results in leaching of the pectins contained in the plant, and a loss of matter, which leads to a smaller frame volume and a greater frame density. These effects are indeed detected by the measurements, as shown by the results presented in Table 7.4, where the retted types of shiv S_B, S_{Bi} and S_C exhibit greater frame densities than the non-retted types of shiv S_A, S_D and S_E.

$$\phi = 1 - \frac{\rho}{\rho_{frame}} \text{ where } \rho_{frame} = \frac{M_{shiv}}{V_{frame}} \qquad [7.12]$$

The open porosity of the types of shiv tested is presented in Figure 7.12. This value is particularly high – generally between 85 and 95%.

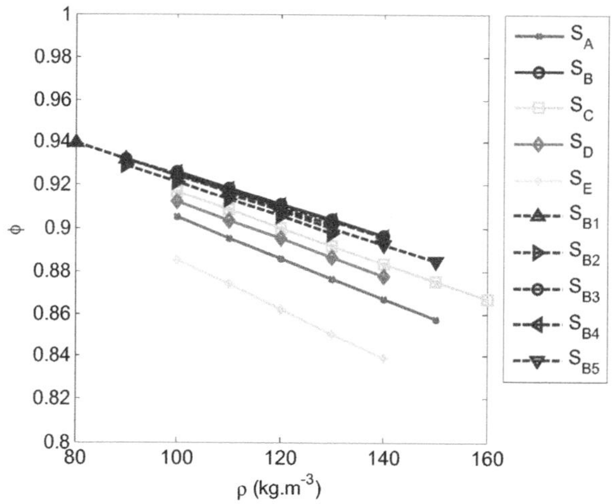

Figure 7.12. *Open porosity of the samples of shiv depending on their apparent density*

7.4.1.2. *Inter-particle porosity and apparent density*

In the case of shiv, the acoustical properties depend more on the inter-particle porosity than on the open porosity [GLE 12]. Similarly to equation [7.12], we can easily get equation [7.13] to describe the inter-particle porosity as a function of the apparent density of the shiv and of the apparent density of the particles $\rho_{particle}$.

$$\phi_{inter} = 1 - \frac{\rho}{\rho_{particle}} \quad [7.13]$$

The apparent density of the particles was evaluated by way of the least squares method in order to minimize the difference between equation [7.13] and inter-particle porosity measured. The measurements were performed on three samples in order to evaluate the reproducibility of the results, and the means and standard deviations are presented for each configuration tested.

The apparent densities estimated are summed up in Table 7.4, and the inter-particle porosities are presented and compared to their theoretical evolution in Figure 7.13. In this figure, we note slight differences (<2%) between the estimations of the porosity and the values projected by the model. However, the predictions are still very satisfactory, and enable us to validate equation [7.13].

The inter-particle porosity is between 65% and 85%, so between 10% and 20% lower than the open porosity of the materials. The apparent densities of the particles can range from 360 to 630 kg.m^{-3}, which is very low in comparison with the frame densities. This is due to the high degree of porosity contained within the particles themselves.

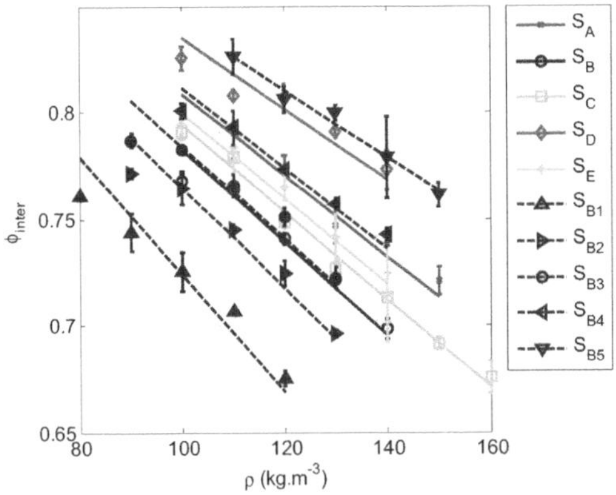

Figure 7.13. *Inter-particle density of the shiv samples depending on their apparent density*

7.4.1.3. *Intra-particle porosity*

The intra-particle porosity $\phi_{particle}$ is evaluated using equation [7.14], and the values are collected in Table 7.4 for the different hemp shivs. The intra-particle porosities are between 43% and 73%, and depend heavily on the PSD. This could be explained by the fact that the smallest particles come from the top of the plant, where the stem is less porous and denser. This effect will be verified with additional analyses in our future endeavors.

$$\phi_{particle} = 1 - \frac{\rho_{particle}}{\rho_{frame}} \qquad [7.14]$$

7.4.1.4. *Average thickness of the particles*

Using the surface mass density ρ_s which is deduced by measuring the mass and analyzing a 2D image of a representative sample of particles (e.g. using the method described by Ceyte [CEY 08]), and the apparent density of the particles $\rho_{particle}$, it is also possible to determine the average thickness of the particles of each type of shiv. The average thickness $\bar{e}$ is therefore calculated by using equation [7.15], and its value is given in Table 7.4 for the types of shiv tested. The thickness of the particles varies between 0.2 and 0.9 mm, which corresponds to a visual estimation.

$$\bar{e} = \frac{\rho_s}{\rho_{particle}} \qquad [7.15]$$

In conclusion, analyzing the porosities of shiv is highly instructive. Indeed, it offers a non-destructive and relatively accessible means to characterize fundamental parameters of the particles: the frame density, the apparent density of the particles, the intra-particle porosity and the thickness of the particles. The other advantage to this acoustic method is its representativeness, as it can be easily applied to sufficiently large samples, which is not necessarily possible with micro-tomography or mercury porosimetry.

Shiv	S_A	S_B	S_C	S_D	S_E	S_{B1}	S_{B2}	S_{B3}	S_{B4}	S_{B5}
ρ_{frame} $(kg.m^{-3})$	1050	1350	1200	1140	870	1330	1270	1320	1340	1300
ρ_s $(kg.m^{-3})$	0.18	0.19	0.23	0.19	0.14	0.33	0.22	0.18	0.15	0.12
$\rho_{particle}$ $(kg.m^{-3})$	520	460	490	610	500	360	430	460	530	630
$\phi_{particle}$	0.50	0.66	0.60	0.47	0.43	0.73	0.67	0.65	0.60	0.51
$\bar{e}$ (mm)	0.35	0.42	0.48	0.32	0.28	0.91	0.52	0.40	0.28	0.18
$R_{particle}$ (mm)	0.28	0.28	0.31	0.28	0.20	0.50	0.36	0.29	0.22	0.15

Table 7.4. *Parameters characterized for all the types of shiv tested*

7.4.2. *Resistivity*

The resistance to the passage of air σ, commonly called resistivity, is an extremely important parameter, which helps characterize the viscous effects in a porous material at low frequencies. The resistivity is defined as the inverse of the static permeability k_0 (equation [7.16]).

$$\sigma = \frac{\eta}{k_0} \qquad [7.16]$$

The measures of resistivity have been compared to various empirical and physical models, describing the air permeability of granular materials as a function of their inter-particle porosity and a characteristic size of the aggregates. All the models developed have the same dependency in $\frac{1}{R_{particle}{}^2}$, where $R_{particle}$ is the equivalent radius of an aggregate, but they are different functions of the porosity ϕ_{inter}. The model identified as being the most pertinent in terms of shiv is a model obtained by self-consistent homogenization based on cubic geometry, frequently used in the case of particles of wood in order to predict variations in pressure. This model was developed by Prieur du Plessis and Woudberg [DUP 08], who validated and extended the well-known Ergun equation to any range of porosity. Its expression is given by equation [7.17].

$$\sigma = \frac{25.4\,\eta}{\left(\frac{4}{3}\Pi\right)^{2/3} R_{particle}{}^2} \frac{(1-\phi_{inter})^{\frac{4}{3}}}{\left(1-(1-\phi_{inter})^{\frac{1}{3}}\right)\left(1-(1-\phi_{inter})^{\frac{2}{3}}\right)^2} \qquad [7.17]$$

Hence, this model was applied to our samples of shiv. The inter-particle porosity was calculated in the various cases using equation [7.13] and the equivalent radius $R_{particle}$ was adjusted by way of the least squares method. The resistivities characterized and their modeled values are presented in Figure 7.14 as a function of the apparent density of the different types of shiv, and the equivalent radii are given in Table 7.4.

The resistivities of the various types of shiv range from 500 to 10000 $N.m^{-4}.s$, which represents a relatively wide range of permeability. Furthermore, we observe a very close correspondence between the experimental results and the model. The equivalent radii are similar in size to the average thickness of the particles, which seems logical in that sound propagation occurs in the direction of the thickness of the particles, with these being posed on their greatest cross-section when the sample is being prepared. A linear regression was carried out between these two parameters, and a strong correlation was shown (R^2 = 0.94). The relation is given in equation [7.18].

$$R_{particle} = 0.462\,\bar{e} + 9.66\,10^{-5} \qquad [7.18]$$

It also proves particularly useful to model the resistivity. Indeed, we were able to demonstrate, on the basis of the above results, that knowing the apparent particle density and an equivalent radius for the shiv enables us to precisely calculate its resistivity, as a function of its apparent density. Figure 7.15 illustrates this well: the normalized permeability $\frac{k_0}{R_{particle}{}^2}$ is shown as a function of ϕ_{inter}, and it shows a master curve.

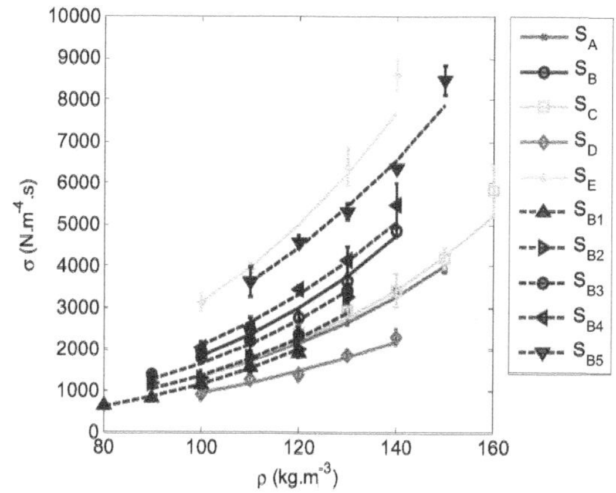

Figure 7.14. *Resistivity of shiv samples as a function of their apparent density*

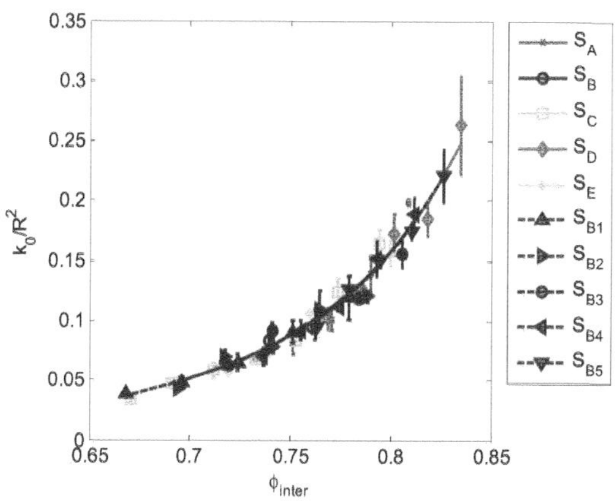

Figure 7.15. *Normalized permeability of shiv samples as a function of their apparent density*

7.5. Conclusion

This chapter discusses the different results obtained as regards the characterization and modeling of the acoustical properties of hemp concretes.

Hemp concrete is a material whose acoustical properties can vary greatly depending on the properties of its components and the way in which it is put to use. In particular, the results presented herein highlight that the parameters of installation have a far stronger influence than the properties of the components, with the apparent density of the material and the binder-to-shiv mass ratio playing an important part. Thus, by combining these different parameters, it is possible to find the optimal formulation of a hemp concrete for any given application.

When modeling the acoustical properties of hemp concretes, we take account of the different scales of porosity existing in the material. It is shown that one of the peculiarities of hemp concrete is that only the inter-particle pores have a part to play in sound dissipation, because of the existence of a high contrast of permeability between these pores and the intra-particle pores. A simulation is performed using semi-phenomenological models based on four acoustic parameters: the porosity, the resistivity, the tortuosity and the viscous characteristic length, and facilitates a very satisfactory prediction of the acoustic properties of the materials.

Finally, it has been shown that by way of acoustic measurements, it is possible to work back to the acoustical parameters, which can then be used to deduce the basic parameters of the materials. Thus it is possible to characterize the apparent density of the particles, their internal porosity and their thickness. This approach enables us, finally, to check the quality of hemp concretes and their raw materials by acoustical measurement.

7.6. Bibliography

[ALL 09] ALLARD J.-F., ATALLA N., *Propagation of sound in porous media – Modelling sound absorbing materials*, Wiley, 2009.

[ARN 01] ARNAUD L., CEREZO V. Qualification physique des matériaux de construction à base de chanvre – Rapport final – Programme de juin 1998 à août 2001, Contrat de recherche CNRS 071462, 2001.

[ARN 12] ARNAUD L., GOURLAY E., "Experimental study of parameters influencing mechanical properties of hemp concretes", *Construction and Building Materials*, 28, p. 50–56, 2012.

[CAS 00] CASTAGNEDE B., AKNINE A., BROUARD B., TARNOW V., "Effects of compression on the sound absorption of fibrous materials", *Applied Acoustics*, 61, p. 173–182, 2000.

[CER 05] CEREZO V. Propriétés mécaniques, thermiques et acoustiques d'un matériau à base de particules végétales: approche expérimentale et modélisation théorique. Doctoral thesis, Ecole doctorale MEGA, Lyon, 2005.

[CEY 08] CEYTE I., Béton de chanvre, définition des caractéristiques mécaniques de la chènevotte, Travail de Fin d'Études, ENTPE 2008 155, 2008.

[DUP 08] DU PLESSIS J. P., WOUDBERG S., "Pore-scale derivation of the Ergun equation to enhance its adaptability and generalization", *Chemical Engineering Science* , 63 (9), p. 2576–2586, 2008.

[GAR 98] GARCIA-JALDON C., DUPEYRE D., VIGNON M., "Fibres from semi-retted hemp bundles by steam explosion treatment", *Biomass and Bioenergy* 14 (3) (1998) 251–260.

[GLÉ 11] GLÉ P., GOURDON E., ARNAUD L., "Acoustical properties of materials made of vegetable particles with several scales of porosity", *Applied Acoustics* 72 (2011) 249–259.

[GLÉ 12] GLÉ P., GOURDON E., ARNAUD L., "Modelling of the acoustical properties of hemp particles", *Construction and Building Materials*, 37, p. 801–811, 2012.

[JOH 87] JOHNSON D.-L., KOPLIK J., DASHEN R., "Theory of dynamic permeability and tortuosity in fluid-saturated porous media", *Fluid Mechanics*, 176, p. 379–402, 1987.

[LAF 97] LAFARGE D., LEMARINIER P., ALLARD J.-F., TARNOW V., "Dynamic compressibility of air in porous structures at audible frequencies", *Journal of the Acoustical Society of America*, 102 (4), p. 995–2006, 1997.

[LEC 03] LECLAIRE P., UMNOVA O., HOROSHENKOV K.-V., MAILLET L., "Porosity measurement by comparison of air volumes", *Review of scientific instruments*, 74 (3), p. 1366–1370, 2003.

[OLN 03] OLNY X., BOUTIN C., "Acoustic wave propagation in double porosity media", *Journal of the Acoustical Society of America*, 114 (1), p. 73–89, 2003.

[PAN 06] PANNETON R., OLNY X., "Acoustical determination of the parameters governing viscous dissipation in porous media", *Journal of the Acoustical Society of America*, 119 (4), p. 2027–2040, 2006.

[SAM 08] SAMRI D. Analyse physique et caractérisation hygrothermique des matériaux de construction: approche expérimentale et modélisation numérique. Doctoral thesis, Ecole doctorale MEGA, Lyon, 2008.

[ZWI 49] ZWIKKER C., KOSTEN C.-W., *Sound Absorbing Materials*, Elsevier, New York, 1949.

Chapter 8

Plant-Based Concretes in Structures: Structural Aspect – Addition of a Wooden Support to Absorb the Strain

8.1. Introduction

At present, the poor mechanical strengths of hemp concrete mean that this material alone is inapt for use in the load-bearing structure of a building. Therefore, it is possible to use it as a filler in a wooden skeleton, for thermal insulation. From a design point of view, the strength of the composite wall is unaffected by the contribution of the hempcrete.

In view of the thermal regulations in force and the characteristics of hempcrete (essentially its conductivity), the thicknesses of hempcrete needing to be used are large, and we may imagine that its presence will be able to absorb some of the mechanical strains in addition to the wooden skeleton. This contribution may, in particular, relate to the absorption of vertical strains but also the actions relating to the effects of wind.

As part of the Agrobat project [FFB 09], a research program on hempcrete installed by projection (i.e. using the "shotcrete" approach), wind-brace tests were carried out in order to evaluate the contribution of hempcrete. The principle behind these tests is presented in Figure 8.1.

Chapter written by Philippe MUNOZ and Didier PIPET.

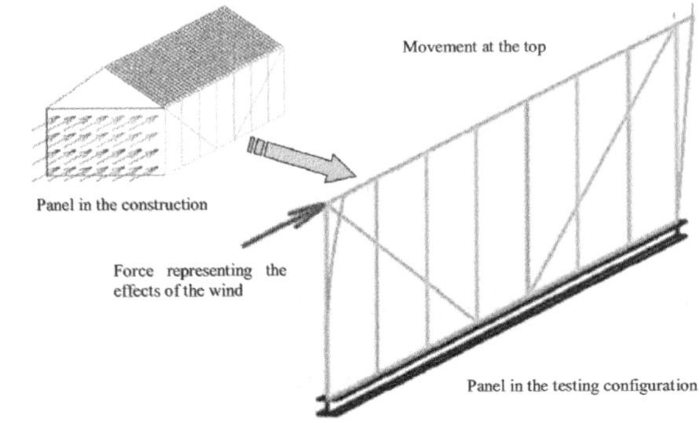

Figure 8.1. *Principle of the wind-brace test*

The standard relating to this test is NF EN 594 [AFN 08], but its domain of application relates to walls braced by a layer of wood-based material. This regulation advocates positioning the panel on a vertical plane, applying vertical loads to each upright, applying continuous horizontal strain to the panel and taking measurements of its movement at three points.

The approach adopted is based on the following principles:

– working on a full-scale panel, approximating those panels commonly used by the construction company "*Le Bâtiment Associé*". The dimensions of these panels would be limited in order to simplify the testing procedure, without this alteration having any effect of the predictable results. Hence, we chose panels 4.56 m in length and 2.48 m in height;

– in that the objective is to perform a comparative study between the wind-bracing behavior of a panel with a wooden framework (with bracing struts) and that of a similar panel in which those struts are replaced with a hempcrete filling, we did not respect the norm to the letter, but every attempt was made to approximate the prescribed conditions when technically and economically possible;

– a few years previously, we had already tested panels comparable to this one, but which were positioned vertically. During the course of these tests, many experimental difficulties were encountered in ensuring friction-free guidance of the lintel. Having learnt from these difficulties, in this case we chose to test the panels on the flat, thereby limiting the friction of the panel in the horizontal plane. In addition, the aim of this type of test is to evaluate the limitations of the strength of a test substance in relation to the effects of the wind, which will be predominant in comparison to the vertical strain.

In a first test, the behavior of a conventional wood-framework braced by struts was examined, in order to validate and possibly adapt the experimental arrangements initially chosen.

A second test was then carried out on an improved test sample (improvements made on the basis of the results of the first stage). The second panel tested was similar to the first, apart from the fact that the bracing struts have been replaced with a hempcrete filling.

Finally, a comparative analysis was performed on the behavior of the two panels in order to evaluate the aptitude of hempcrete to take part in the bracing of the wooden structure.

8.2. Preliminary test

8.2.1. *Description of the panel*

The panel tested comprises a wooden skeleton 145 mm thick, braced by two 145 × 45 mm struts. In its upright view, the overall dimensions of the panel are 2.48 m in height by 4.56 m in length. Figure 8.2 gives a fuller picture of the makeup of the wooden skeleton. The different elements of the wooden framework are nailed together.

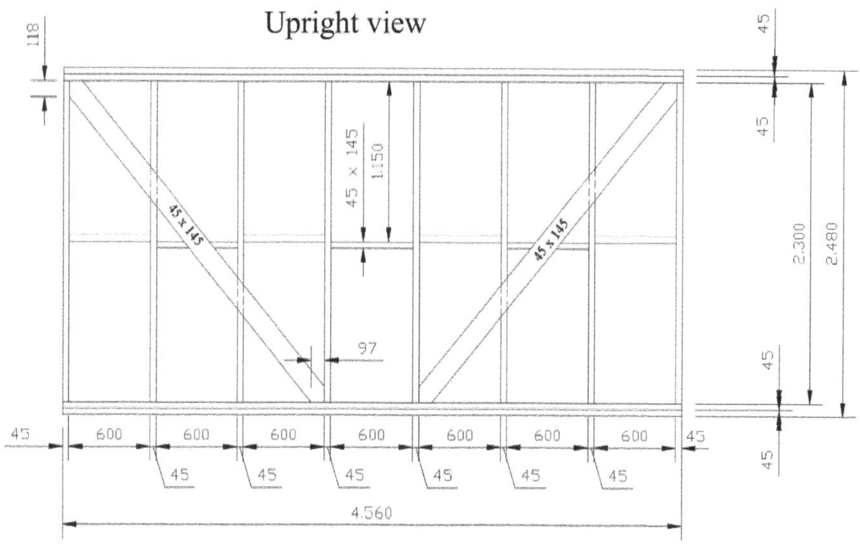

Figure 8.2. *Structure of the wooden frame*

8.2.2. *Putting the panel in place on the bracing bank*

The panel is positioned flat, and rests on eight concrete blocks through two sliding plates – one made of Teflon, the other of polished stainless steel, with a friction coefficient of 0.04 – as shown in Figure 8.3.

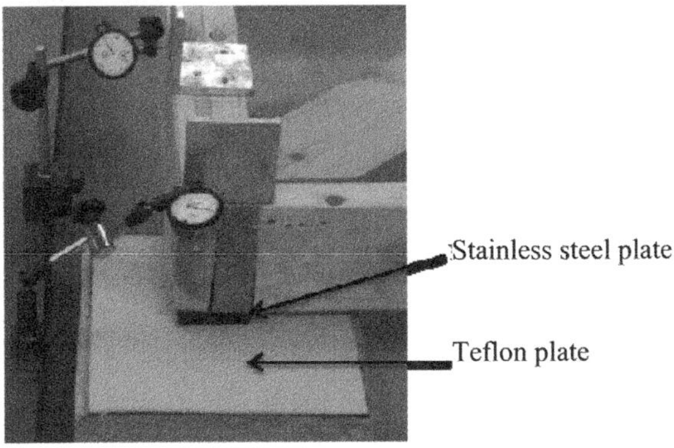

Figure 8.3. *Supports of the panel*

The lower crosspiece is connected to the lower "HEB" H-shaped girder by 5×70 VBA screws spaced 300 mm apart. The girder is held in place by threaded rods anchored in the plaster using a "Spit Grip" lip anchor, and is propped up on the horizontal plan by 100×100 angle irons fixed in the plaster in the same way. The arrangement of all of these construction devices is illustrated in Figure 8.4.

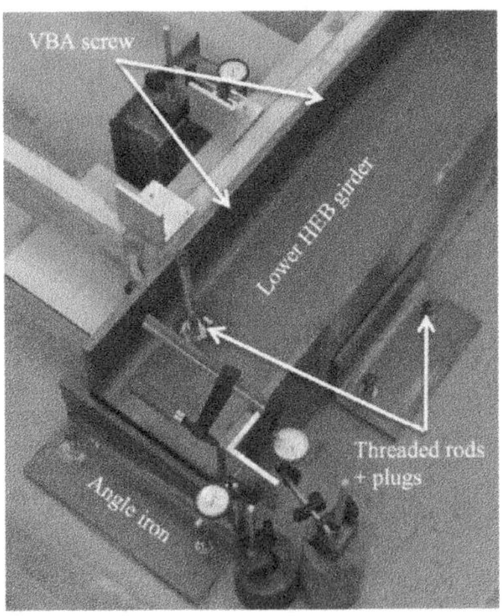

Figure 8.4. *Emplacement of the panel on the lower HEB*

On the upper crosspiece, a flat, no-friction support is made to counteract the rotation of the panel caused by the action of the jack. This flat support is put in place on the upwind upper HEB, comprising a Teflon plate and a stainless steel plate (see Figure 8.5).

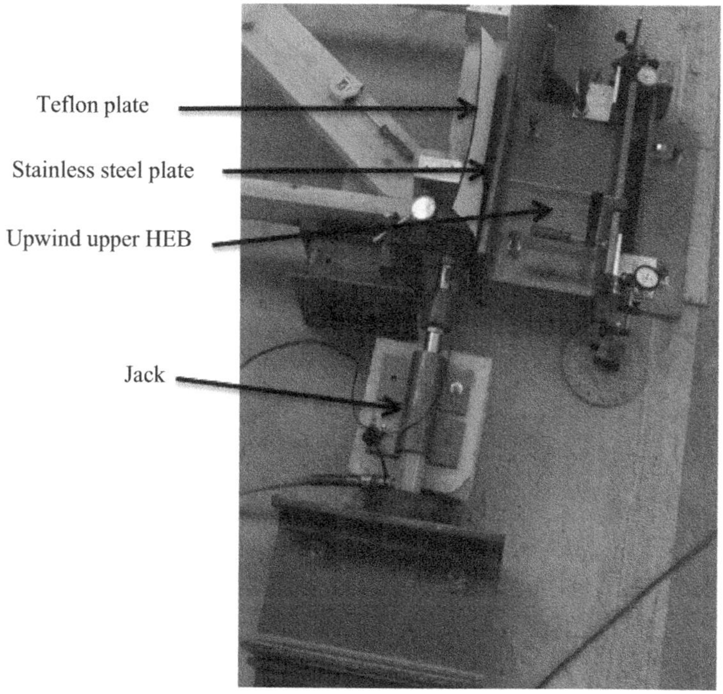

Figure 8.5. *Countering of the rotation of the top of the panel upwind*

8.2.3. *Longitudinal loading and measurement of the movements*

The longitudinal force is exerted by a 50 kN manual jack, whose action is measured by a strain-gauge sensor. The norm prescribes a velocity of loading such that the movement of the top of the panel is between 2 and 4 mm/min. We shall apply increasing strain in steps of 100 daN (in order to be able to take measurements of the data).

The movements are measured by mechanical comparators at 1/100 mm. As Figures 8.6 and 8.7 show (bird's-eye view of the test configuration), 13 points of measurement were used. They were chosen so as to evaluate any movement of the HEBs forming the test bank, and so as to reveal the modalities of the panel's behavior under strain.

In this configuration, the effect of the vertical strains on the uprights is not modeled. However, the upwind upper HEB applies a localized vertical force on the panel when it tends to rise up (under the influence of the longitudinal strain) but with the configuration of the experimental system as it is, we cannot measure the intensity of this force.

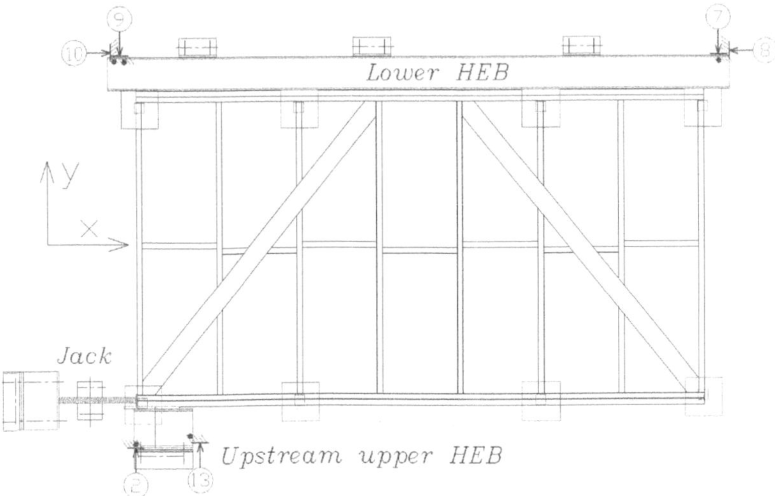

Figure 8.6. *Meaurement of the movements of the test bank*

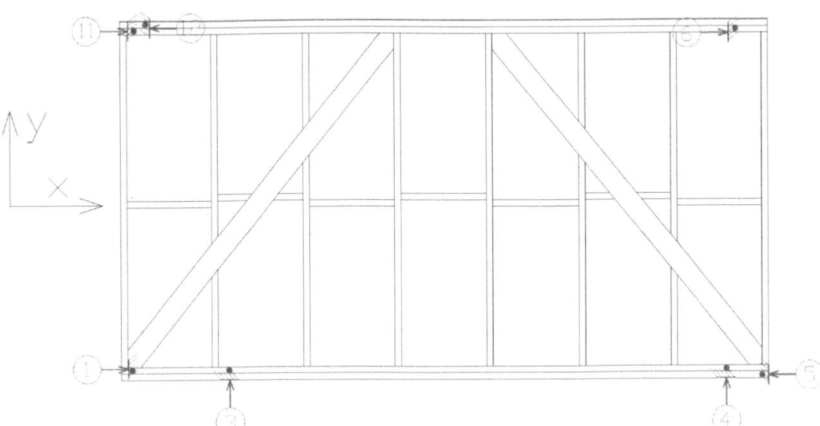

Figure 8.7. *Measurement of the movements of the braced wooden panel*

8.2.4. *Behavior of the test bank*

When strain was applied to the wooden panel, we took a regular measurement of the values of the movements up to 1935 daN. In terms of the behavior of the test bank, our results show the following trends:

– at each stage of strain, it takes a long time (around 8 minutes) to read and manually record all the data. During this operation, the panel continues to deform, (by structural creeping and redistribution of forces), which leads to a 6-10% drop in the load applied. A quicker, automated acquisition of data will need to be implemented;

– the blocking of the lower HEB along the "*x*" axis proves relatively effective, in that the movements recorded at the level of points 8 and 10 are similar and are less than 0.4 mm. The results are less satisfactory, however, in terms of movement along the "*y*" axis. There is a progressive displacement downwind of the lower HEB (measuring point 7 in Figure 8.6), which moves back by 1.5 mm. On the upwind side, the lower HEB is first dragged by the panel (negative displacement at measuring point 9, which reaches a value of 0.5 mm for a load of 1000 daN) before moving back by 1 mm at 1935 daN. Thus, the fixing of the lower HEB to the ground is insufficient, although the displacements remain small in relation to those recorded at the level of the panel being tested;

– the blocking of the upwind upper HEB (see Figure 8.5) is evaluated on the basis of measuring points 2 and 13, the movements of which along the "*y*" axis are presented in Figure 8.8. As before, we note that the fixing is insufficient, but the existence of movements attests to the mobilization of a "vertical" load (in relation to the configuration of the panel in a construction) which we are unable to evaluate on this first arrangement. In addition, this load is only applied on the upwind side of the crosspiece, whereas in reality, the strains of the upper levels of the construction limit the value of this support reaction, and they are applied to all the upright pieces. This connection will therefore be reviewed in the second test. Finally, the difference of the displacements between points 2 and 13 shows that there is a slight rotation of the upwind upper HEB between 800 and 1800 daN. This movement has an influence on the deformations of the upper crosspiece and skews the interpretation of the shifts of measuring points 3 and 4.

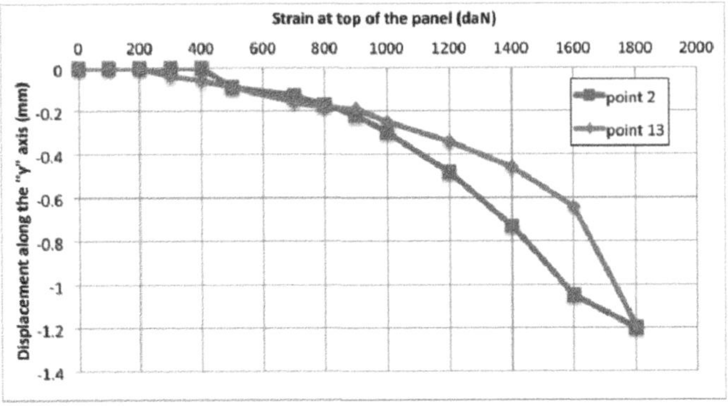

Figure 8.8. *Displacements along the "y" axis of the upwind upper HEB*

8.2.5. *Behavior of the wooden panel*

When strain is applied to the panel, the upwind strut is subject to a compression force, and the downwind strut works in traction. When the movement of the upper crosspiece is great enough, there is rotation of the intermediary uprights, leading the struts to lock with these uprights. This locking will remain until the panel ruptures, so the uprights are able to play a part in bracing the structure. This strain causes significant bending of the uprights. The strain borne by the foot of the fourth upright causes it to twist. The deformation of the panel is illustrated in Figure 8.9.

Figure 8.9. *Deformation of the panel at the upwind and downwind parts*

During this test, no rupture of the wooden elements occurred. The progressive damage done to the panel is localized at the nailed joints by gradual creeping of the nails in relation to the wood.

When the maximum strain (2200 daN) was reached, the joint between the stressed strut and the foot of the fifth upright gave way, as shown in Figure 8.10.

Thus, all the strain was placed on the compromised strut. The bracing force then fell to 1850 daN and the displacement along the "x" axis of the upper crosspiece continued with a constant force. We halted the test with a displacement of 130 mm, because the compromised strut was absorbing a large amount of elastic deformation energy and becoming highly unstable. The norm NF EN 594 stipulates that the test be aborted when the horizontal displacement of the top of the upwind upright reaches 100 mm (whether or not the panel has actually ruptured).

Figure 8.10. *Rupture of the joint of the stressed strut at the foot of the fifth upright*

As regards the horizontal movements of the lower crosspieces, measuring points 11 and 6 give the upwind and downwind displacements of the crosspiece fixed to the uprights, while measuring point 12 gives the displacement of the crosspiece fixed to the lower HEB.

The displacements of points 6 and 11 remain identical throughout the test, which shows that there is no change in the length of the crosspiece attached to the uprights. Conversely, if we compare the data from points 11 and 12, we can see that there is a gradual slippage between the two crosspieces, reaching 2.5 mm at the end of the test. In addition, because of the difference between the displacements of points 12 and 10, the crosspiece fixed to the HEB slips in relation to it by 0.6 mm.

At the level of the upper crosspieces, the movements of the measuring points 1 and 5 are similar (relative difference between the two points below 10%). Hence, there is no significant change in the length nor relative slippage of the upper crosspieces.

8.3. Test on a composite panel of a wooden skeleton and hempcrete

8.3.1. *Description of the panel*

The panel tested is formed of a non-braced skeleton 145 mm in thickness, filled with banked hempcrete. Its total thickness is 280 mm and the skeleton protrudes from the hempcrete on one surface. In an upright view, the overall dimensions of the panel are 2.48 m high by 4.56 m long. Figure 8.11 offers a detailed view of the makeup of the wooden skeleton.

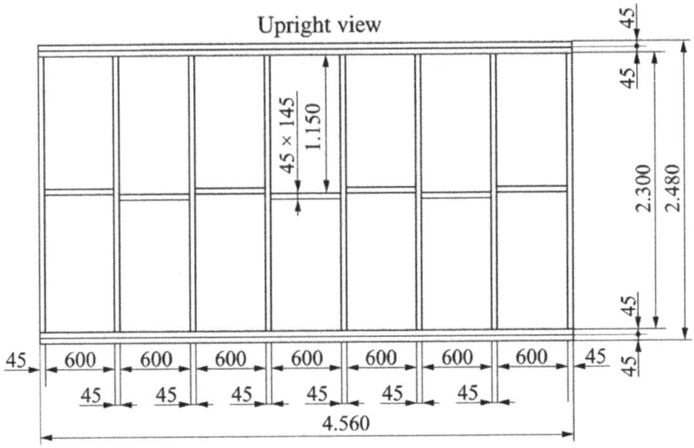

Figure 8.11. *Makeup of the wooden skeleton*

At the time of the test, the hempcrete had been in place for 1 year. The formula used is that employed for an application in "walling", and is characterized by a binder content of 220 kg/m^3 and an apparent density of 300-350 kg/m^3.

8.3.2. *Emplacement of the panel on the bracing bank*

Preliminary note: given that the panel was positioned flat on the bracing bank, the true vertical loads applied by the upper floors or by the roof structure become horizontal loads to be applied to the panel in the laboratory. However, in the report that follows, we shall keep the term "vertical loads" to denote these actions.

Similarly, the term "upper crosspieces" denotes the crosspieces situated in line with the jack. The "lower crosspieces" are those on the opposite side to the upper crosspieces. The configuration of the panel installed on the bracing bank is illustrated by Figure 8.12.

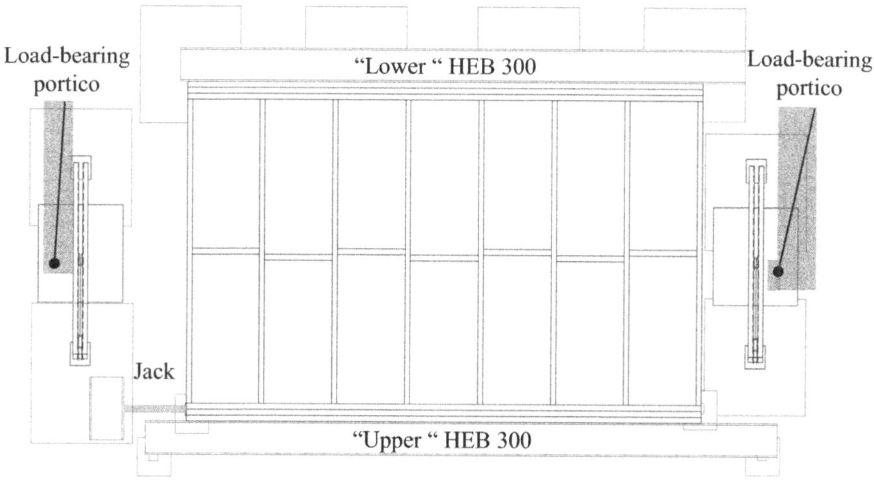

Figure 8.12. *Bird's-eye view of the panel installed on the bracing bank*

The panel rests on the floor on five supports, comprising a polished, greased plate of stainless steel and a Teflon plate. In relation to the test described above, a silicone grease was added between the Teflon and stainless steel plates. This limits the friction coefficient between the supports and the panel to 0.01. The principle of the supports is shown in Figure 8.13.

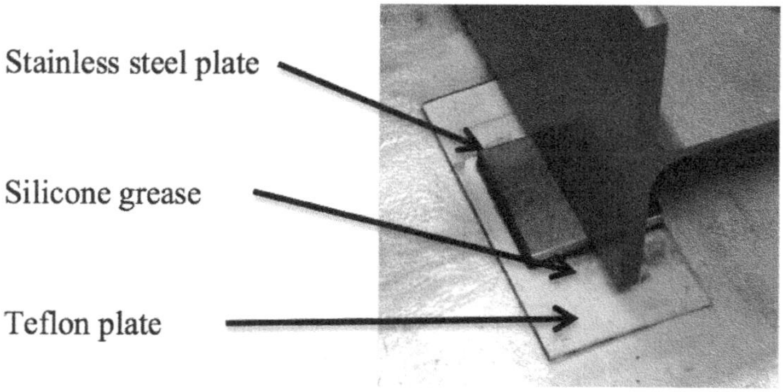

Figure 8.13. *Typical sliding support*

Two lintels (145 mm × 45 mm) are positioned "above" and "below" the panel. The upper lintel is nailed to the upper crosspieces. The lower lintel is simply bolted between the lower crosspieces and the lower HEB 300. These devices are defined by the norm and are presented in Figure 8.14.

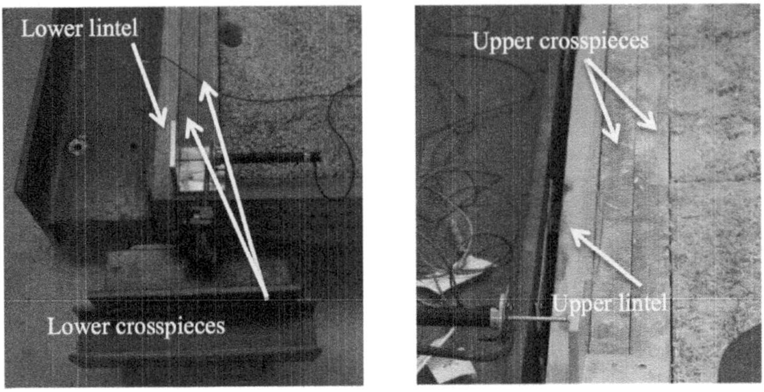

Figure 8.14. *Positioning of the lintels at the upper and lower parts of the panel*

The join between the panel tested and the lower panels in the construction or the foundation bedplates of the building is formed by the lower HEB 300 affixed in four reinforced concrete foundation bedplates. The panel tested is fixed to this HEB by 7 threaded rods 10 mm in diameter distributed from the center of each hempcrete panel, as shown in Figure 8.15.

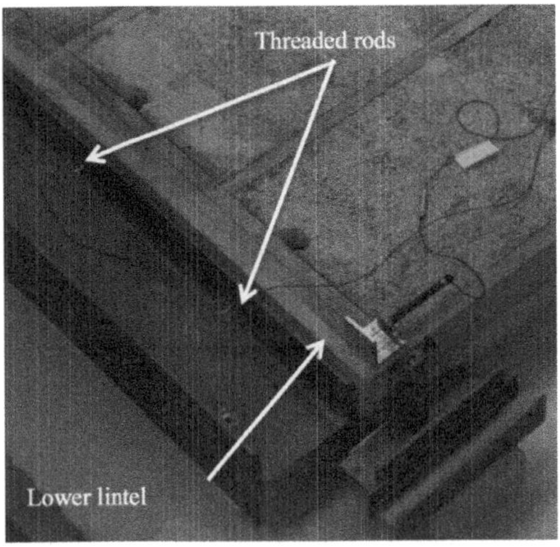

Figure 8.15. *Joint between the panel and the lower HEB*

8.3.3. *Vertical loading*

Figure 8.16 illustrates the "vertical" loading of the panel, which is exerted by the upper HEB 300 pressing on the panel through eight supports (100 mm × 100 mm) of the same combination of greased stainless steel and Teflon plates as described above. These supports are positioned in line with the uprights. They will not prevent the longitudinal movement (i.e. in the direction of the pressure from the jack) of the upper lintel of the panel.

The two beds of concrete, each weighing 1000 daN, through the pulley cables and the HEB, create a 2000 daN force distributed locally around the 8 uprights. Thus, a vertical load of 250 daN is applied to each upright, which corresponds to a distributed load of 440 daN/m on the upper lintel.

As prescribed by the norm NF EN 594, the vertical load on the upwind upright is shifted by 100 mm in relation to the side of the panel (for details see Figure 8.16).

The upper HEB 300 rests on the ground through three greased stainless steel/ Teflon supports as described above. Thereby, we are assured of nearly all the weight of the concrete blocks' being transferred to the panel.

This system for strain application, which might be described as "supple" in that the strain applied remains constant (even when the upper crosspiece moves vertically during the test), is a fairly close approximation of genuine conditions of use, and conforms to the norm.

Figure 8.16. *Vertical loading and a detailed view at the level of the upper lintel*

8.3.4. *Longitudinal loading and measurement of the movements*

The longitudinal strain is exerted by a 50 kN manual jack, whose action is measured by a deformation-gage strain sensor (see the detailed view in Figure 8.16).

The movements are measured by 50 mm resistive sensors and a 150 mm inductive sensor. As illustrated in Figure 8.17, 5 measuring points were chosen. The norm requires that the movements of points 113, 117 and 120 be measured. Points 118 and 119 were added to verify the pertinence of the "vertical" loading.

All the sensors are linked to a recording center which takes a measurement every 5 seconds from each one of the instruments.

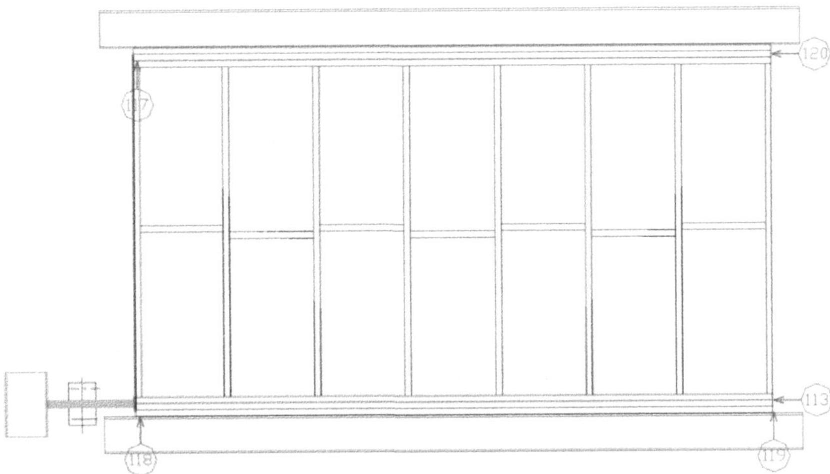

Figure 8.17. *Position and direction of the movement measuring points*

The norm prescribes longitudinal loading which refers to the probable rupture stress estimated by the previous tests. As this is the first test we have performed on this type of structure, we hypothesized that the resistance of the panel would be similar to that of the wood-framed panel from the previous tests (i.e. 2000 daN). The cycle of strain application imposed by the norm resulting from this hypothesis is illustrated by Figure 8.18.

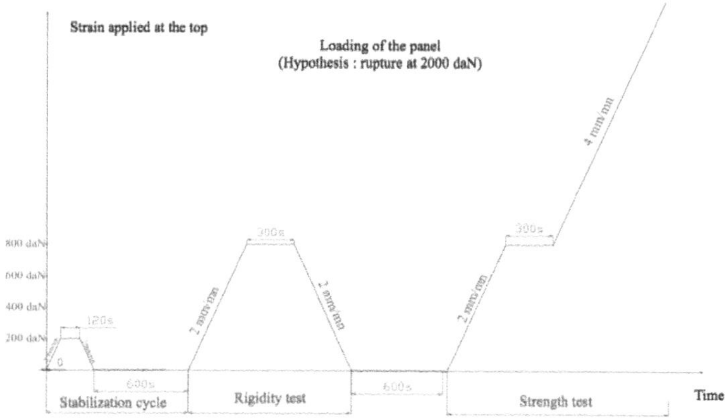

Figure 8.18. *Longitudinal loading of the panel as prescribed by the norm*

The stress levels recorded during the test is illustrated in Figure 8.19. It differs from the simulated curve as regards the speeds of loading and unloading because our strain application was not controlled in terms of movement, and the strain relief was controlled by the rate of retreat of the jack. However, the values of loading at each stage are respected.

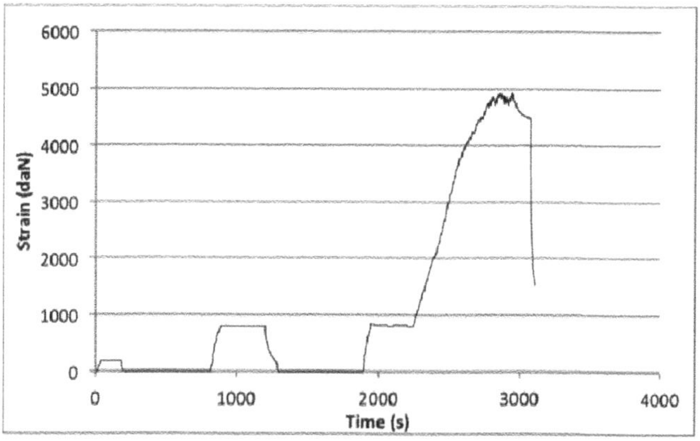

Figure 8.19. *Longitudinal loading applied*

8.3.5. *Running of the test*

The shape of the curve showing the progression of the horizontal movements of the crosspiece (sensor 113 as shown in Figure 8.17) is similar to the change in the strain applied.

During the rigidity test, this movement reaches a value of 1.4 mm (under a strain of 808 daN) and we observe a slight creep of 0.11 mm during the "plateau" of this strain level. When the strain is relieved, the movement instantly falls back to 0.32 mm, and then we observe a recovery by 0.1 mm.

During the strength test, the first application of strain (up to 854 daN) causes a movement of 1.35 mm, which is comparable to the value obtained during the rigidity test.

After the plateau of strain, for which we observe a creep by 0.06 mm, the movements increase up to 80.57 mm for a maximum recorded strain of 4915 daN.

The other movements measured are more difficult to interpret, because the sensors lack stability and the movements do not evolve steadily. In fact, significant jumps are observed, because there are phases of blocking followed by sudden relaxation. Besides the uncertainties intrinsic to the sensors being used, this is probably due to the transfer of strain between the wooden skeleton and the hempcrete, coupled with phenomena of shearing of the hempcrete. In addition, it is likely that the friction between the base of the panel and the lower lintel, as well as partial separation by slippage of the join between the panel and the HEB, skews the measurement (particularly at sensors 117 and 120 as shown in Figure 8.17).

Sensor 117 shows a separation between the lower lintel and the lower crosspiece by the upwind upright (the value of which reaches 20 mm by the end of the test) illustrated in Figure 8.20.

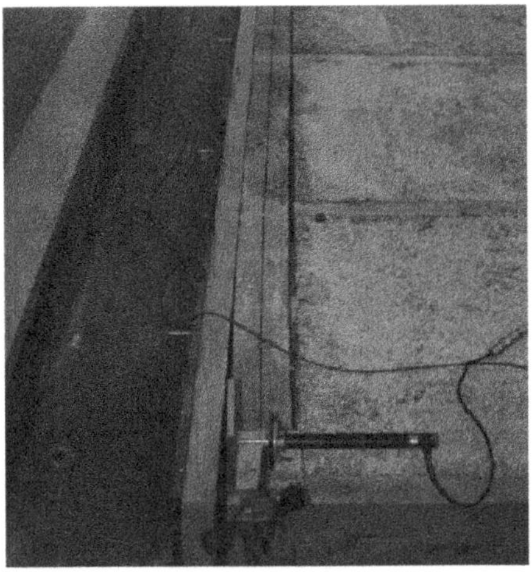

Figure 8.20. *Separation of the lower crosspiece at the end of the test*

The similar values obtained by sensor 118 show that there is no significant variation in the length of the upwind upright (i.e. on the side of the jack).

The variations in amplitude at sensor 119 are no greater than 5 mm, and therefore are more limited: there is no significant vertical movement of the top of the downwind upright.

As regards sensor 120, which measures the horizontal movement (in the direction of the jack) from the foot of the downwind upright, there are too many variations to calculate the relative displacement between the top and the bottom of the upright as the norm stipulates. However, the movements at the bottom are relatively small in comparison to the movements at the top (less than 10 mm for sensor 120 at the end of the strength test, compared to 80 mm given by sensor 113). The information from this sensor will therefore not be taken into account (which leads to an underestimation of the rigidity of the panel because we take account of the absolute horizontal displacement of the top of the downwind upright instead of looking at its relative displacement in relation to its bottom).

8.3.6. *Feature of the ruin of the panel*

At the end of the strength test, we observe that the hempcrete has developed a network of "stairwell-shaped" cracks, which forms a diagonal slab of compressed hempcrete as shown in Figure 8.21. The mechanism of formation of this diagonal is related to the shear stress within the hempcrete caused by the field of displacements imposed by the bending of the vertical uprights in the skeleton.

Figure 8.21. *Cracking of the hempcrete at the end of the test (photo taken from the point of horizontal loading)*

In addition, we noted separation of the upwind upright, which suffers a significant amount of bending in its top half (see Figure 8.22). This upright compresses the hempcrete behind it, but the separation seen in the hempcrete shows that it is not subject to this strain throughout the entirety of its thickness (280 mm) at the local level. We can therefore hypothesize that the overall strength of the panel would be greater if the upwind upright were wider than 145 mm, thereby directly mobilizing a larger thickness of hempcrete.

Figure 8.22. *Shearing along the hempcrete at the level of the upwind upright*

Finally, as Figure 8.23 shows, we can also see separation, at various points, between the uprights and the hempcrete, in the areas of contact subjected to the traction force.

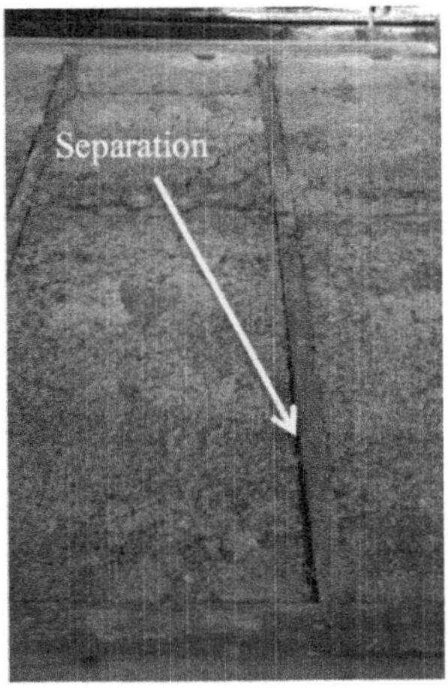

Figure 8.23. *Separation between the hempcrete and the vertical uprights*

Finally, let us point out that the ruination of the sample being tested was never truly attained by the time the test was stopped, because the panel was still resisting a 5000 daN force (the maximum capacity of the jack). In addition, the horizontal movement of the upper crosspiece is then 80 mm, so less than the 100 mm limit set out by the norm NF EN 594.

8.4. Results and comparative analysis

In terms of the panel filled with hempcrete, we can calculate the rigidity of bracing from the gradient of the curve representing the horizontal strain depending on the horizontal movement of the top of the panel, shown in Figure 8.24.

In terms of the rigidity test, we obtain a value of 577 daN/m when stressed and 667 daN/m when relaxed. This value appears again in the first phase of the strength test, for which we obtain a gradient of 632 daN/m. These values are calculated for a 17% degree of loading, corresponding to a horizontal strain of 800 daN.

It should, however, be noted that the loading/unloading cycle reveals behavior which is not perfectly elastic (we observe non-reversibility between the phases of loading and unloading) and we register a slight residual displacement at the end of the relaxation phase. These observations can likely be put down to the plasticization of the joints of the wooden structure and/or the properties of the wood/hempcrete interface.

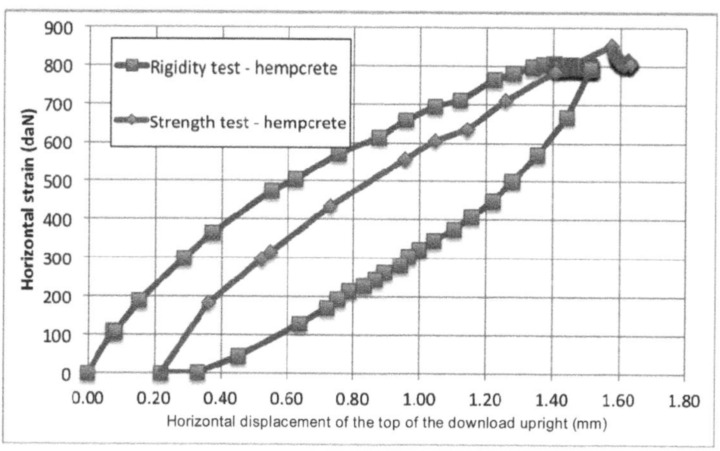

Figure 8.24. *Bracing rigidity of the hempcrete-filled panel*

Finally, the behavior exhibits a slight degree of non-linearity, which tends to reduce during the strength test. This phenomenon is similar to what we observe when carrying out tests to measure instantaneous compression moduli in test tubes of hempcrete.

If we apply the same approach for the braced wooden panel, we obtain a bracing rigidity measurement of 68 daN/m at 800 daN, and 74 daN/m for a degree of loading of 17%, comparable to that of the hempcrete-filled panel.

Let us point out that according to the norm, the rigidity is calculated by taking account of the difference in horizontal displacement between the top and the bottom of the panel. In these conditions, the braced wooden panel would exhibit a rigidity of 74 daN/m at 800 daN and 80 daN/m at 17%.

In Figure 8.25, we have shown the changes in the horizontal strain in comparison with the horizontal movement of the top of the downwind upright for the two panels. We note that the behavior of the hempcrete-filled panel is more satisfactory than that of the braced wooden panel:

– the rigidity of the wood/hempcrete panel is greater than that of the braced wooden structure;

– the peak of strain is decidedly higher;

– finally, the deformation at the level of that peak is more significant.

Furthermore, for a given level of strain, the hempcrete-filled panel deforms far less (which is entirely logical, given the rigidity values calculated).

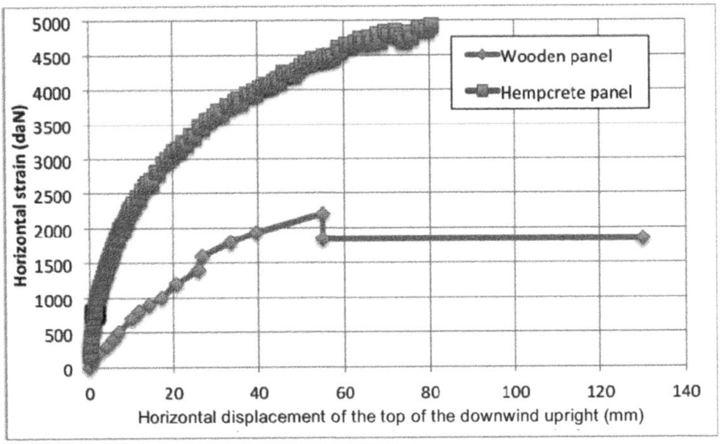

Figure 8.25. *Bracing behavior of the "braced wood" and "hempcrete-filled" panels*

8.5. Conclusions and reflections

To begin with, a test bank was designed and built in order to carry out a full-scale bracing test on a panel with a wooden skeleton braced by struts. Apart from the flat positioning of the panel, in many ways it is in line with a norm currently in force.

The results obtained enabled us to validate the experimental setups, while giving indications as to ways in which the experiment might be optimized. These indications were followed and the improvements applied when a second test bank was constructed.

A second test on a wooden panel – not braced this time, but rather filled with banked hempcrete – was then carried out.

The results, when analyzed, demonstrated that the presence of the hempcrete as a filler of the panel is able not only to substitute the bracing struts (which, incidentally, are the pieces which require most work when the skeleton of the panel is being built), but also to greatly improve the mechanical behavior in terms of bracing. Indeed, the peak charge is 2.2 times higher, but in particular, the bracing rigidity is increased almost tenfold.

Remember that in the two tests, the wooden skeleton of the panels was rigorously made identical apart from the presence of the bracing struts.

Finally, the hempcrete provides a degree of ductility which is advantageous as the point of rupture approaches, in that it limits any risk of instability at the level of the elements of the wooden skeleton.

Let us point out that since these tests were carried out, we have continued to make alterations to the testing bank to be used, replacing the strain application system and the sensors in order to integrate them into an acquisition and control system whereby the test can be conducted more rigorously.

While these tests still need to be repeated on a greater number of test substances in order to confirm these trends, and a statistical analysis of the measurements recorded needs to be carried out, we can conclude that the development of a computation model that is able to take account of the contribution of hempcrete in absorbing the strain in bracing seems to present real potential.

This approach should enable us to optimize or simplify the panels with wooden skeletons in order to enhance the economic relevance of the construction system.

Besides the intrinsic characteristics of the skeleton, to feed into this model, we need a realistic idea of the mechanical characteristics of the hempcrete (particularly its compressive strength, but above all the deformation moduli and the shear strength).

8.6. Acknowledgements

The CRDA, where both authors of this chapter work, is a structure supported by the *Conseil Régional de Champagne-Ardenne* (Champagne-Ardenne Regional Council) and *Délégation Régionale à la Recherche et à la Technologie* (Regional Delegation for Research and Technology). The Agrobat project was financed by the FFB Champagne-Ardenne, with the contribution of Judith Lego and Didier Valem from the *Direction des Affaires Techniques de la FFB*, Sonia Guldener from the *FFB Champagne-Ardenne* and the company *Le Bâtiment Associé*.

8.7. Bibliography

[FFB 09] FFB CHAMPAGNE ARDENNE, Projet Béton de chanvre AGROBAT, Summary report 128 pages, February 2009.

[AFN 08] AFNOR, NF EN 594 – Essai de raideur et résistance au contreventement des murs à ossature bois – Méthodes d'essais, AFNOR, November 2008.

Chapter 9

Examination of the Environmental Characteristics of a Banked Hempcrete Wall on a Wooden Skeleton, by Lifecycle Analysis: Feedback on the LCA Experiment from 2005

The aim of this chapter is not to update our lifecycle analysis (LCA) performed on hempcrete in 2005. Rather, it is to re-contextualize that study, and to give the reader certain indications to help understand the working hypotheses made at the time and the results of the investigation. In addition, this chapter affords us the opportunity to place our 2005 LCA in perspective in relation to the present-day context in the domain of hempcrete.[1]

9.1. Introduction

Besides their intrinsic properties – which confer upon them performances recognized as good in terms of thermal, acoustic, hygrometric and mechanical behavior as well as fire resistance – bio-sourced materials (essentially drawn from plant resources) seem to exhibit environmental advantages, which make them very

Chapter written by Marie-Pierre BOUTIN and Cyril FLAMIN.

1 In 2005, the same team performed a lifecycle analysis on thermoplastic compounds impregnated with hemp fibers, and on a banked hempcrete wall with a wooden skeleton, at INRA – the *Centre de Lille-Estrées-Mons*.

valuable in the context of sustainable construction in today's world. LCA, a method which was developed in the 1960s to evaluate the environmental impacts relating to the use of a product or service, has been applied to many such innovative construction materials. In particular, one might cite the recent study on straw bales used in a wall over a wooden skeleton and on a similar construction coupling a wooden skeleton with miscanthus concrete [LOP 11], and on a flax concrete block [TOU 12]. Hence, by using such materials, we would be able to limit the use of non-renewable fossil resources and decrease the greenhouse gas (GHG) balance by absorption of CO_2, which plants need in order to grow.

In 2005, the Direction des Politiques Economique et Internationale (DPEI) (Dept. Of Economic and International Policy) in France's Agriculture and Fisheries Ministry (MAP), in partnership with hemp-growers, asked the Institut National de Recherche Agronomique (INRA – National Institute for Agricultural Research) to carry out the first lifecycle analyses on thermoplastic compounds impregnated with hemp fibers, and on a banked hempcrete wall with a wooden skeleton, in conformance with the ISO 14040 series of standards.

In the early 2000s, France was the largest producer of hemp in Europe. This market sector then sought to diversify its outlets. Thus, in addition to paper-making, attention turned to the new markets of plasturgy (thermoplastic compounds impregnated with hemp fibers) and construction (banked hempcrete over a wooden skeleton in walling). We need to be aware of the potential environmental impacts due to the manufacture of these two types of products in order to discover which stages of their lifecycle are most harmful to the environment. In the context of a quest for continual improvement, such knowledge enables us to further decrease the potential negative impacts of these products for the environment by focusing our efforts on these particular stages.

In addition, the hemp lobby deems that, in terms of the prospect of substituting mineral fibers in composite plastics and alternatives in conventional construction methods, there is no better promotion than to demonstrate the environmental gains facilitated by the presence of hemp in these new products. However, at the time, these environmental characteristics had not yet been totally proven.

Figure 9.1 shows the outline of the study as it was defined in 2005 by the committee in charge of the project: to perform LCAs on banked hempcrete on a wooden skeleton (the building market) and thermoplastic compounds impregnated with hemp fibers (the plasturgy market), with emphasis on the agronomic part in order to highlight the role and environmental advantage to hemp in these materials. As these two materials each used one of the two products gained by the primary transformation of hemp straw (hemp shiv and fibers), their use as examples offered a nearly complete view of the plant's value.

The governing committee, in charge of defining the boundaries of the study, guiding and monitoring the LCA, contained members of institutions (the agriculture ministry, ADEME, the ecology ministry), industrial players in the hemp trade (BCB-Lhoist, AFT Plasturgie, UTC, Plasticana) and members of technical and R&D institutes (FNPC, INRA, Construire en chanvre).

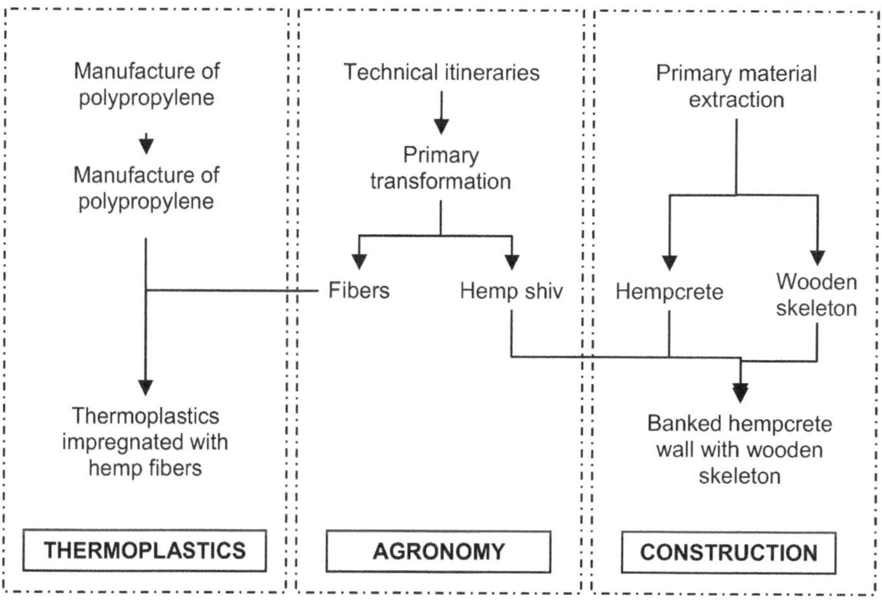

Figure 9.1. *Field of the LCA study (source courtesy of FNPC, LCDA, PDM Industrie, Eurochanvre, INRA, AFT Plasturgie, BCB-Lhoist, Construire en Chanvre)*

It is our intention, in the rest of this chapter, to revisit the main results of the LCA on the banked hempcrete wall on a wooden skeleton, and the reflection carried out in order to perform this (at the time, groundbreaking) work on the evaluation of the environmental benefits of hempcrete.

9.2. Description of the products studied

Hemp is used as a starter crop in crop rotation systems.

It provides two main types of products which are cortical fiber and hemp shiv.

– Made from hemp straw, the cortical fiber previously used for the textile industry was, in 2005, used primarily in paper-making. However, thanks to its mechanical properties, it was also used in the making of products such as composites, insulation used for construction and in the production of reinforced thermoplastics to be used in the automobile industry.

– Made from the marrow of the stalk of the hemp plant, hemp shiv in 2005 was mainly used to make animal litter, because of its capacity for absorption. Yet its properties of low density and its high insulating capability meant it was becoming more and more widely used in the composition of certain construction materials. This trend is now confirmed by the shift in the use of hemp shiv by the building industry.

When mixed with an aerated-lime-based binder and ready-made hydraulic binder (for the purposes of the study, Tradical 70® was used), hemp shiv can be used in the making of concretes which, as they are not load-bearing themselves, are then mechanically shot onto a load-bearing wooden structure. The concrete is shot onto a banking panel placed behind the wooden skeleton, until a thickness sufficient to cover the load-bearing structure is attained. The wood-skeleton/hempcrete wall created thereby exhibits specific and excellent technical characteristics, such as a high insulating capacity, good sound correction and permeability to water vapor.

9.3. Method for environmental evaluation of bio-sourced materials

Let us here give a little more detail about the lifecycle analysis (LCA) method; LCA is a standardized method (ISO 14 040 series) which can quantify the impacts of a "product" (be it a manufactured good, a service or a procedure), from the extraction of the primary materials that make it, right up until its disposal at the end of its life, encompassing the phases of distribution and use along the way – i.e. it offers a "cradle to the grave" view of a product's environmental impact [BOE 05].

An LCA is built around four phases:

– definition of the objectives and scope of the study:

- the system under study (its sub-systems, and the stages within those sub-systems);

- the functional unit (i.e. the common unit that serves as a reference point to express the environmental balance of a "product". The functional unit enables us to quantify all the results of an LCA in relation to the service rendered by the "product").

– Analysis of the inventory of flows of matter and energy associated with the stages of the lifecycle, boiled down to the functional unit chosen.

– Evaluation of the potential impacts resulting from the flows of matter and energy listed. For this purpose, coefficients are used to calculate the contribution of each flow to the various environmental impacts under study. These coefficients are determined by scientific experts in each domain (e.g. the coefficients for calculating the greenhouse effect were provided by the IPCC[2]).

– Interpretation of the results obtained through the lens of the objectives chosen: this interpretation is performed iteratively, i.e. after each of the three previous steps, in order to verify that the results obtained are able to satisfy the objectives of the investigation. It is also at this stage that the robustness of the results is evaluated (e.g. by sensitivity analysis).

When communicating the results of an LCA, the testing conditions and the results must be presented in a detailed and transparent manner. The working methodology must be clearly recapped, because the results will only be meaningful in view of the working hypotheses made. When the results of the study are intended to be published, a summary write-up clearly presenting the objectives and scope of the study, the main limitations and the hypotheses, etc., must be drafted and made available to all.

With results that invite a comparison between products or procedures, this report must necessarily include a critical review – that is, an examination of the study by an independent expert. This expert may act alone or as part of a critical review board bringing together specialists in the sector in question and the main interested parties, with the essential aim being to bring the combined expertise of the LCA team and the sector professionals to bear, and of course, ensure total impartiality. Comments and responses to the recommendations issued by the critical review must be included in the summary report when it is distributed [BOE 05]. It should be noted that, because it was not possible in the context at the time, the responses to the critical review of the hemp LCA in 2005 were not included in the summary report, which led to the removal from the report of the comparison between the results for the banked hempcrete wall on a wooden skeleton and those for a wall built of double polystyrene concrete blocks with the same thermal resistance. Without calling into question the methodology applied and the results obtained, this critical review also listed a number of points which could be improved in the study so as to conform more closely to the stipulations of the ISO 14040 series of standards.

---

2 Intergovernmental Panel on Climate Change.

There are many advantages to LCA. It can be a decision-support tool when designing or renovating a building or when choosing a material to use in these operations, with the aim being to select the best technology available and minimize the impacts on the environment. In addition, it can help increase our familiarity with a product in order to be better able to target improvements to it, or facilitate and lend credibility to any communication about that product [LOP 11].

The results of an LCA on a material may also be useful in the writing of the *Fiche de Déclaration Environnementale et Sanitaire* (FDES – Environmental and Sanitary Declaration [Certificate]) used in France (see Chapter 1). This certificate is one of the tools used in the voluntary HQE® (high environmental quality) approach (a method intended to help project managers monitor and control the impacts of a building on the outside environment and create an indoor environment which is safe and comfortable for its occupants). It is based on the French norm NF P 01-010 (which defines the principles applicable to the provision of information about a product's environmental and sanitary characteristics). The INIES database housed by the CSTB references all the FDESs of construction products which conform to this standard.

9.4. Lifecycle analysis on hempcrete – methodology, working hypotheses and results

The LCA on banked hempcrete on a wooden skeleton was thus carried out in accordance with the methodology set out in the ISO 14 040 series of standards, at each stage respecting the definition of the parameters crucial for its realization. The results which the study yielded cannot be discussed here without first reviewing the methodology employed and the working hypotheses used.

9.4.1. *Delimitation of the system under study*

The lifecycle tree drawn up by the working groups devoted to the "agronomy" and "construction" parts of the lifecycle can be used to identify the main stages of the lifecycle of the wall, and the main inputs and outputs, whether in terms of energy flow or materials.

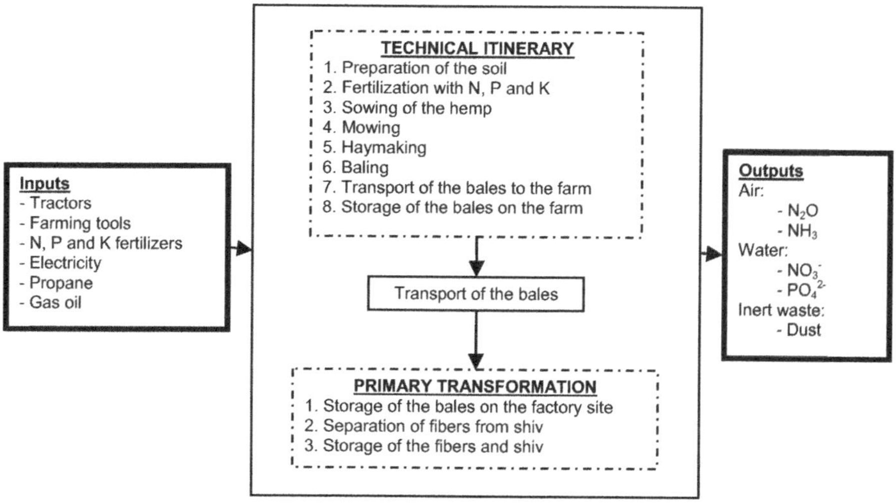

Figure 9.2. *Stages considered in the study of hemp shiv production (Source courtesy of FNPC, LCDA, PDM Industrie, Eurochanvre, INRA)*

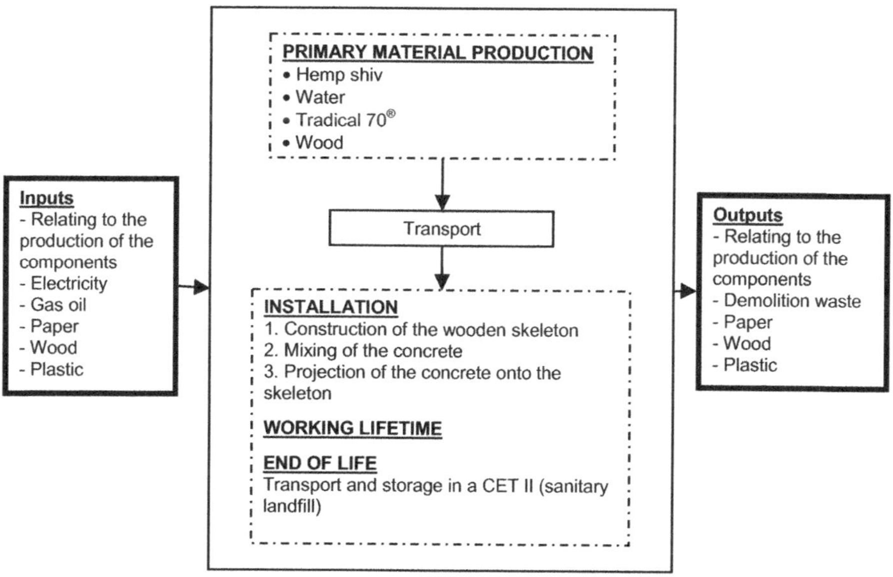

Figure 9.3. *Simplified lifecycle tree for the banked hempcrete wall on a wooden skeleton (Source courtesy of BCB-Lhoist, Construire en chanvre)*

The elaboration of such a system leads us to make a certain number of hypotheses and choices, which must always be clearly stated because they have a bearing on the interpretation of the results of the LCA. If exclusion is argued for, certain stages of the cycle may not be taken into account if they fall under the remit of the cutoff rule authorized in the ISO 14041 standard. This cutoff rule sets the criterion for the inclusion of the inputs and outputs of the process. In 2005, the choice was made to follow the cutoff criterion set out in the standard governing the environmental and sanitary quality of construction materials (NF P01 010), which sets the threshold at 98%. Occasionally, the environmental impacts of certain stages or items cannot be taken into account, for want of available data; such is the case, for instance, with stages spread out over the whole of the lifecycle, such as the behavior with sudden changes in loading. By no means can this exclusion apply to key stages of the lifecycle.

Here, we shall not go into details about the motives for excluding each of the stages which are given in the LCA study, but we list them here for easier understanding and interpretation of the results.

In terms of the "agronomy" part of the cycle, the system studied does not take account of:

– the stage of production of seeds and production of hempseed sometimes associated with the production of straw;

– the waste relating to the moderate use of synthetic fertilizers (packaging, rinsing products for the agricultural material…);

– the impact relating to the manufacture of the machines that perform the primary transformation, the storage sites and the conditioning of the fiber and the shiv.

In addition, the dust caused by the process of separation of the fibers and the shiv was taken into account as a waste product because at the time, there was no avenue by which this by-product could be used and given value.

In terms of the "construction" part of the cycle, the system studied does not take account of:

– the electrical consumption needed to build the wooden skeleton;

– the stages of sudden changes in the loading and the material's behavior at these times.

– the reuse and/or recycling of the hempcrete material were not taken into account either (the hempcrete lost during projection onto the wall is mainly put back into the concrete-mixer).

For the entire cycle, the cutoff criterion prescribed by the NF P01 010 norm was applied; hence, inputs which account for less than 2% of the total mass of the inputs were not taken into account.

The *functional unit* chosen for the study of the banked hempcrete wall on a wooden skeleton was to perform the function of a load-bearing wall over a 1 m^2 surface area with a thermal resistance of 2.36 m^2K/W for a year. This value for the thermal resistance was chosen to be equivalent to that of a conventional wall built of double-polystyrene concrete blocks.

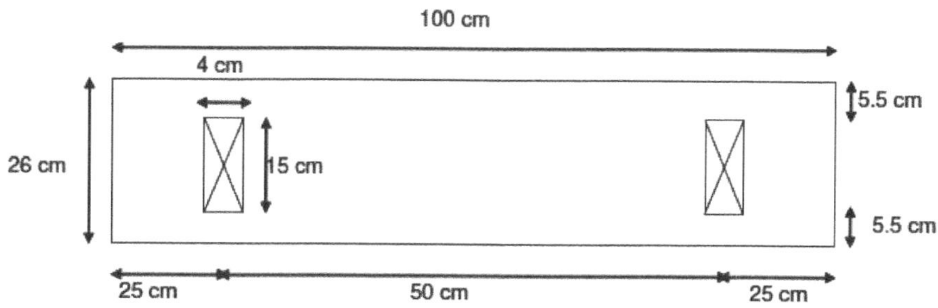

Figure 9.4. *Transversal cross-section of the functional unit (Source courtesy of BCB-Lhoist, Construire en chanvre) representing two wooden uprights embedded in the hempcrete matrix (typical cross-section)*

The *typical lifetime* chosen for the wall in question was estimated by the group of experts to be 100 years, on the basis of the lifetimes of other materials such as autoclaved concrete or terracotta bricks. Maintenance of the wall is assumed to relate essentially to the flagging of the surface of the wall and not its core. This parameter is crucial in an LCA, because it is used to define the *reference flow* which represents the mass of materials required to ensure the defined function of the functional unit. In this study, it was calculated on the basis of the quantities of materials needed to construct 1 m^2 of wall as defined in the functional unit, divided by the lifetime. In fact, the longer the lifetime, the smaller is the reference flow, and therefore the lesser are the environmental impacts associated with the quantities of materials involved.

Component	Necessary amount (kg/m^2 of wall)	Typical lifetime (years)	Reference flow (kg/func. unit)
Tradical 70®	54.5	100	0.545
Hemp shiv	24.8	100	0.248
Water	37.2	100	0.372
Wood	5.5	100	0.055

Table 9.1. *Reference flow (Source: BCB-Lhoist, Construire en chanvre)*

Given that the oldest buildings to use hempcrete are less than 30 years old, the lack of feedback about the behavior of these constructions over time means it is impossible to actually verify the hypothesis of a 100-year working life. However, an aging test carried out since by the CEBTP has shown that hempcretes, if they respect the crucial requirements, can withstand 20 freeze-thaw cycles, which is similar to the performances of a soft sandstone [CEB 05].

9.4.2. *Inventory analysis*

To perform an inventory analysis we need, firstly, to define a reference scenario for the procedure under investigation, and secondly, to gather data regarding the nature and amount of inputs and outputs involved during the different stages of that scenario.

Indeed, for the production of the same good, there may be procedures that are similar but where the nature or the amount of inputs or outputs involved are noticeably different. Such is the case, in particular, for agricultural production, which depends heavily on the soil – a factor which varies from one place to another. Thus, depending on the soil types, the quality and quantities of nitrogenous fertilizers used for a crop for a given yield may vary.

The *reference scenario* drawn up by the working group comprising experts in the field of hempcrete (BCB-Lhoist and Construire en chanvre) is intended to include the characteristics of the trade at the time. This scenario is the foundation for the inventory of flows of matter and energy in and out of the system (inventory analysis), and therefore for the evaluation of the environmental impacts, so it needs to reflect the organization of the domain as accurately as possible (see Figure 9.5).

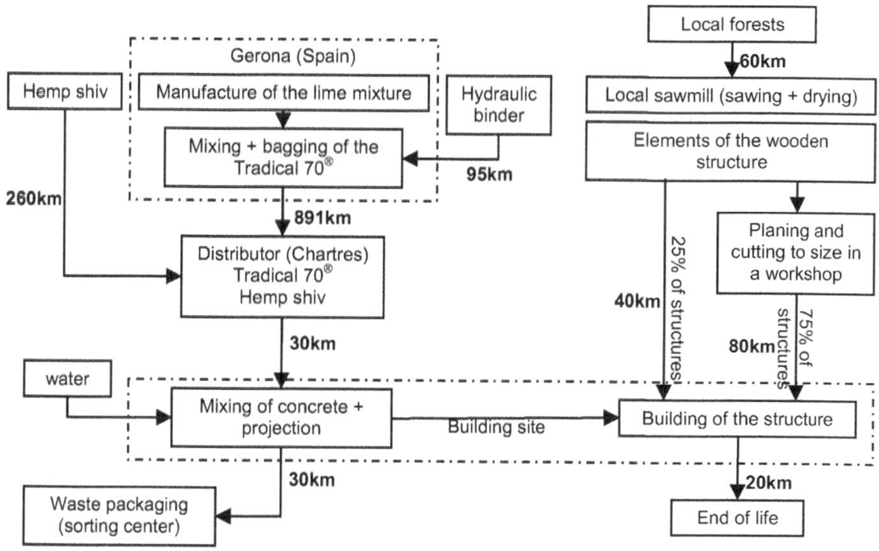

Figure 9.5. *Reference scenario for a lifecycle analysis on a banked hempcrete wall on a wooden skeleton (Source courtesy of BCB-Lhoist, Construire en chanvre)*

The drawing up of a reference scenario includes a task of reflection which, again, leads us to make a certain number of working hypotheses to simplify a system which is always complex, whilst still remaining faithful to the ground truth. The main bias which creeps in is an under-estimation of the major stages in terms of environmental impacts.

Two main hypotheses were used in 2005 when constructing the reference scenario for the lifecycle of the banked hempcrete wall on a wooden skeleton. These hypotheses relate to the distances over which the materials need to be transported (by road) and the end of the natural life of the wall.

Thus, in 2005, the Tradical 70® was manufactured in Gerona in Spain, and then distributed over hundreds of kilometers to the three main distribution depots – in Brittany, Alsace and Isere. In order to simplify the system without minimizing this significant phase of transport, the average distance of distribution was estimated by taking the barycenter of distribution as a function of the volumes transported. As this barycenter was located in Chartres, the distance over which the Tradical 70® was transported was estimated at 891 km. The distance of distribution of the hemp shiv in question in the 2005 study was evaluated at 260 km because it was made and conditioned in Bar sur Aube, under the brand Chanvribat®.

The distances of transport relating to the construction of the wooden skeleton were estimated on the advice of an expert party (Construire en chanvre). They may seem slight in comparison with the distances travelled by the Tradical 70® or the shiv, but wood remains an important resource in France, and this domain is not subject to the same supply constraints. The distances are estimated on the basis of past experience, demonstrating that the use of alternative building materials such as hempcrete often goes hand-in-hand with a desire to stimulate local development, which is often accompanied by a quest to reduce the impact of the construction on the environment.

The second working hypothesis relates to the end of life of the wall in question, which is defined as storage in the nearest *centre d'enfouissement technique de classe II* (CET II – sanitary landfill) to the site. Indeed, as no data are available from tests on the degradation of hempcrete (rate of mineralization of the hemp shiv, nature and quantities of atmospheric emissions and lixiviates), there is no proof that would enable us to consider hempcrete to be an inert waste product that can be stored on a CET III site. Therefore, it had to be treated in the same way as raw wood, for which there were no data available either in 2005 about the nature and quantities of atmospheric emissions. Thus, as per the suggestion of the norm P 01-010, the end-of-life phase in the 2005 study takes account only of the mass of waste stored and the emissions relating to the transport by truck to the nearest CET II.

With these working hypotheses lain down and the reference scenario established, *the inventory analysis of the flows of matter and energy in and out of the system* is performed for each identified stage. The inventories of production of the Tradical 70® (lime, hydraulic binder), the wood and the other components of the wall and of the multiple stages identified in the scenario are performed using the software Simapro®, which includes numerous bibliographical references and databases such as Ecoinvent®.

The inventory of the amounts of substances likely to be emitted or consumed during the production of the hemp shiv was calculated using the software Biofit® [CAL 00], on the basis of a reference scenario drawn up by the group of experts involved in the study, which reflects a typical itinerary of hemp production (the result of inquiries made to hemp producers). The value of the inputs and the transport distances attributed to the production of hemp shiv was calculated on the basis of the values provided by experts in the domain (FNPC, LCDA, Eurochanvre and PDM Industrie) and the value of the outputs (nitrates, phosphates, etc.) on the basis of formulae taken from the existing body of literature, the details of which are given in the 2005 publication.

Initially calculated for the production of a kilogram of hemp straw, the quantities of substances involved are then distributed by a mass-based allocation between the different commercializable parts of the plant (in this case hemp shiv and fibers) in order to evaluate their specific contribution in terms of environmental impact.

In the 2005 LCA, the committee in charge estimated this mass distribution to be 60% for shiv and 40% for fiber; the coefficients of *mass allocation* were calculated as follows:

$$CmS = Ms/(Ms+Mf) \text{ and } CmF=Mf/(Ms+Mf)$$

– Ms: mass of shiv produced per ton of hemp straw (kg/t). The value of Ms is around 480 kg/t.

– Mf: mass of fiber produced per ton of hemp straw (kg/t). The value of Mf is around 350 kg/t.

– CmS: coefficient of mass allocation of the shiv. The value chosen in the study was rounded off at 60% by the committee in charge.

– CmF: coefficient of mass allocation of the fiber. The value chosen in the study was rounded off at 40% by the committee in charge.

The 2005 LCA also presents the results in terms of an economic allocation, taking account of the market value of both products. At the time, this economic allocation was estimated by the committee to be 32% for the shiv and 68% for the fiber. Note that ADEME, in its study of a simplified methodology for the conducting of LCAs of bioproducts in 2009, suggests using the economic allocation on the co-products created by separation of plant fibers if the differences between the mass allocation and the economic allocation are too great (the threshold is set at 10 points) as is the case here.

This system of allocation between shiv and fiber of the environmental impact relating to the manufacture of a kilo of hemp straw involves two factors. To begin with, it recalls an important working hypothesis in this study – i.e. the fact that the dust, which may account for up to 15% of the total weight, is considered to be a waste product. If there were clearly-identified outlets for the dust, it would then be considered as a co-product and would have its own portion of the environmental impact assigned to it, which would naturally decrease the estimation of the environmental impact of the production of hemp shiv. This allocation then implies that the fiber produced concomitantly with the shiv will in turn have its integral value increased (in the production of thermoplastic compounds in terms of the 2005 study). Otherwise, the proportion of environmental impact assigned to the shiv would have to be reviewed and increased.

The final working hypothesis relating to the inventory analysis relates to the storage of carbon in the product. Indeed, it is at this level of the study that the main asset of the banked hempcrete wall on a wooden skeleton – a biosourced material – is taken into account. It is a question of the *carbon absorber* which is manifested at the level of three components of the wall by two different mechanisms: photosynthesis (wood and shiv) and recarbonation of the lime.

Agricultural products are formed by using carbon dioxide (CO_2) from the atmosphere to drive the process of *photosynthesis*. As these products break down, the CO_2 is returned to the atmosphere. The quantity of CO_2 needed to create the dry hemp material is calculated as follows:

$$Qco_2 = (QDM \times PCDM) \times (MMco_2/MMc)$$

– Qco_2: mass of CO_2 needed to create a mass QDM of dry material.

– QDM: mass of dry material considered (g).

– PCDM: Proportion of carbon in the dry material. According to ADEME (1998), dry hemp material is composed of 45.9% carbon.

– $MMco_2$: molar mass of CO_2, equal to 44 $g.mol^{-1}$.

– MMc: molar mass of carbon, equal to 12 $g.mol^{-1}$.

By the above formula, 1.7 g of CO_2 are needed to create 1 g of dry material. This quantity of CO_2 taken up by the plant must therefore be taken away from the carbon balance of the hemp industry. Inventory analysis of the lifecycle of the banked hempcrete wall on a wooden skeleton shows that 52.2 kg equivalents (eq.) of CO_2/m^2 of wall are stored in the hemp shiv.

The same equation can be used to evaluate the amount of CO_2 stored in the wood; thus, when we consider the percentage of carbon (49.4%) in dry wood [WER 03], 9.9 kg eq. of CO_2/m^2 of wall are stored in the wooden skeleton. Hence, 62.1 kg eq. CO_2/m^2 of wall in total are stored by the organic components of that wall, by way of photosynthesis.

The phenomenon of *recarbonation* of the lime, for its part, can be explained as the absorption of atmospheric CO_2 by the slaked lime contained in the hempcrete throughout the operational lifetime of the wall. Indeed, the slaked lime is the product of the hydration of quicklime, which itself is made by decarbonation of limestone under the effect of heat (kiln firing at 900°C):

– Decarbonation:

$CaCO_3 \rightarrow CaO$ (quicklime) + CO_2 (emitted and taken into account in the study among the outputs from the lime-production process).

– Hydration:

$CaO + 2H_2O \rightarrow Ca(OH)_2$ (slaked lime) + H_2O.

– Recarbonation:

$Ca(OH)_2 + CO_2 \rightarrow CaCO_3 + H_2O$.

The proportion of CO_2 thus absorbed is estimated by the experts in lime from the working group (BCB-Lhoist) as 249 kg/ton of Tradical 70®, so 13.6 kg eq. CO_2/m^2 of wall. The rate of recarbonation is faster when the phenomenon is not slowed down by the presence of an outer flagging on the surface of the wall.

In the final analysis, 75.7 kg eq. CO_2 are stored by photosynthesis and recarbonation in 1 m^2 of the walling under study, which counterbalance the GHG emitted during the lifecycle and, therefore, decrease the balance in terms of the greenhouse effect of the banked hempcrete wall on a wooden skeleton.

The interested reader can refer to the 2005 study for further details on the construction of the reference scenarios and the inventory analysis which, finally, contained some 530 substances.

9.4.3. *Impact evaluation*

The aim of this part of the chapter is to quantify the potential environmental impacts that may be caused by the procedure under study here. Given that hempcrete is a construction material, the impacts considered in the study were chosen in line with the recommendations formulated in the norm NF P01-010, governing the environmental and sanitary quality of construction products. Nine potential environmental impacts were chosen:

– resource exhaustion (in kg Sb eq.). This indicator, expressed in kilograms per antimony equivalent (abbreviated to "kg Sb eq.") estimates the amount of resources consumed;

– atmospheric acidification (in kg SO_2 eq.). This indicator expresses the quantities of sulfur dioxide equivalent (notated kg SO_2 eq.) given off into the atmosphere;

– greenhouse effect over 100 years (in kg CO_2 eq.). This indicator expresses the amounts of equivalent carbon dioxide, the reference GHG (notated kg CO_2 eq.) given off into the atmosphere;

– destruction of the ozone layer (in kg CFC-11 eq.). This indicator expresses the quantities of equivalent CFCs, the gases responsible for the destruction of the ozone layer (notated kg CFC-11 eq.) given off into the atmosphere. In the CML referential framework, protoxide of nitrogen (N_2O), a gas resulting from nitrogenous fertilizers and which has an indirect effect on the destruction of the ozone layer is not expressed in CFC-11 eq. The destruction of the ozone layer caused by this gas was therefore not taken into account in the 2005 LCA;

– formation of photochemical ozone (in kg C_2H_4 eq.). This indicator expresses the quantities of equivalent ethylene (notated kg C_2H_4 eq.) given off into the atmosphere;

– consumption of non-renewable energy (in MJ). This indicator expresses the amount of non-renewable energy (in MJ) consumed;

– waste production (in kg). This indicator expresses the amounts of waste products (kg) produced;

– air pollution (in m^3);

– water pollution (in m^3).

For each substance listed during the course of the inventory analysis, there is an equivalence factor for a given mass of substances in a given potential environmental impact. These factors are known as characterizing factors. The values calculated for each substance are then converted into impacts using these factors. The sum is then calculated for each of the potential impacts examined in the LCA.

Impacts	Production of primary materials			Installation	Operational lifetime	End of life	Transport (total)	Total	
	Hemp shiv		Other prim. mat.					Mass	Economic
	Mass	Economic							
Resource exhaustion (kg Sb eq.)	2.8×10^{-2}	1.5×10^{-2}	7.7×10^{-2}	1.2×10^{-3}	0	0	2.6×10^{-2}	1.3×10^{-1}	1.2×10^{-1}
Atmospheric acidification (kg SO_2 eq.)	5.1×10^{-2}	2.7×10^{-2}	4.8×10^{-2}	1.3×10^{-3}	0	0	5.1×10^{-3}	1.0×10^{-1}	8.2×10^{-2}
Greenhouse effect (100yrs) *including carbon storage (kg CO_2 eq.)*	-45.9 *Shiv = -52.2*	-24.5 *Shiv = -27.9*	23.1 *Wood = -9.9*	0.2 *0*	-13.6 *Lime = -13.6*	0 *0*	6.7×10^{-1} *0*	-35.5 *-75.7*	-14.1 *-51.4*
Destruction of the ozone layer (kg CFC-11 eq.)	7.1×10^{-7}	3.8×10^{-7}	3.3×10^{-6}	3.4×10^{-7}	0	0	5.7×10^{-6}	9.9×10^{-6}	9.7×10^{-6}
Formation of photochemical ozone (kg C_2H_4 eq.)	7.1×10^{-4}	3.8×10^{-4}	4.2×10^{-3}	5.0×10^{-5}	0	0	3.8×10^{-4}	5.4×10^{-3}	5.0×10^{-3}
Non-renewable energy (MJ)	52.3	27.9	265.8	19.9	0	0	56.3	394.2	369.9
Air pollution (m^3)	674	360	207.2	14.6	0	0	128.2	1024	709.9
Water pollution (m^3)	4.3	2.3	2.2	6.1×10^{-2}	0	0	1.1×10^{-1}	6.7	4.7
Waste production (kg)	6	3.2	No data	0.9	0	98	No data	104.9	102.1

Table 9.2. *Possible environmental impacts over 100 years relating to the manufacture of m² of a wall of banked hempcrete on a wooden skeleton (Source courtesy of BCB-Lhoist, [KEL 03; WER 03], Construire en chanvre, CTBA, ADEME, INRA)*

Table 9.2 presents the environmental impacts of each phase of the lifecycle of 1 m² of a wall made of banked hempcrete on a wooden skeleton.

In order to read this table correctly, you will need the following information:

– in order to be taken down to the functional unit, all these results must be divided by 100 (a typical lifetime);

– because the change in type of allocation affects only the "production of primary materials" part, it is presented separately to the other materials (Tradical 70®, wood and water);

– as regards the "greenhouse effect over 100 years", the table gives the values of the specific measurements for the three carbon absorbers in the hempcrete wall on a wooden skeleton – i.e. hemp shiv, wood and lime.

9.4.4. *Results and interpretation of the lifecycle*

As the results provided by LCAs are complex, they require a certain amount of interpretation. This interpretation is done by identifying those stages of production which may have a major impact on the environment, and if necessary, altering the value of certain sensitive parameters in the reference scenario in order to test their role in the potential environmental impact of the product.

Like any activity, the lifecycle of a banked hempcrete wall on a wooden skeleton has an impact on the environment, as attested by the results shown in Table 9.2, with the exception of the 100-year greenhouse effect. Indeed, the wall under investigation has a favorable effect on this impact: it stores more carbon by photosynthesis and recarbonation than is emitted by its lifecycle. Thus, 67 kg of equivalent CO_2 are stored per m^2 of wall, as opposed to 33 kg of equivalent CO_2 emitted, which represents a net balance of 35 kg CO_2 eq. stored over 100 years per m^2 of hempcrete wall for mass allocation (14 kg CO_2 eq. stored for economic allocation). As shown in Figure 9.6, the main absorber of carbon remains the hemp shiv.

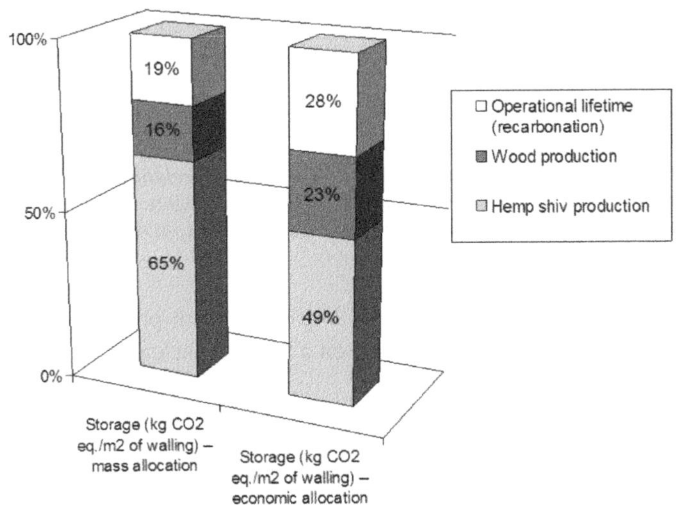

Figure 9.6. *Distribution of the storage of CO_2 between the three carbon absorbers in the banked hempcrete wall on a wooden skeleton*

9.5. Interpretations of the lifecycle, conclusions and reflections

The results (shown in Figure 9.7) demonstrate that the differences between the economic and mass allocation are small in terms of most of the potential impacts considered, with the exception of the greenhouse effect and air and water pollution. This phenomenon can be accounted for by the fact that the value of these impacts is greatly influenced by the production of the hemp shiv. In the knowledge that the impact of shiv production represents 60% (mass allocation) or 32% (economic allocation) of the potential impact of straw production, the differences observed vary greatly.

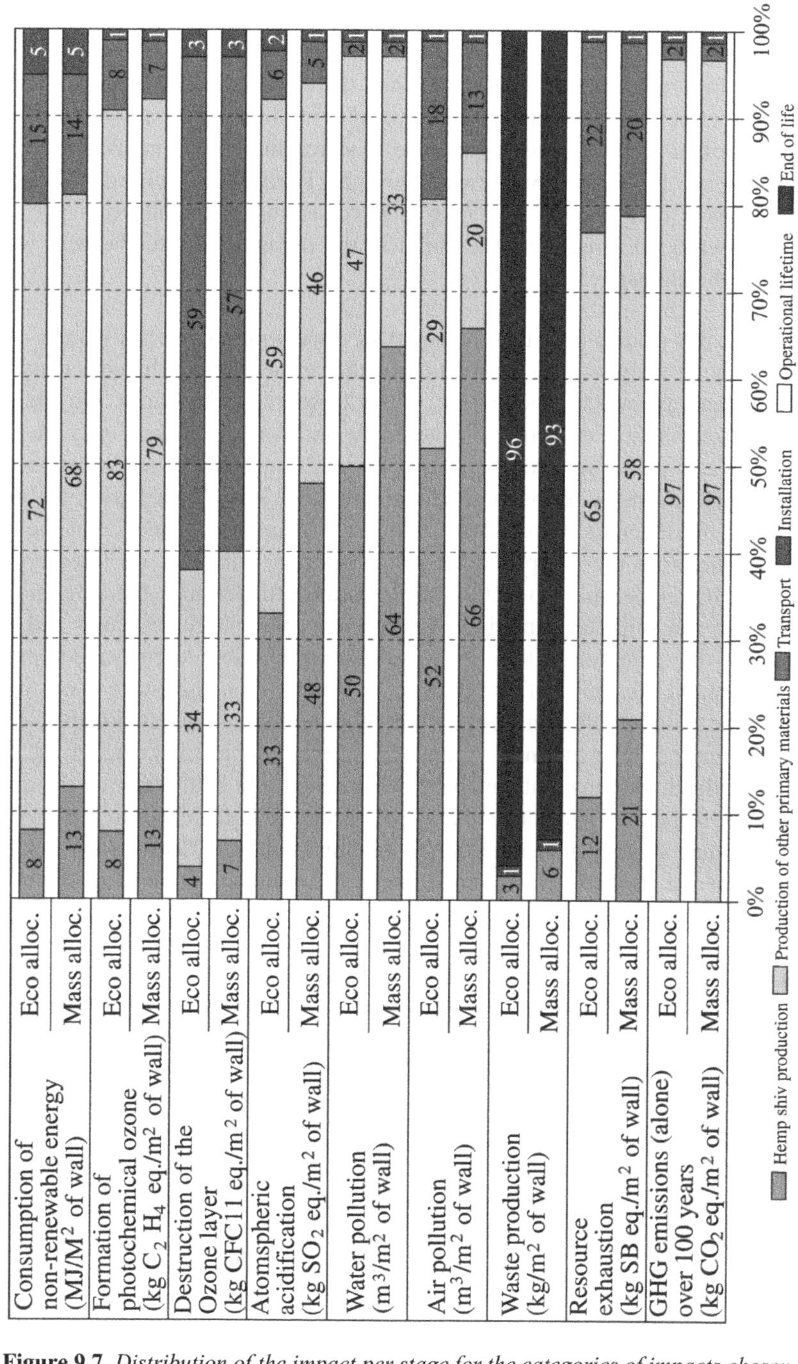

Figure 9.7. *Distribution of the impact per stage for the categories of impacts chosen*

In the context of the working hypotheses chosen, the results of the lifecycle analysis carried out on the wall of banked hempcrete on a wooden skeleton (see Table 9.2, Figures 9.6 and 9.7) show that there is a favorable impact on the greenhouse effect; the hempcrete wall constitutes an interesting carbon absorber for a duration of at least 100 years, because it stores more CO_2 in its three carbon absorbers – i.e. the hemp shiv, the wood and the Tradical® (lime) – than is emitted by its lifecycle (net balance of 35 kg CO_2 eq. stored over 100 years per m^2 of hempcrete wall). In time, the CO_2 stored in the organic parts of the wall will be restored to the atmosphere.

Besides, this study shows that the hemp's contribution to potentially unfavorable impacts may be marginal (for destruction of the ozone layer, GHG emissions, non-renewable energy, waste production); minor (resource exhaustion). On the other hand, the contribution of hemp to potentially unfavorable impacts may be major (air- and water pollution and, to a lesser extent, atmospheric acidification). The emphasis placed on the agronomic part in the 2005 study reveals that the use of nitrogenous fertilizers is a crucial stage in the environmental balance of the hemp plant and therefore the hemp shiv. It has a bearing, primarily, on three types of impact: GHG emission (production of mineral fertilizer and the fate of these fertilizers in the soil), consumption of non-renewable energy resources and water pollution by the emission of nitrates. It is therefore possible to envisage improving the environmental balance of hemp straw by working on this stage – on the one hand, on the dose of fertilizer applied, which should be as low as possible, and on the other hand on the creation of varieties of fertilizer which make better use of the nitrogen. Indeed, the sensitivity analysis performed in the 2005 study on the dose of nitrogen for the same yield shows that a 20% decrease in the doses of nitrogen used would facilitate an assuagement of all the potential impacts of around 10%, with the exception of the impacts "waste production", "destruction of the ozone layer" and "water pollution".

In addition, developing outlets for the dust, which is considered as a waste product in the study but which can represent up to 15% of the total weight of the plant, would indirectly decrease the environmental impact attributed to the shiv, because the overall impact of the plant would be distributed among three co-products instead of two. At present, the dust is still not valued as a co-product; its particle size distribution and its highly irregular quality (a sometimes high mineral content) means that the dust can only be commercialized in terms of a heating product, compressed into pellets for individual heaters, or as an organic soil-enriching product (tests currently being carried out).

The results of the study also show that the manufacture of the Tradical 70® is the stage which contributes most the GHG emissions (97% of the total emissions), the formation of photochemical ozone, the consumption of non-renewable energy and the exhaustion of resources, because it is composed primarily of lime-based minerals kiln-fired at a high temperature (900°C).

Transport (over the whole of the lifecycle) is the stage with the single greatest impact in terms of destruction of the ozone layer. It is also in second place as regards the impact on the consumption of non-renewable energy and the greenhouse effect. At the time of our earlier study, one of the recommendations in terms of improving the environmental balance of the hempcrete wall was to promote the development of the supply chain to decrease the distances of transport. This is currently being done, because other manufacturers of hydraulic binder have entered into the market, with a resulting increase in the number of production sites, including sites on French territory. For instance, Tradical 70® is now also produced in Besançon. The number of sites for hemp shiv production is also expanding. In parallel to the four or five main industrial producers who devote 10-20% of their hemp shiv production to the construction market, we are witnessing the development of a local production network in a short circuit, by groups of smaller hemp growers who devote 90-100% of their production to the construction industry (Construire en chanvre). The development of this network ensures better division of the territory, the effect of which is a decrease in the distances of transport.

In summary, we must note here that certain hypotheses made for lack of available data, do have an effect on the results of the study. For instance, at the end of its life, the wood is likely to be disposed of by being burnt. This release of CO_2 would then partly counterbalance the carbon advantage of the three absorbers. In the case of a potential methanization in an anaerobic environment (burial), the methane released would have an even greater impact, because its global warming potential (GWP) is 25 times greater than that of CO_2 (the GWP of methane over 100 years is 25 times according to the fourth report issued by the IPCC). The balance of the LCA could be conserved or refined by developing the chains of recycling or value creation which would ensure a restitution of the carbon stored in the form of CO_2 rather than CH_4 (methane). Thus, the wooden part could be reused or given value in energetic terms whereas the hempcrete part could be used in the area of composting, in ballasting or in agricultural soil-enrichment. A sensitivity analysis could evaluate the impact of this type of end-of-life disposal. It should be noted that recent LCAs are still asking questions about the different scenarios for the end of life of agromaterials and their environmental impact. The study performed by Lopez *et al* [LOP 11] also attempts to look more closely at the decomposition of bio-sourced materials when they are dumped and the recuperation of the methane emissions, and/or to envisage other scenarios for the end of life which could help to get more value out of the products: careful spreading of the straw (in order to favor aerobic rather than anaerobic decomposition), energy value creation (wood), etc.

Beyond the results of the LCA, the main improvement for the field put forward by the 2005 study was technological evolutions – e.g. the development of large-scale production of hempcrete bricks that are easier to install, particularly in the context of a new build. These new blocks, industrially manufactured, would enable construction contractors to use less binder and less water. In terms of load-bearing bricks, the wood structure would be removed, and consequently so would the danger at the end of its natural life (re-release in the form of CO_2 or methane) while conserving the main carbon absorber – the shiv; thus, the potential environmental impact of the hempcrete would probably be diminished. In 2012, many industrialists offer hempcrete bricks made using a hydraulic binder, but these cannot be used in load-bearing structures, and an evaluation of their environmental impact still needs to be carried out. Such a study could be based on the recent works at CoDEM [TOU 12], analyzing the environmental impacts of flax concrete bricks, with the main contributor to these impacts still being the cement used.

Banked hempcrete, for its part, is particularly well adapted for building renovation; quite apart from its thermal resistance, which satisfies the thermal regulations in force (RT 2005) with walls of less than 30 centimeters in thickness, "its hygrothermic performance, which combines mass transfer and phase change with the other thermal phenomena, means it is able to greatly exceed the expected energy performances. In addition, users have noted that they attain a satisfactory level of thermal comfort at relatively low ambient temperatures (17 or 18°C, or sometimes less), which leads to a significant decrease in energy consumption" (*Construire en Chanvre*).

Besides the initiatives by actors in the field as regards standardization/labeling, intended to characterize the different types of hemp shiv and certify hemp/lime combinations in order to buoy the field up in the construction sector, a collective sanitary and environmental declaration certificate (FDES) on hempcrete is currently being drafted under the aegis of the ministry in charge of ecology. In line with the norm P 01 010, this FDES will enable professionals in the sector to display information in a rational manner about the environmental and sanitary aspects of their products. It will also enable data relating to hempcrete to be used for the overall evaluation of the performances of buildings.

9.6. Bibliography

[AFN 04] AFNOR (2004) Norme NF P01 010 Qualité environnementale des produits de construction.

[BOE 05] BOEGLIN N., VEUILLET D., "Introduction à l'Analyse de Cycle de Vie", note de synthèse externe, ADEME, May 2005.

[BOU 05] BOUTIN M.P., FLAMIN C., QUINTON S., GOSSE G., Analyse du cycle de vie de : mur en béton chanvre banché sur ossature bois, INRA, Lille, France, 2005

[CAL 00] CALZONI J., CASPERSEN N., DERCAS N., GAILLARD G., GOSSE G. HANEGRAAF M., HEINZER L., JUNGK N., KOOL A., KORSUIZE G., LECHNER M., LEVIEL B, NEUMAYR R., NIELSEN A.M., NIELSEN P.H., NIKOLAOU A.., PANOUTSOU C. PANVINI A., PATYK A.,RATHBAUER J., REINHARDT G.A., RIVA G., SMEDILE E STETTLER C., PEDERSON WEIDEMA B., WORGETTER M.., VAN ZEIJTS H., Bioenergy for Europe: which ones fit best? A comparative analysis for the community, 178 pages, 2000

[CEB 05] CEBTP 2005, Utilisation de matériaux renouvelables en parois de maisons à ossature bois, 132 pages, October 2005

[CON] Construire en Chanvre. URL, http://www.construction-chanvre.asso.fr

[INI] INIES,URL: http://www.inies.fr

[ISO 06] ISO 14040 management environnemental – analyse de cycle de vie – principes et cadre, 2006.

[KEL 03] KELLENBERGER D., ALTHAUS H-J., KÜNNIGER T., JUNGBLUTH N. - 2003 – Life Cycle Inventories of Building Products – Ecoinvent report No. 7 – Dübendorf, December 2003.

[LAU 09] LAURANSON R., SCHULTZE A., TOULLEC M., Etude d'une méthodologie simplifiée pour la réalisation des ACV des bioproduits, synthèse de l'étude Etude réalisée pour le compte de l'ADEME par BIO Intelligence Service, December 2009

[LOP 11] LOPEZ A., ALTMEYER S., PONS M.-N., BARTHÉLÉMY M., NICOLAS A., ACV de quatre matériaux de construction de construction et isolation biosourcés, PROGEPI, France, June 2011.

[TOU 12] TOUBON L., LAIDOUDI B., CRIGNY A., AMONY S., DUPRÉ B., Blocs de construction en béton de lin : analyse du cycle de vie, CoDEM Picardie, Dury, France, 2012

[WER 03] WERNER F., ALTHAUS H-J., KÜNNIGER T., RICHTER K., JUNGBLUTH N. – Life Cycle Inventories of Wood as Fuel and Construction Material - Ecoinvent report No. 9 – Dübendorf, December 2003.

List of Authors

Sofiane AMZIANE
Blaise Pascal University
Clermont-Ferrand
France

Laurent ARNAUD
ENTPE
Lyon
France

Marie-Pierre BOUTIN
BRGM
France

Florence COLLET
University of Rennes 1
France

Gilles ESCADEILLAS
LMDC
Toulouse
France

Cyril FLAMIN
INRA
France

Philippe GLÉ
ENTPE
Lyon
France

Emmanuel GOURDON
ENTPE
Lyon
France

Étienne GOURLAY
ENTPE
Lyon
France

Yves HUSTACHE
Eco-Innovation & Stratégie
France

Christophe LANOS
University of Rennes 1
France

Gérard LENAIN
SOC INNOVANTE CONSTRUC
CHANVRE (SI2C)
France

Camille MAGNIONT
IUT Toulouse
France

Philippe MUNOZ
CRDA
Lycée ARAGO
Reims
France

Vincent NOZAHIC
Blaise Pascal University
Clermond-Ferrand
France

Vincent PICANDET
University of South Brittany
France

Didier PIPET
CRDA
Lycée ARAGO
Reims
France

Driss SAMRI
ENTPE
Lyon
France

Henri VAN DAMME
IFSTTAR
France

Index

A

Absorption, 68, 69, 249, 250
ACERMI, 9
Acoustical, 243-263
Aerated lime, 86-92, 120, 125, 136, 159, 164, 247
Agro-concretes, 1-19
Arithmetic mean, 58, 60, 61
Ash, 34, 35
Average elongation, 57, 58

B

Blast furnace slag, 103-106
Bracing tests, 287

C

Cambium, 28
Capillary absorption coefficient, 92
Casting, 28, 65, 66, 68, 69, 118, 120, 123, 244, 249
Cellulose, 15, 32, 33
Chemical composition, 15, 16
Clinker, 76-78, 84, 105, 168
Compacting, 28, 66, 67, 121, 124, 125, 131, 139, 144, 154, 157, 158, 162- 164, 170, 174-176
Confinement stress, 35
Construction details, 150
Convexity, 42, 43, 52, 53, 66
Convexity ratio, 42, 52, 53
Cortex, 29

D

Defibration, 30, 34-37, 39, 49, 52, 65, 259
Density, 35, 36
Distribution model, 59-65
Dynamic bulk modulus, 253, 254, 257, 258
Dynamic density, 253, 256-258

E

Eco-materials, 11, 12
Elasticity modulus, 155, 158, 163, 174
Ellipse, 44-47, 51-54
Elongation, 43, 45- 47, 50, 57-58, 65, 175
Environmental impacts, 2-3, 84, 92, 100, 101, 103, 106, 110
Epidermis, 28

F

Factor of 4, 4
Féret, 42-44, 54, 58, 65
FHS, 35
Fibers, 10, 13-15, 17, 27-29, 33-45, 52, 54, 65, 75, 93, 109, 175, 289, 290, 296, 301
Flatness, 58
Fly ash, 101-103
Formulation, 117-152
Frame density, 258, 259, 261

G

Geometric mean, 58, 60, 63, 65, 66
Grenelle, 3, 4

H

Hemicelluloses, 15, 33
Hemp shiv, 28
HQE, 6, 294
HS, 35
Hydration, 78-81, 96
Hydraulic limes, 89-92
Hydric, 83, 88, 92, 96, 100, 103, 105, 110
Hydric properties, 83, 88, 92, 96, 100, 103, 105, 110

I

Image analysis, 39-47
Installation, 143-145
Intra-particle porosity, 164, 258, 261

K

Kaolinite, 97

L

LCA (Life-Cycle Analysis), 4, 8
Lignin, 33, 34
Lime-pozzolan binders, 92-106
Lipids, 31, 34
Loading cycle, 156, 285
Longitudinal loading, 271, 272, 280, 281

M

Major axis, 43, 44, 54
Mechanical properties, 81-82, 88, 91-92, 99-100, 102, 105, 109
Mechanical strength(s), 81, 87, 88, 91-93, 99, 102-105, 110, 111, 128, 137, 153, 154, 167, 170, 174-176, 267
Meristem, 29
Metakaolin, 96
Minor axis, 43, 54

N

Natural hydraulic limes (NHLs), 89-92
Natural pozzolans, 93-96

P

Particle size distribution (PSD), 36, 37
Pectins, 15, 16, 18, 31, 33, 34
Phloem, 17, 28
Plant cell walls, 32, 33
Plaster, 106-110
Porosity, 30
Portland cements, 75-84
Pozzolanic admixtures, 75, 80, 81, 93, 101
Pozzolanic reaction, 11, 80, 81, 92, 95, 98, 100, 102
PREBAT, 5, 167
Principle, 146-152
Professional regulations, 119, 146, 147, 151, 152, 154, 159
Projected area, 41-52, 54, 55

Projection, 38, 145, 170-173, 238, 267, 297, 299

R

Resistivity, 262, 263
Resolution, 40, 41, 53
Rigidity, 15, 33, 120, 175, 176, 282, 283, 285, 286
Rigidity test, 282, 285
RT 2012, 4, 7, 238

S

Sieving, 38, 39
Sound absorption, 245-251, 252, 256, 258
Standard deviation, 46, 58-61, 63, 65, 66, 260
Strength test, 282, 283, 285
Sustainability, 6, 12
Sustainable development, 1-3

T

Thermal properties, 75, 83, 100, 103, 105, 111, 145, 179
Thresholding, 41
Tortuosity, 212, 256, 257, 264

V, W, X

Vertical loading, 279
Viscous characteristic length, 256
Waxes, 15, 31, 34
Xylem, 15, 17, 28, 29, 34

Printed and bound by CPI Group (UK) Ltd, Croydon, CR0 4YY

13/09/2026

14961075-0005